# LEWIS AND CLARK EXPEDITION
VOLUME I.

*Meriwether Lewis.*

# HISTORY
## OF
# THE EXPEDITION
## OF
## CAPTAINS LEWIS AND CLARK
### 1804-5-6

REPRINTED FROM THE EDITION OF 1814

WITH INTRODUCTION AND INDEX
BY
JAMES K. HOSMER, LL.D.,
AUTHOR OF "A SHORT HISTORY OF THE MISSISSIPPI VALLEY,"
"THE STORY OF THE LOUISIANA PURCHASE," ETC.;
PRESIDENT OF THE AMERICAN
LIBRARY ASSOCIATION.

In Two Volumes, with Portraits and Maps
VOLUME I.

SECOND EDITION

CHICAGO
A. C. McCLURG & CO.
1903

Copyright
A. C. McCLURG & CO.
1902
Published Oct. 15, 1902
Second Edition published January 15, 1903

*Composition by The Dial Press, Chicago.*
*Presswork by John Wilson & Son,*
*University Press, Cambridge, Mass., U. S. A.*

# HISTORY OF THE EXPEDITION OF CAPTAINS LEWIS AND CLARK

## 1804 - 5 - 6

## VOLUME I

Vol. 1 Trade Paperback ISBN: 1-58218-697-9
Vol. 1 Hardcover ISBN: 1-58218-698-7
Vol 2 Trade Paperback ISBN: 1-58218-702-9
Vol 2 Hardcover ISBN: 1-58218-703-7

## As Published in 1903

All rights reserved, which includes the right to reproduce this book or portions thereof in any form whatsoever except as provided by the U.S. Copyright Laws. For information address Digital Scanning, Inc.

Digital Scanning and Publishing is a leader in the electronic republication of historical books and documents. We publish many of our titles as eBooks, as well as hardcover and trade paper editions. DSI is committed to bringing many traditional little known books back to life, retaining the look and feel of the original work.

Published by DIGITAL SCANNING, INC. Scituate, MA 02066
781-545-2100  www.digitalscanning.com

©2002 DSI Digital Reproduction
First DSI Printing: 2002

# HISTORY

OF

# THE EXPEDITION

UNDER THE COMMAND OF

*CAPTAINS LEWIS AND CLARK.*

TO

THE SOURCES OF THE MISSOURI,

THENCE

ACROSS THE ROCKY MOUNTAINS

AND DOWN THE

RIVER COLUMBIA TO THE PACIFIC OCEAN

PERFORMED DURING THE YEARS 1804—5—6.

By order of the

GOVERNMENT OF THE UNITED STATES.

PREPARED FOR THE PRESS

BY PAUL ALLEN, ESQUIRE.

IN TWO VOLUMES.

VOL. I.

*PHILADELPHIA:*
PUBLISHED BY BRADFORD AND INSKEEP; AND
ABM: H. INSKEEP, NEWYORK.
J. Maxwell, Pduer.
1814.

*Facsimile title-page 1814 edition*

DISTRICT OF PENNSYLVANIA, to wit:

BE IT REMEMBERED, That on the twenty-second day of January, in the thirty-eight year of the independence of the United States of America, A. D. 1814, Bradford & Inskeep, of the said district, have deposited in this office the title of a book, the right whereof they claim as proprietors, in the words following, to wit:

"History of the Expedition under the Command of Captains Lewis and Clark, to the Sources of the Missouri, thence across the Rocky Mountains, and down the River Columbia to the Pacific Ocean. Performed during the Years 1804–5–6, by order of the Government of the United States. Prepared for the press by Paul Allen, Esquire."

In conformity to the act of Congress of the United States, entitled "An act for "the encouragements of learning, by securing the copies of maps, charts, and books, "to the authors and proprietors of such copies during the times therein mentioned." And also the act, entitled, "An act supplementary to an act, entitled, "An "act for the encouragement of learning, by securing the copies of maps, charts,and "books, to the authors and proprietors of such copies during the times therein men- "tioned," and extending the benefits thereof to the arts of designing, engraving, and etching historical and other prints."

DAVID CALDWELL,
Clerk of the District of Pennsylvania.

*Facsimile of original copyright notice*

# PUBLISHERS' NOTE.

NOTWITHSTANDING that in America few names are more familiar upon the tongue than Lewis and Clark, it is a singular fact that a full and adequate account of what they did has long been almost unattainable. The published work of 1814 has quite disappeared from the market. The fragmentary accounts that preceded this edition have become very rare, as also the various foreign presentments, English, Dutch, German, and French. The noble reprint which appeared in 1893, prefaced, annotated, and supplemented in every needful way by Dr. Elliott Coues, a work of great bulk and cost, is entirely out of print.

The lack thus existing, felt now more and more as the centenary of the great exploration draws near, the present edition has been planned to fill. The text used is that of the 1814 edition, which must hold its place as the only account approaching adequacy. On its appearance, it at once superseded the broken narratives that had gone before; and Dr. Coues, the latest and most eminent of the editors of Lewis and Clark, after listing in an exhaustive bibliography of the subject a large number of works, pronounces it the only text worthy of attention.

This new edition, then, is a reprint of the Biddle text of 1814 complete. Its bulk is not increased by annotation, but it has been carefully supervised by Dr. James K. Hosmer, a specialist in matters pertaining to Western history. Dr. Hosmer has prepared a careful analytic Index, a feature which the original edition lacked, and furnished an Introduction giving the events which led up to the great expedition, and the vast development that has flowed from it, in a way to make plain the profound significance of the achievement. Fac-similes of all the maps of the 1814 edition are given, together with new reproductions of the best extant portraits of the heroes of the enterprise.

The publishers take pleasure in offering this work to the reading public, in the hope that it may fill all requirements and become the standard popular edition of this great American classic.

*September 1, 1902.*

# CONTENTS OF VOLUME I.

|  | PAGE |
|---|---|
| PUBLISHERS' NOTE | vii. |
| INTRODUCTION | xvii. |
| PREFACE TO EDITION OF 1814 | xxxvii. |
| LIFE OF CAPTAIN LEWIS BY THOMAS JEFFERSON | xli. |

### CHAPTER I.

The party set out on the expedition and pass Wood river—Description of the town of St. Charles—Osage Woman river—Gasconade and Osage rivers described—Character of the Osage Indians—Curious traditionary account of their origin—The party proceed, and pass the Mine river—The two Charitons—The Kanzas, Nodawa, Newahaw, Neeshnabatona, Little Nemahar, each of which are particularly described—They encamp at the mouth of the river Platte—A particular description of the surrounding country—The various creeks, bays, islands, prairies, etc., given in the course of the route  1

### CHAPTER II.

Some account of the Pawnee Indians—Council held with the Ottoe and Missouri Indians—Council held with another party of the Ottoes—Death of Sergeant Floyd—The party encamp near the mouth of Whitestone river—The character of the Missouri, with the rivers that enter it—The surrounding country—The various islands, bays, creeks, etc., given in the course of the expedition  34

### CHAPTER III.

Whimsical instance of superstition of the Sioux Indians—Council held with the Sioux—Character of that tribe, their manners, etc.—A ridiculous instance of their heroism—Ancient fortifications—Quicurre river described—Vast herds of Buffaloe—Account of the Petit Chien or Little Dog—Narrow escape of George Shannon—Description of White river—Surprising fleetness of the Antelope—Pass the river of the Sioux—Description of the Grand Le Tour, or Great Bend—Encamp on the Teton river  56

x.                          *Contents*

CHAPTER IV.

Council held with the Tetons—Their manners, dances, etc.—Chayenne river—Council held with the Ricara Indians—Their manners and habits—Strange instance of Ricara idolatry—Another instance—Cannonball river—Arrival among the Mandans—Character of the surrounding country, and of the creeks, islands, etc.                                      88

CHAPTER V.

Council held with the Mandans—A prairie on fire, and a singular instance of preservation—Peace established between the Mandans and Ricaras—The party encamp for the winter—Indian mode of catching goats—Beautiful appearance of northern lights—Friendly character of the Indians—Some account of the Mandans—The Ahnahaways and the Minnetarees—The party acquire the confidence of the Mandans by taking part in their controversy with the Sioux—Religion of the Mandans, and their singular conception of the term medicine—Their tradition—The sufferings of the party from the severity of the season—Indian game of billiards described—Character of the Missouri, of the surrounding country, and of the rivers, creeks, islands, etc. 126

CHAPTER VI.

The party increase in the favour of the Mandans—Description of a buffaloe dance—Medicine dance—The fortitude with which the Indians bear the severity of the season—Distress of the party for want of provisions—The great importance of the blacksmith in procuring it—Depredations of the Sioux—The homage paid to the medicine stone—Summary act of justice among the Minnetarees—The process by which the Mandans and Ricaras make beads—Character of the Missouri, of the surrounding country, and of the rivers, creeks, islands, etc.                                        158

CHAPTER VII.

Indian method of attacking the buffaloe on the ice—An enumeration of the presents sent to the President of the United States—The party are visited by a Ricara chief—They leave their encampment, and proceed on their journey—Description of the Little Missouri—Some account of the Assiniboins—Their mode of burying the dead—Whiteearth river described—Great quantity of salt discovered on its banks—Yellowstone river described—A particular account of the country at the confluence of the Yellowstone and Missouri—Description of the Missouri, the surrounding country, and of the rivers, creeks, islands, etc.                                       186

## Contents

### CHAPTER VIII.

Unusual appearance of salt—The formidable character of the white bear—Porcupine river described—Beautiful appearance of the surrounding country—Immense quantities of game—Milk river described—Extraordinary character of Bigdry river—An instance of uncommon tenacity of life in a white bear—Narrow escape of one of the party from that animal—A still more remarkable instance—Muscleshell river described  213

### CHAPTER IX.

The party continue their route—Description of Judith river—Indian mode of taking the buffaloe—Slaughter river described—Phenomena of nature—Of walls on the banks of the Missouri—The party encamp on the banks of the river to ascertain which of the streams constitute the Missouri—Captain Lewis leaves the party to explore the northern fork, and Captain Clark explores the southern—The surrounding country described in the route of Captain Lewis—Narrow escape of one of his party  241

### CHAPTER X.

Return of Captain Lewis—Account of Captain Clark's researches with his exploring party—Perilous situation of one of his party—Tansy river described—The party still believing the southern fork the Missouri, Captain Lewis resolves to ascend it—Mode of making a place to deposit provisions, called *cache*—Captain Lewis explores the southern fork—Falls of the Missouri discovered, which ascertains the question—Romantic scenery of the surrounding country—Narrow escape of Captain Lewis—The main body, under Captain Clark, approach within five miles of the falls, and prepare for making a portage over the rapids  269

### CHAPTER XI.

Description and romantic appearance of the Missouri at the junction of the Medicine river—The difficulty of transporting the baggage at the falls—The party employed in the construction of a boat of skins—The embarrassments they had to encounter for want of proper materials—During the work the party much troubled by white bears—Violent hail-storm, and providential escape of Captain Clark and his party—Description of a remarkable fountain—Singular explosion heard from the Black mountains—The boat found to be insufficient, and the serious disappointment of the party—Captain Clark undertakes to repair the damage by building canoes, and accomplishes the task  295

*Contents*

CHAPTER XII.                                                    PAGE

The party embark on board the canoes—Description of Smith's river—Character of the country, etc.—Dearborne's river described—Captain Clark precedes the party for the purpose of discovering the Indians of the Rocky mountains—Magnificent rocky appearances on the borders of the river denominated the Gates of the Rocky mountains—Captain Clark arrives at the three forks of the Missouri without overtaking the Indians—The party arrive at the three forks, of which a particular and interesting description is given        323

CHAPTER XIII.

The name of the Missouri changed, as the river now divides itself into three forks, one of which is called after Jefferson, the other after Madison, and the other after Gallatin—Their general character—The party ascend the Jefferson branch—Description of the river Philosophy, which enters into the Jefferson—Captain Lewis and a small party go in advance in search of the Shoshonees—Description of the country, etc., bordering on the river—Captain Lewis still preceding the main party in quest of the Shoshonees—A singular accident which prevented Captain Clark from following Captain Lewis's advice, and ascending the middle fork of the river—Description of Philanthropy river, another stream running into the Jefferson—Captain Lewis and a small party, having been unsuccessful in their first attempt, set off a second time in quest of the Shoshonees        351

CHAPTER XIV.

Captain Lewis proceeds before the main body in search of the Shoshonees—His ill success on the first interview—The party with Captain Lewis at length discover the source of the Missouri—Captain Clark with the main body still employed in ascending the Missouri or Jefferson river—Captain Lewis's second interview with the Shoshonees attended with success—The interesting ceremonies of his first introduction to the natives detailed at large—Their hospitality—Their mode of hunting the antelope—The difficulties encountered by Captain Clark and the main body in ascending the river—The suspicions entertained of Captain Lewis by the Shoshonees, and his mode of allaying them—The ravenous appetites of the savages illustrated by a singular adventure—The Indians still jealous, and the great pains taken by Captain Lewis to preserve their confidence—Captain Clark arrives with the main body, exhausted by the difficulties which they underwent        378

# Contents

xiii.

### CHAPTER XV.

PAGE

Affecting interview between the wife of Chaboneau and the chief of the Shoshonees—Council held with that nation, and favourable result—The extreme navigable point of the Missouri mentioned—General character of the river and of the country through which it passes—Captain Clark in exploring the source of the Columbia falls in company with another party of Shoshonees—The geographical information acquired from one of that party—Their manner of catching fish—The party reach Lewis river—The difficulties which Captain Clark had to encounter in his route—Friendship and hospitality of the Shoshonees—The party with Captain Lewis employed in making saddles, and preparing for the journey   406

### CHAPTER XVI.

Contest between Drewyer and a Shoshonee—The fidelity and honour of that tribe—The party set out on their journey—The conduct of Cameahwait reproved, and himself reconciled—The easy parturition of the Shoshonee women—History of this nation—Their terror of the Pawkees—Their government and family economy in their treatment of their women—Their complaints of Spanish treachery—Description of their weapons of warfare—Their curious mode of making a shield—The caparison of their horses—The dress of the men and of the women particularly described—Their mode of acquiring new names   433

### CHAPTER XVII.

The party, after procuring horses from the Shoshonees, proceed on their journey through the mountains—The difficulties and dangers of the route—A council held with another band of the Shoshonees, of whom some account is given—They are reduced to the necessity of killing their horses for food—Captain Clark, with a small party, precedes the main body in quest of food, and is hospitably received by the Pierced-nose Indians—Arrival of the main body amongst this tribe, with whom a council is held—They resolve to perform the remainder of their journey in canoes—Sickness of the party—They descend the Kooskooskee to its junction with Lewis's river, after passing several dangerous rapids—Short description of the manners and dress of the Pierced-nose Indians   463

# LIST OF PORTRAITS AND MAPS.

### VOLUME I.

PORTRAIT OF MERIWETHER LEWIS      *Frontispiece*
From new photograph of the original painting in Independence Hall, Philadelphia.

MAP OF LEWIS AND CLARK'S TRACK      facing 124
Facsimile of the original drawing by William Clark.

PLAN OF ANCIENT FORTIFICATION ON THE MISSOURI RIVER      facing 276
Facsimile from the edition of 1814.

MAP OF THE FALLS OF THE MISSOURI RIVER      facing 400
Facsimile from the edition of 1814.

# INTRODUCTION

LEWIS AND CLARK were known in their time as the explorers of Louisiana, and to understand what they achieved, it must be known what the Louisiana of the beginning of the nineteenth century was, and how it came to be. In the period of Columbus, when the world was full of colonizing zeal, the French were not behind the Spaniards, Portuguese, Dutch, or English. It was felt then, as it is felt now, that the greatness, perhaps the existence, of France depended upon the possession of a domain outside of Europe, and it is said that during the sixteenth century eighteen separate attempts were made by France to establish a foothold in America. French colonization has always proceeded in this way: brilliant adventurers, full of initiative, vividly imaginative, able and intrepid, have led the way; all that it was possible for individuals to accomplish has been accomplished again and again. Why is it that in spite of the splendid achievement of her sons France has always as a colonizing nation stood behind her contemporaries and rivals? Because her pioneer sons have never received from the government or from the nation adequate support; and in addition to that, because among themselves her path-breakers have too often found it impossible to harmonize. Call it vanity, or call it by the more honourable name ambition, this foible of powerful spirits has seemed in the case of Frenchmen to exist inordinately. These causes have repeatedly brought it about that enterprises which began full of promise have ended at last in failure. In the middle of the eighteenth century France seemed more likely than England to become the ruler of India. Rarely in any age has there been a more impressive manifestation of ability and courage than that made by those splendid sons of France, Duplaix, La Bourdonnais, Bussy, Lally-Tollendal, and Suffren. It came to nothing because they had no proper support from king or nation,—still more, perhaps, because they could not agree among themselves, wasting in internal jealous bickering energy which if directed against the obstacles that confronted them might have won success. To-day in India France is a forgotten name, while England is sovereign. In America, too, where at first the prospects of French dominion seemed almost certain, utter failure was the ultimate result,—and for the same reason as in the Orient.*

Though Jacques Cartier sailed up the St. Lawrence as early as 1534, New France was not founded until the coming in 1603 of Champlain, perhaps the noblest type in the whole impressive series of colonizers that went forth into the world under the banner of the lilies. In resource and courage he was matchless: in imagination he conceived such projects as occupying America with a French empire and even connecting the Pacific with the Atlantic by an Isthmian canal. In temper he seems to have been, moreover, unlike his class in general, sweet and disposed to coöperation. If properly sustained, as has been said, he would have colonized from Quebec to Florida, or westward from the mouth of the St. Lawrence to the Mississippi. After thirty years of heroic striving, however, he died in his little fortress at Quebec, the great wilderness about him as yet scarcely touched by any impress, his fine energy throughout his life crippled and deadened by want of sympathy and support from home. When his life went out it was like the extinction of a torch; and for a

* See W. Frewen Lord's "Lost Empires of the Modern World" on this topic.

generation a darkness brooded which seemed too heavy to lift. But the story of New France was to have other chapters.

In 1666 landed in Canada Robert Cavalier, Sieur de La Salle, a scion from a noble Norman stock of Rouen, as well endowed perhaps with great qualities as his predecessor, who, because his lot fell in the time of Louis XIV., a prince who with all his weaknesses could yet appreciate a brave man and give him some support, worked out a magnificent achievement. He might have done still more than he did; but hampered by an unamiability which made it difficult for him to work with others, his bold and well-planned schemes encountered constantly a spirit of mutiny in the followers upon whom he relied to carry them out,—until at last, shipwrecked on the coast of Texas, deserted by the admiral whose attachment he had failed to win, and balked by the companions whom his haughtiness had estranged, he fell under the weapons of his own men, in the wilderness, in 1687. Had fortune been a little more favourable, or had he himself been less harsh and imperious, how much more he might have done! and yet scarcely any other path-breaker of the old time or the new has ever accomplished so much. Pressing on to find the road to China from the rapid at Montreal where his *seigneurie* was located, he did not indeed reach Cathay; but he discovered the Ohio; first of white men, in his bark the "Griffin," he traversed the great lakes, and first of white men he sailed down the Mississippi to its mouth. First of men he mapped the vast country intelligently from the great river on the west to the Atlantic, from the St. Lawrence to the Gulf of Mexico, and projected and partially built a chain of posts running from Canada to the city he meant to build at the mouth of the Mississippi. Seldom indeed has pioneer let such light in upon a wilderness as did La Salle, tramping and paddling through leagues by the thousand, undaunted though forests hung dark, though torrents dashed, and though the red-handed Iroquois haunted his trail. It was this most intrepid of explorers, who, standing not far from the Gulf, hoisting the banner of France, and entertaining in his heart chivalrous loyalty to his king, conferred the name *Louisiana* in 1682 upon the far-stretching territory both east and west of the river whose secrets no man knew, though he had broken a way to them.

    The colony which La Salle did not live to found was nevertheless established in 1699 at Biloxi, on the Gulf, by Iberville, a Canadian seigneur, who, after showing much prowess in Hudson's Bay and the West Indies, took up in Louisiana the work which La Salle left incomplete. Iberville is but a transitory figure upon our stage; but his young brother Bienville, who in 1717 founded New Orleans and long guided its destinies, steadfastly upheld the fortunes of the colony. Matters went as they have commonly gone in the history of French colonization. The king and the great people at home had small interest in the dependency except as a means of enriching themselves, It was turned over at first to the monopolist Crozat, and afterwards to John Law, to be exploited in the Mississippi Bubble. For a long time few settlers came, and these had little thought of home-making and permanent life. They wandered off in the hunt for gold, or sought the savages to trade for furs. The Mississippi Bubble, though so disastrous in one way, and so marked by iniquity, did work a certain advantage to Louisiana. Through misrepresentation, sometimes indeed through kidnapping, some thousands of settlers, fairly respectable in character, were transferred thither. Once there they could not return, and they slowly adapted themselves to their exile. Men preponderating largely, wives were afforded by the picturesque expedient of *filles à la cassette,* "girls with little trunks,"—enterprising young women of respectable character, induced for a small dowry and trousseau to cross the ocean to unknown husbands waiting for them on the other side. In this rough and ready match-making, which was thoroughly practical and turned out successfully, the excellent Ursuline nuns often played a great part—recruiting the girls in the provinces of France, chaperoning them on the long voyage, and sheltering them in the convent on their arrival until properly mated under sanction of the Church.

But the prosperity of the colony was never otherwise than very languid. Above Pointe Coupée population practically ceased. To the right and left of the river the vast stretches were uninvaded. But if little can be said about the townsmen or the farmers, not so as to the explorers: in that field the French have never been wanting, nor were they slothful here. Bourgmont, Du Tisné, the brothers Mallet, are the names of men of action in the Southwest, who traced far the courses of the Red and the Arkansas, penetrated the desert toward Santa Fé, and brought to the world's knowledge the Pawnees, the Comanches, and many another remote tribe. But these pathfinders of the South were less interesting than those of the far North. In the days of the great Louis, the son of an officer of the regiment Carignan-Salières, a body of regular troops that did good service in Canada in the early time, went to France, became a soldier, and under the name La Vérendrye fought at Malplaquet, where, encountering a trooper of Marlborough, he received six sabre-cuts and was left for dead on the field. But he lived to return to New France, where, establishing himself in Manitoba, on and beyond the utmost verge of Louisiana, he dwelt in the forest and brought up sons of a like spirit with himself. These ranged the desert far and wide, until the younger La Vérendrye, in 1742, striking into the track to be followed sixty years later by Lewis and Clark, sighted the Rocky Mountains. How near France came to possessing North America, and how utterly she failed in the end! When Braddock was destroyed in 1755, and the establishment of Fort Duquesne seemed to make sure to France her hold upon the Ohio and Mississippi Valleys; when Montcalm, a great soldier, by seizing Oswego, made Ontario a French lake, and presently after, by his victories at Fort William Henry and Ticonderoga, put himself into a position to thrust from the rear at the unprotected heart of New England, the continent seemed about to fall to the lilies. But a great minister, Pitt, arose in England, and a capable general, Wolfe, appeared to lead in America. In 1759 came the victory of the Plains of Abraham, and for France all was over. She resigned at once to England not only Canada, but all that part of La Salle's old Louisiana which lay east of the Mississippi except New Orleans. Broken in spirit, and fearing to lose that also to England, in 1762 she ceded that to Spain,—resigning at the same time to Spain her title to the half of the Mississippi Valley west of the river to which the name Louisiana now became restricted. Thus in North America France became completely dispossessed.

But nothing is so remarkable, as we follow in history the fortunes of France, as the buoyancy with which again and again, after apparent submergence, we presently find her once more on the top of the wave. To the humiliations of Crécy and Agincourt succeed ere long the triumphs of Joan of Arc. The apparent disintegration through civil strife of the days of the Fronde is followed by the concentrated might of the monarchy of Louis XIV. So at the crisis we are considering, though for the moment dragged in the dust by English success, it is but for a moment: in a few years by her well-timed aid the Thirteen Colonies go free, and it is England's turn

*Introduction* xxiii.

to sink into what appears to be the abyss of ruin. In less than a generation Napoleon is at the front with the world at his feet.

We understand well that Napoleon profoundly affected Europe; but do we make it real to ourselves that that most portentous of apparitions in the field of human affairs affected no less profoundly America? Washington was the father of our country: Lincoln was the preserver of the union of our country: Napoleon doubled the area of our country. The conjunction of the name of the Corsican with those of Washington and Lincoln seems grotesque, but indeed he was the instrument employed by fate or providence to enlarge our bounds to their present amplitude. Here is the story in brief: Napoleon, rising into power like every great Frenchman of the past and the present, desired to be a colonizer. He had a scheme on the one hand for penetrating to India; on the other hand for regaining and restoring the magnificent New France in America, to which in the dark days of Louis XV. had come such catastrophe,— and at this latter scheme we must glance. Spain, to whom in 1762 Louisiana, shorn of its eastern half, had been ceded, was at the close of the century sinking more and more rapidly in the decadence which had so long been in progress. Napoleon, now First Consul, eager and confident through mighty successes, resolved to take back from her the great province obtained from France in the day of disgrace. According to his wont with the weak, he proceeded with little ceremony. Giving to Spain a small Italian possession, he took in exchange Louisiana, and prepared with all energy a fleet and army to confirm his hold against all comers. Victor, soon to be one of the most brilliant of the marshals, was to be the leader. The United States was to be treated very cavalierly, as Spain had already been: New France was to be restored with all the promise it had had before the death of Montcalm. But there were delays which even the First Consul's energy could not abridge. Spain, though weak, found ways to baffle with diplomacy: San Domingo, where the blacks under Toussaint l'Ouverture had revolted, must be reconquered as a preliminary; by the hands of the negroes, and their ghastly ally the yellow fever, many thousands perished, though they had before been victors in Egypt and at Marengo. When at last all was ready the sky over Europe became dark with impending war. Could the First Consul afford to engage his hands in an American complication when every resource might soon be demanded nearer home? A weaker character might have felt he was committed to the American enterprise and pushed it through at every risk. Not so Napoleon. He measured accurately the crisis, gave up the plans so long studied and laboured for, and plunged at once into a new path. If he could not dominate two continents, in Europe at any rate he would be master.

How Napoleon withdrew from his enterprise is for us a story very memorable. The United States when its third President, Thomas Jefferson, assumed office in 1801 was still a feeble nation. The Fathers, in 1787, building wiser than they knew, had indeed laid the foundation of what was to become perhaps the mightiest power of the world; in 1801, however, the frame was ill-knit and the step still faltered. Jefferson, thoroughly a man of peace, wise, humane, and of noble moral courage, found at his coming difficulties to cope with greater even than those that confront most Presidents. Among them was the necessity of quieting the South and West, sections that had helped powerfully to elect him, and which were exasperated because their path to the sea, the Mississippi River, was blocked by the Spaniards. When it became known that Louisiana had been retroceded by Spain to France excitement increased, for it was believed that the

## Introduction

more vigorous policy upon which Spain had entered had been inspired by France, which expected soon to be in possession and would certainly embarrass the Union. The air was full of talk of war, and Congress, under Federalist instigation, considered a plan for appropriating $5,000,000, and raising 50,000 men to seize the mouth of the Mississippi before the French could establish themselves there. Jefferson had loved the French, and could not believe in their ill-will to America. That confidence, and his liking for peace, caused him to seek another way of escape than by force of arms. He had sent soon after his inauguration Livingston to Paris to arrange for an amicable settlement of difficulties; and now as the tumult increased he sent besides James Monroe, giving him $2,000,000 in hand, and definitely empowering him to treat for the purchase of New Orleans and the Floridas. Of gaining the great country west of the river he had no thought. It had been in his mind indeed to make some exploration of it, though it was nominally Spanish territory: it was, however, scarcely less a no-man's land than is to-day the Antarctic continent; and the President had talked over with an enterprising young man, his private secretary, Meriwether Lewis, a plan for an expedition into the untracked desert.

Suddenly Jefferson's two negotiators in Paris, Livingston, who had labored in vain in his mission for many months, and Monroe, who had just stepped off his ship, found themselves confronted by an astounding proposition. The policy of the French government had completely changed. The old friend of the negotiators, Barbé Marbois, minister of the Treasury, who had lived in America, came to them from the First Consul demanding that they should buy not simply New Orleans and the river's mouth but the whole vast Louisiana, that the purchase-money should be not $2,000,000 but $15,000,000, and that the bargain should be signed and sealed without delay. How the envoys dared to take the responsibility one can hardly see: in their instructions there was no shadow of authorization for it. Take it, however, they did, yielding to the most imperious of human wills. Napoleon, secure now that Louisiana would not fall to England, strode off to threaten his island enemy across the Channel from Boulogne, and presently after to cripple Austria at Ulm. In due time news reached Jefferson of what had been done. At that moment in his career Jefferson is really a pathetic figure. He believed fully that the constitution gave the President no authority to take any such step; moreover, he and most of the better Americans believed that what had been bought was valueless, and worse. That emigration should pass across the river into the new country would be a positive detriment: in the words of Madison, Secretary of State, it would "dilute the population" disastrously and probably lead to a dissolution of the Union. While the administration and its friends were thus sick at heart, from the opposition came a roar of denunciation and ridicule before which even the stoutest might quail.

But as months went on the confusion settled. However the East might storm, the West and South were well content. The river-mouth was gained, the Great West was open for the freest ranging. The people settled into the belief that the price paid had been none too large. When Congress met in October, 1803, a great majority sustained the Louisiana Purchase; and though for many years a vigorous minority rebelled against it, the country accommodated itself to it until the transaction came to be recognized for what it is, the greatest piece of good fortune that befell the country between the adoption of the constitution and the saving of the Union in '65.

Louisiana was gained: what lay concealed within that vast area which foot of white man had rarely pressed, it became

*Introduction*

imperative to know, and Jefferson proceeded promptly and wisely. He published presently a message imparting to the country such information as could be gleaned from the stories of hunters and rovers, a document laughed at at the time for its extravagance, but which is really remarkable as falling so far short of the truth. More important than this, he commissioned Meriwether Lewis, the young secretary with whom before the purchase he talked over a plan to explore the West, to organize an expedition as soon as possible and break a path through the wilderness. Lewis associated with himself William Clark, a friend a few years older than himself with whom he had served under Wayne. These gathered a force of picked men, two score or so in number, most of whom were mustered into the service of the United States; and on the 14th of May, 1804, they set out from near St. Louis to cross the continent.

In classic mythology no figures are more attractive than the Dioscuri, Castor and Pollux, twin sons of Leda and Jupiter, who sailed in the Argo under Jason for Colchis, and who were the especial patrons of the sailor and the pathfinder. Though full of virile power, their fame has about it little touch of blood. Castor is the tamer of horses, Pollux the boxer and wrestler. Both are skilful with weapons, but seldom shedders of blood. Their function is that of guides and protectors, until, raised to the skies, they become the constellation Gemini. Lewis and Clark are the Dioscuri of American tradition. Sons of Virginia in the same generation they stand forth as twins. Though manful in the highest degree, they are never sanguinary. Bold sailors, matchless with the rifle, tamers of the wild horses of the desert, the result of their striving was purely beneficent. They fought against no enemy but the forces of nature.

Their boat, fifty-five feet long, rowed by twenty-two oars,

decked fore and aft, with an arrangement amidships whereby the bulwarks could in a moment be heightened to ward off an attack, must have resembled the Argo, if that craft ever existed outside of fable. If a Jason must be found for the expedition, it is no great stretch to see him in Jefferson; for although not present in body, his mild, wise, energetic, courageous spirit thoroughly imbued both leaders and men. Pushing into the mouth of the Missouri where the turbid current thrusts aside the purer Mississippi, a great square sail trimmed to catch the easterly breeze aiding their efforts, the twenty-two stout oarsmen bent to their work, the forest walls on either hand echoing the rhythmic stroke. Throughout the summer their task was laborious but scarcely perilous. Thus far the wilderness was sometimes traversed by hunters and trappers from St. Louis, and as they advanced northward they began to encounter Hudson's Bay men. Before winter, however, they had to deal with the fierce Sioux, escaping battle only by marvellous tact and the steadfast courage which always appals wild beasts and wild men. In dealing with the Indians, it may be supposed that Captain Clark was especially skilful. He was a younger brother of George Rogers Clark, the conqueror of the Northwest, the paleface beyond all others held in respect by the wild tribes, and perhaps equalled his elder brother as a manager of savages. Through all their contests they came unscathed themselves and without inflicting harm, the only exception being that Lewis, isolated with three followers among the Blackfeet, was obliged to slay two, to save himself and his party.

The expedition spent its first winter among the Mandans, the interesting tribe afterwards blotted out by small-pox, which was much in advance of Indians in general on the way toward civilization. In the early spring of 1805, abandoning their large boat for which the upper Missouri was too shallow,

*Introduction* xxix.

and embarking in canoes, they pushed on westward into a region absolutely untenanted. No Indians were seen for months, the first appearing as they were penetrating the Rocky Mountains. The conditions of the explorer's life have never been more ideal than those of the expedition during its second summer. About them stretched grassy plains teeming with deer and buffalo, till each stretch of prairie seemed like the sheet in the vision of Peter, let down by the four corners and charged with life of every kind. It is pleasant to read in the record that they shot no creature except as their necessity required it. They had their annoyances from the prickly pear which pierced their moccasins, from the rapids which threatened to overturn the canoes, above all from the "musquetoes." They encountered perils, especially from rattlesnakes and grizzly bears. They bore all lightly, and as the summer drew toward its close, passing the Gates of the Mountains, they reached the spot where the canoes must be forsaken: the great continental divide must in some way be crossed, whence they hoped to descend by some stream on the other side to the Pacific. They had counted upon meeting long before this Indians of the plains and mountains from whom horses might be procured, to bear the burdens which the canoes had hitherto conveyed. But since leaving the Mandans there had been neither sight nor sound of man; and though the two Captains ranged far and wide ahead of their men, spying out from every point that offered a wide view, the land appeared to be swept of inhabitants. The embarrassment was becoming grave.

At this point in the story, had Lewis and Clark really been the Dioscuri, it would have been proper to immolate victims and wet the earth with libations; after which some favoring goddess might have been expected to appear, Venus, Juno, or Minerva, to help the heroes out of their quandary. Help came, and from a feminine hand; afforded by no goddess but by one who deserves to be remembered with honor and gratitude. The only woman who figures in the story of Lewis and Clark is the young squaw Sacajawea, the Bird-woman, of the Snake or Shoshone tribe, who, having been captured in her childhood from her people in the mountains by the Minnetarees, had been discovered by Lewis and Clark among the latter, who dwelt near the Mandans. The Indian girl had grown to womanhood and was the slave and wife of Chaboneau, a French waif, who like many another *voyageur* had sunk into absolute savagery. The Bird-woman had not forgotten the tongue of her people; and the Captains, thinking she might be useful as an interpreter when they reached the mountains, had persuaded her and her husband to join them. Though carrying at her back her pappoose, but a few months old, she bore all that her male companions did, and quickly made her way to their respect by her efficiency and kindly nature. Near the Three Forks of the Missouri the Bird-woman recognized the spot where in childhood she had been taken captive. When Indians were at last encountered, lo, they were her friends! The first squaw they met it happened had been captured at the same time with Sacajawea; together they had endured captivity some years, until the friend, at last escaping, returned to her people. The two women met with affection; and when presently after, at the council, the Bird-woman set out to interpret, lo, it became revealed that the mountain-chief whose words she was rendering was her own brother! The happy beginning was at once followed up. Gifts were bestowed, and a cordial friendship was plighted. The Shoshones furnished guides and horses, and smoothed the way with the tribes that lived still farther west. But for the fortunate chance the expedition might have been blocked then and there.

*Introduction* xxxi.

The Captains had found friends, but not until now did they encounter real hardships. The abundant game of the plains here failed them; in the scarcity they came near starvation, saving their lives only by eating their horses. The trails were passable only with the severest struggle; in the bleak and tangled wilderness even the guides were at fault. But no heart in the command failed. They met the hordes of the uplands with frank good nature in their faces, extending with their hands presents,—flags, medals, scarlet cloth, beads, and rarely failed to make good impression. They speculated sometimes, amused, over what kind of a message it was that finally filtered into the savage mind, as they smoked the pipe of peace and pow-wowed about the councilfire. The speech of Lewis or Clark, done into French for Chaboneau, was set over into Minnetaree for Sacajawea, whose Shoshone forthwith rendered into Ootlashoot, Chopunnish, or whatever the barbarian dialect might be, at length became intelligible to those who sat waiting. But some kind of a message was conveyed: the greatness of Uncle Sam was announced, the good disposition of the mighty ruler Jefferson under whose sway they had come. Somewhat naively the tribes were adjured to remain at peace among themselves and to practice virtues, possible only in a very different stage of society. Mr. Olin D. Wheeler, in his book "Wonderland,"* tells an interesting Indian tradition, as to how the strangers came among them. A Catholic priest, long settled among the Flatheads, gives the story of an old chief, which he got from his father, the Chief Three Eagles. Scouting out one day from his tepees, Three Eagles suddenly descried such a train as he had never seen before, —a company of white men, without blankets, leading packhorses heavily laden, in front of whom rode two chiefs, well

\* Northern Pacific publication for 1900.

in advance and spying out the land on every side. His first thought was that they had been robbed; and fearing that they might rob in return, he hurried back to his people and ordered that the horses should all be brought in. From an ambush Three Eagles and his people kept watch of the strangers, lost in wonder as to whether the apparitions were men or spirits. When at last the Indians were espied the two captains galloped boldly toward them, extending their hands and shouting cheerful greeting. Then came the council at which the pipe-of-peace was smoked.

In good time, thanks to their own energy and also to the good friends they made, the divide was crossed. For such woodmen as they it was a short task to make once more canoes: embarked in these, they passed through rapids and torrents into the Columbia River. In the gorges of the Cascades there were perils to surmount, but neither rock nor cataract had longer terrors for them. First of civilized men they beheld the Multnomah Fall; Mount Hood, Mount Adams, Mount St. Helens, lifted aloft their snows before them. Suddenly there was a pulse in the stream as of mighty forces thrusting in: it was the tidal throb of the near-at-hand sea, and presently they beached their canoes within sight and sound of the great ocean's breakers. There was time before winter fell to provide a shelter. Their task was performed: the continent was crossed. Whether or not they could return was by no means sure, and Lewis and Clark took means to get to the world knowledge of what they had done. If they perished the world should know of their achievement.

With the spring of 1806, they started homeward. As far as the divide the way was plain, but among those peaks and labyrinthine gorges they lost themselves as before. The savages, however, were friendly as ever: the kindred of the

*Introduction* xxxiii.

Bird-woman were faithful as before with guides and horses. Sacajawea herself when the trail was lost seemed to have the instinct of a wild migrating bird, and could orient herself better than it could be done with the compass. They came out safely on the head-streams of the Missouri, and found the powder and food they had buried undisturbed in the *caches*. The party divided, Lewis following the Missouri, while Clark traced the course of the Yellowstone. Now it was that Lewis ran his greatest danger. Caught with three companions by a company of Blackfeet, their horses were stolen, and they were on the point of losing their rifles. Then for the only time war was declared. In self-defense two of the savages were killed, and the Dioscuri themselves never had a wilder ride than the gallop over the plains upon their unbroken steeds, through which they at last reached safety. Where the Yellowstone and Missouri unite the two parties came together again. Through the closing weeks of summer the boats drifted rapidly down, and one day in September, 1806, saluting the flag they had carried so far with a parting volley, the Captains and their men stepped ashore at St. Louis.

Never was success more complete. From first to last all went smoothly, not at all because the dangers and difficulties were small, but because the skill and courage with which they were confronted were consummate. Lewis and Clark were never found wanting, and in all the effort they coöperated without a touch of jealousy. From first to last among the men there was scarcely a trace of insubordination: each worked to his full capacity, yielding to the guidance of the leaders whose natural ascendency they thoroughly recognized. The student of Lewis and Clark learns to respect them all,— the stout sergeants, Pryor, Ordway, and Patrick Gass, the latter of whom in his quaint diary supplements nobly the record of the chiefs;—the blacksmith Shields, York the negro slave, whom the Indians thought great "medicine," the half-breed Drewyer, past-master of woodcraft, the Frenchman Cruzat whose fiddle resounded night after night in the desolate camps while the men danced off their pains and fears. But most of all the lone woman Sacajawea is an object of interest. Her figure in the story of Lewis and Clark is very pathetic and engaging, and in Indian story few characters appear whose desert was greater. A captive and a slave she followed the trail or worked with the men in forcing on the canoes. Her husband Chaboneau soon proved to be inefficient and cowardly; but as dangers and hardships gathered, the heart and head of the squaw showed ever new resources. It is doubtful if the expedition could have pushed its way through without her. It is exasperating to read that at the end, while Chaboneau received $500, his slave-wife got nothing, apparently not even her freedom. Years afterward a traveller who encountered her reports that she was ill and had apparently not long to live. In her manners and dress she followed civilized ways. Evidently her experience with the pale-faces had aroused an aspiration which the poor creature would have followed if she could. When and where she died is unknown and there is no memorial of her.

Since the days of Lewis and Clark many other bold explorers have been busy opening to the day the wilderness through which they first broke a path. Zebulon Montgomery Pike at the same time with them followed the Mississippi as they did the Missouri, reaching Leech Lake, which to this day lies remote. Returning, he plunged westward until he sighted at last the peak which bears his name. Though less fortunate, he was perhaps the equal of the two Captains in skill and manly worth. As the 19th century advanced scientists and soldiers pressed on together to open the country,—

## Introduction

Long, Simpson, Frémont, Stevens, Clarence King, Hayden, and many another. When the locomotive appeared exploration was stimulated, and presently it became plain that the wonderful machine was such a transformer as the world had never before seen. The great New West which the world beholds as having come to pass in the expanse traversed by Lewis and Clark may rightly be called the child of the locomotive. Never before when men have occupied new lands has the occupation been so rapid. At the present moment there remains not a fragment of territory in the vast region pierced by their track which has not been admitted to statehood. Commonwealths shoulder to shoulder, populous, politically complete, socially organized according to advanced standards, are ranged along that old pathway. It is not strange that some feel we have gone quite too rapidly; and our grandchildren may wish their forbears had been slower in the exploitation of the resources of the fine domain. It is not in place to discuss here this question, or to deal with the other problems which beset and threaten the present moment.

What Lewis and Clark laid open is a world upon which nature has lavished her bounties. It is already occupied by many million of English-speaking people,—a race formed by the assimilation into a strong Anglo-Saxon stock of elements from a number of the better breeds of men. The appliances of the highest civilization abound among them. The principles of the best polity ever evolved in the progress of the human race are established for their government. The present fruition is scarcely calculable: the hope for the future is boundless.

JAMES K. HOSMER.

*Minneapolis Public Library,*
*August 21, 1902.*

# PREFACE
## TO THE EDITION OF 1814.

IN presenting these volumes to the public, the editor owes equally to himself and to others to state the circumstances which have preceded the publication, and to explain his own share in compiling them.

It was the original design of Captain Lewis to have been himself the editor of his own travels, and he was on his way towards Philadelphia for that purpose when his sudden death frustrated these intentions. After a considerable and unavoidable delay, the papers connected with the expedition were deposited with another gentleman, who, in order to render the lapse of time as little injurious as possible, proceeded immediately to collect and investigate all the materials within his reach.

Of the incidents of each day during the expedition a minute journal was kept by Captain Lewis or Captain Clark, and sometimes by both, which was afterwards revised and enlarged at the different periods of leisure which occurred on the route. These were carefully perused in conjunction with Captain Clark himself, who was able from his own recollection of the journey, as well as from a constant residence in Louisiana since his return, to supply a great mass of explanations, and much additional information with regard to part of the route which has been more recently explored. Besides these, recourse was had to the manuscript journals kept by two of the serjeants, one of which, the least minute and valuable, has already been published. That nothing might be wanting to the accuracy of these details, a very intelligent and active member of the party, Mr. George Shannon, was sent to contribute whatever his memory might add to this accumulated fund of information.

From these copious materials the narrative was sketched nearly in its present form, when other pursuits diverted the attention of the writer, and compelled him to transfer his manuscript, in its unfinished state, with all the documents connected with it, to the present editor, to prepare them for the press and superintend the publication. That he may not seem to arrogate anything from the exertions of others, he should therefore state that, although the whole work was thus submitted to his entire discretion, he found but little to change, and that his labour has been principally confined to revising the manuscript, comparing it with the original papers, and inserting such additional matter as appears to have been intentionally deferred by the writer till the period of a more mature revisal. These circumstances, which would otherwise be indifferent to the public, are mentioned merely to account for imperfections, which are in some degree inseparable from any book of travels not written by the traveller. In a work of pure description, indeed, like the present, where the incidents themselves are the sole objects of attraction, the part of an editor is necessarily subordinate, nor can his humble pretensions aspire beyond the merit of rigid adherence to facts as they are stated to him. This has been very diligently attempted, and for this, in its full extent, the editor deems himself responsible.

The present volumes, it will be perceived, comprise only the narrative of the journey. Those parts of the work which relate to the various objects of natural history, observed or collected during the journey, as well as the alphabets of the

## Preface to the Edition of 1814

Indian languages, are in the hands of Professor Barton, and will, it is understood, shortly appear.

To give still further interest to the work, the editor addressed a letter to Mr. Jefferson, requesting some authentic memoirs of Captain Lewis. For the very curious and valuable information contained in his answer, the public, as well as the editor himself, owe great obligations to the politeness and knowledge of that distinguished gentleman.

PAUL ALLEN.

PHILADELPHIA, January 1, 1814.

# LIFE OF CAPTAIN LEWIS

[*Contributed to the Edition of 1814 by Thomas Jefferson.*]

MONTICELLO, August 18, 1813.

SIR: In compliance with the request conveyed in your letter of May 25, I have endeavoured to obtain, from the relations and friends of the late Governor Lewis, information of such incidents of his life as might be not unacceptable to those who may read the narrative of his western discoveries. The ordinary occurrences of a private life, and those also while acting in a subordinate sphere in the army, in a time of peace, are not deemed sufficiently interesting to occupy the public attention; but a general account of his parentage, with such smaller incidents as marked his early character, are briefly noted; and to these are added, as being peculiarly within my own knowledge, whatever related to the public mission, of which an account is now to be published. The result of my inquiries and recollections shall now be offered, to be enlarged or abridged as you may think best; or otherwise to be used with the materials you may have collected from other sources.

Meriwether Lewis, late Governor of Louisiana, was born on the eighteenth of August, 1774, near the town of Charlottesville, in the county of Albemarle, in Virginia, of one of the distinguished families of that state. John Lewis, one of his father's uncles, was a member of the king's council, before the Revolution. Another of them, Fielding Lewis, married a sister of General Washington. His father, William Lewis, was the youngest of five sons of Colonel Robert Lewis, of

Albemarle, the fourth of whom, Charles, was one of the early patriots who stepped forward in the commencement of the Revolution, and commanded one of the regiments first raised in Virginia and placed on continental establishment. Happily situated at home, with a wife and young family, and a fortune placing him at ease, he left all to aid in the liberation of his country from foreign usurpations, then first unmasking their ultimate end and aim. His good sense, integrity, bravery, enterprise, and remarkable bodily powers, marked him as an officer of great promise; but he unfortunately died early in the Revolution. Nicholas Lewis, the second of his father's brothers, commanded a regiment of militia in the successful expedition of 1776 against the Cherokee Indians, who, seduced by the agents of the British government to take up the hatchet against us, had committed great havoc on our southern frontier by murdering and scalping helpless women and children, according to their cruel and cowardly principles of warfare. The chastisement they then received closed the history of their wars, and prepared them for receiving the elements of civilization, which, zealously inculcated by the present government of the United States, have rendered them an industrious, peaceable, and happy people. This member of the family of Lewises, whose bravery was so usefully proved on this occasion, was endeared to all who knew him by his inflexible probity, courteous disposition, benevolent heart, and engaging modesty and manners. He was the umpire of all the private differences of his county—selected always by both parties. He was also the guardian of Meriwether Lewis, of whom we are now to speak, and who had lost his father at an early age. He continued some years under the fostering care of a tender mother, of the respectable family of Meriwethers, of the same county; and was remarkable even in infancy for enterprise, boldness, and

## *Life of Captain Lewis*

discretion. When only eight years of age he habitually went out, in the dead of night, alone with his dogs, into the forest to hunt the raccoon and opossum, which, seeking their food in the night, can then only be taken. In this exercise, no season or circumstance could obstruct his purpose—plunging through the winter's snows and frozen streams in pursuit of his object. At thirteen he was put to the Latin school, and continued at that until eighteen, when he returned to his mother, and entered on the cares of his farm; having, as well as a younger brother, been left by his father with a competency for all the correct and comfortable purposes of temperate life. His talent for observation, which had led him to an accurate knowledge of the plants and animals of his own country, would have distinguished him as a farmer; but at the age of twenty, yielding to the ardour of youth and a passion for more dazzling pursuits, he engaged as a volunteer in the body of militia which were called out by General Washington, on occasion of the discontents produced by the excise taxes in the western parts of the United States; and from that situation he was removed to the regular service as a lieutenant in the line. At twenty-three he was promoted to a captaincy; and, always attracting the first attention where punctuality and fidelity were requisite, he was appointed paymaster to his regiment. About this time a circumstance occurred which, leading to the transaction which is the subject of this book, will justify a recurrence to its original idea. While I resided in Paris, John Ledyard, of Connecticut, arrived there, well known in the United States for energy of body and mind. He had accompanied Captain Cook on his voyage to the Pacific ocean and distinguished himself on that voyage by his intrepidity. Being of a roaming disposition, he was now panting for some new enterprise. His immediate object at Paris was to engage a mercantile company in the fur-trade of the western coast of America, in which, however, he failed. I then proposed to him to go by land to Kamschatka, cross in some of the Russian vessels to Nootka Sound, fall down into the latitude of the Missouri, and penetrate to, and through, that to the United States. He eagerly seized the idea, and only asked to be assured of the permission of the Russian government. I interested, in obtaining that, M. de Simoulin, minister plenipotentiary of the empress at Paris, but more especially the Baron de Grimm, minister plenipotentiary of Saxe-Gotha, her more special agent and correspondent there in matters not immediately diplomatic. Her permission was obtained, and an assurance of protection while the course of the voyage should be through her territories. Ledyard set out from Paris, and arrived at St. Petersburgh after the empress had left that place to pass the winter, I think, at Moscow. His finances not permitting him to make unnecessary stay at St. Petersburgh, he left it with a passport from one of the ministers; and at two hundred miles from Kamschatka, was obliged to take up his winter quarters. He was preparing, in the spring, to resume his journey, when he was arrested by an officer of the empress, who by this time had changed her mind and forbidden his proceeding. He was put into a close carriage, and conveyed day and night, without ever stopping, till they reached Poland, where he was set down and left to himself. The fatigue of this journey broke down his constitution, and when he returned to Paris his bodily strength was much impaired. His mind, however, remained firm, and he after this undertook the journey to Egypt. I received a letter from him, full of sanguine hopes, dated at Cairo, the fifteenth of November, 1788, the day before he was to set out for the head of the Nile; on which day, however, he ended his career and life,—and thus failed the first attempt to explore the western part of our northern continent.

In 1792, I proposed to the American Philosophical Society that we should set on foot a subscription to engage some competent person to explore that region in the opposite direction; that is, by ascending the Missouri, crossing the Stony mountains, and descending the nearest river to the Pacific. Captain Lewis, being then stationed at Charlottesville, on the recruiting service, warmly solicited me to obtain for him the execution of that object. I told him it was proposed that the person engaged should be attended by a single companion only, to avoid exciting alarm among the Indians. This did not deter him; but Mr. Andre Michaux, a professed botanist, author of the "Flora Boreali-Americana" and of the "Histoire des Chesnes d'Amerique," offering his services, they were accepted. He received his instructions, and when he had reached Kentucky in the prosecution of his journey he was overtaken by an order from the minister of France, then at Philadelphia, to relinquish the expedition, and to pursue elsewhere the botanical inquiries on which he was employed by that government,—and thus failed the second attempt for exploring that region.

In 1803, the act for establishing trading houses with the Indian tribes being about to expire, some modifications of it were recommended to Congress by a confidential message of January 18, and an extension of its views to the Indians on the Missouri. In order to prepare the way, the message proposed the sending an exploring party to trace the Missouri to its source, to cross the Highlands, and follow the best water communication which offered itself from thence to the Pacific ocean. Congress approved the proposition, and voted a sum of money for carrying it into execution. Captain Lewis, who had then been near two years with me as private secretary, immediately renewed his solicitations to have the direction of the party. I had now had opportunities of knowing him intimately. Of courage undaunted; possessing a firmness and perseverance of purpose which nothing but impossibilities could divert from its direction; careful as a father of those committed to his charge, yet steady in the maintenance of order and discipline; intimate with the Indian character, customs, and principles; habituated to the hunting life; guarded, by exact observation of the vegetables and animals of his own country, against losing time in the description of objects already possessed; honest, disinterested, liberal, of sound understanding, and a fidelity to truth so scrupulous that whatever he should report would be as certain as if seen by ourselves; with all these qualifications, as if selected and implanted by nature in one body for this express purpose, I could have no hesitation in confiding the enterprise to him. To fill up the measure desired, he wanted nothing but a greater familiarity with the technical language of the natural sciences, and readiness in the astronomical observations necessary for the geography of his route. To acquire these he repaired immediately to Philadelphia, and placed himself under the tutorage of the distinguished professors of that place, who, with a zeal and emulation enkindled by an ardent devotion to science, communicated to him freely the information requisite for the purposes of the journey. While attending, too, at Lancaster, the fabrication of the arms with which he chose that his men should be provided, he had the benefit of daily communication with Mr. Andrew Ellicot, whose experience in astronomical observation, and practice of it in the woods, enabled him to apprise Captain Lewis of the wants and difficulties he would encounter, and of the substitutes and resources offered by a woodland and uninhabited country.

Deeming it necessary he should have some person with him of known competence to the direction of the enterprise, in the event of accident to himself, he proposed William Clark, brother of General George Rogers Clark, who was approved, and, with that view, received a commission of captain.

In April, 1803, a draught of his instructions was sent to Captain Lewis, and on the 20th of June they were signed in the following form:

"To MERIWETHER LEWIS, Esquire, captain of the first regiment of infantry of the United States of America:

"Your situation as secretary of the President of the United States has made you acquainted with the objects of my confidential message of January 18, 1803, to the legislature; you have seen the act they passed, which, though expressed in general terms, was meant to sanction those objects, and you are appointed to carry them into execution.

"Instruments for ascertaining, by celestial observations, the geography of the country through which you will pass have been already provided. Light articles for barter and presents among the Indians, arms for your attendants, say for from ten to twelve men, boats, tents, and other travelling apparatus, with ammunition, medicine, surgical instruments, and provisions, you will have prepared, with such aids as the secretary at war can yield in his department; and from him also you will receive authority to engage among our troops, by voluntary agreement, the number of attendants above mentioned; over whom you, as their commanding officer, are invested with all the powers the laws give in such a case.

"As your movements, while within the limits of the United States, will be better directed by occasional communications, adapted to circumstances as they arise, they will not be noticed here. What follows will respect your proceedings after your departure from the United States.

"Your mission has been communicated to the ministers here from France, Spain, and Great Britain, and through them to their governments; and such assurances given them, as to its objects, as we trust will satisfy them. The country of Louisiana having been ceded by Spain to France, the passport you have from the minister of France, the representative of the present sovereign of the country, will be a protection with all its subjects; and that from the minister of England will entitle you to the friendly aid of any traders of that allegiance with whom you may happen to meet.

"The object of your mission is to explore the Missouri river, and such principal streams of it as, by its course and communication with the waters of the Pacific ocean, whether the Columbia, Oregon, Colorado, or any other river, may offer the most direct and practicable water communication across the continent, for the purposes of commerce.

"Beginning at the mouth of the Missouri, you will take observations of latitude and longitude, at all remarkable points on the river, and especially at the mouths of rivers, at rapids, at islands, and other places and objects distinguished by such natural marks and characters, of a durable kind, as that they may with certainty be recognised hereafter. The courses of the river between these points of observation may be supplied by the compass, the log-line, and by time, corrected by the observations themselves. The variations of the needle, too, in different places, should be noticed.

"The interesting points of the portage between the heads of the Missouri, and of the water offering the best communication with the Pacific ocean, should also be fixed by observation; and the course of that water to the ocean, in the same manner as that of the Missouri.

"Your observations are to be taken with great pains and accuracy; to be entered distinctly and intelligibly for others as well as yourself; to comprehend all the elements necessary, with the aid of the usual tables, to fix the latitude and longitude of the places at which they were taken; and are to be rendered to the war-office, for the purpose of having the calculations made concurrently by proper persons within the United States. Several copies of these, as well as of your other notes, should be made at leisure times, and put into the care of the most trustworthy of your attendants to guard, by multiplying them, against the accidental losses to which they will be exposed. A further guard would be, that one of these copies be on the cuticular membranes of the paper-birch, as less liable to injury from damp than common paper.

"The commerce which may be carried on with the people inhabiting the line you will pursue renders a knowledge of those people important. You will therefore endeavour to make yourself acquainted, as far as a diligent pursuit of your journey shall admit, with the names of the nations and their numbers;

"The extent and limits of their possessions;

"Their relations with other tribes or nations;

"Their language, traditions, monuments;

"Their ordinary occupations in agriculture, fishing, hunting, war, arts, and the implements for these;

"Their food, clothing, and domestic accommodations;

"The diseases prevalent among them, and the remedies they use;

"Moral and physical circumstances which distinguish them from the tribes we know;

"Peculiarities in their laws, customs, and dispositions;

"And articles of commerce they may need or furnish, and to what extent.

"And, considering the interest which every nation has in extending and strengthening the authority of reason and justice among the people around them, it will be useful to acquire what knowledge you can of the state of morality, religion, and information among them; as it may better enable those who may endeavour to civilize and instruct them to adapt their measures to the existing notions and practices of those on whom they are to operate.

"Other objects worthy of notice will be—

"The soil and face of the country, its growth and vegetable productions, especially those not of the United States;

"The animals of the country generally, and especially those not known in the United States;

"The remains and accounts of any which may be deemed rare or extinct;

"The mineral productions of every kind, but more particularly metals, lime-stone, pit-coal, and saltpetre; salines and mineral waters, noting the temperature of the last, and such circumstances as may indicate their character;

"Volcanic appearances;

"Climate, as characterized by the thermometer, by the proportion of rainy, cloudy, and clear days; by lightning, hail, snow, ice; by the access and recess of frost; by the winds prevailing at different seasons; the dates at which particular plants put forth, or lose their flower or leaf; times of appearance of particular birds, reptiles, or insects.

## 1.  *Life of Captain Lewis*

"Although your route will be along the channel of the Missouri, yet you will endeavour to inform yourself, by inquiry, of the character and extent of the country watered by its branches, and especially on its southern side. The North river, or Rio Bravo, which runs into the gulf of Mexico, and the North River, or Rio Colorado, which runs into the gulf of California, are understood to be the principal streams heading opposite to the waters of the Missouri, and running southwardly. Whether the dividing grounds between the Missouri and them are mountains or flat lands, what are their distance from the Missouri, the character of the intermediate country, and the people inhabiting it, are worthy of particular inquiry. The northern waters of the Missouri are less to be inquired after, because they have been ascertained to a considerable degree, and are still in a course of ascertainment by English traders and travellers; but if you can learn anything certain of the most northern source of the Missisipi, and of its position relatively to the Lake of the Woods, it will be interesting to us. Some account, too, of the path of the Canadian traders from the Missisipi, at the mouth of the Ouisconsing* to where it strikes the Missouri, and of the soil and rivers in its course, is desirable.

"In all your intercourse with the natives, treat them in the most friendly and conciliatory manner which their own conduct will admit; allay all jealousies as to the object of your journey; satisfy them of its innocence; make them acquainted with the position, extent, character, peaceable and commercial dispositions of the United States; of our wish to be neighbourly, friendly, and useful to them, and of our dispositions to a commercial intercourse with them; confer with them on the points most convenient as mutual emporiums, and the articles of most desirable interchange for them and us. If a few of their influential chiefs, within practicable distance, wish to visit us, arrange such a visit with them, and furnish them with authority to call on our officers on their entering the United States, to have them conveyed to this place at the public expense. If any of them should wish to have some of their young people brought up with us, and taught such arts as may be useful to them, we will receive, instruct, and take care of them. Such a mission, whether of influential chiefs, or of young people, would give some security to your own party. Carry with you some matter of the kine-pox; inform those of them with whom you may be of its efficacy as a preservative from the small-

* Wisconsin

pox, and instruct and encourage them in the use of it. This may be especially done wherever you winter.

"As it is impossible for us to foresee in what manner you will be received by those people, whether with hospitality or hostility, so is it impossible to prescribe the exact degree of perseverance with which you are to pursue your journey. We value too much the lives of citizens to offer them to probable destruction. Your numbers will be sufficient to secure you against the unauthorized opposition of individuals, or of small parties; but if a superior force, authorized, or not authorized, by a nation, should be arrayed against your further passage, and inflexibly determined to arrest it, you must decline its further pursuit and return. In the loss of yourselves we should lose also the information you will have acquired. By returning safely with that, you may enable us to renew the essay with better calculated means. To your own discretion, therefore, must be left the degree of danger you may risk, and the point at which you should decline, only saying, we wish you to err on the side of your safety, and to bring back your party safe, even if it be with less information.

"As far up the Missouri as the white settlements extend, an intercourse will probably be found to exist between them and the Spanish posts of St. Louis opposite Cahokia, or St. Genevieve opposite Kaskaskia. From still further up the river the traders may furnish a conveyance for letters. Beyond that you may perhaps be able to engage Indians to bring letters for the government to Cahokia, or Kaskaskia, on promising that they shall there receive such special compensation as you shall have stipulated with them. Avail yourself of these means to communicate to us, at seasonable intervals, a copy of your journal, notes, and observations of every kind, putting into cypher whatever might do injury if betrayed.

"Should you reach the Pacific ocean, inform yourself of the circumstances which may decide whether the furs of those parts may not be collected as advantageously at the head of the Missouri (convenient as is supposed to the waters of the Colorado and Oregon or Columbia) as at Nootka sound, or any other point of that coast; and that trade be consequently conducted through the Missouri and United States more beneficially than by the circumnavigation now practised.

"On your arrival on that coast, endeavour to learn if there be any port within your reach frequented by the sea vessels of any nation, and to send two of your trusty people back by sea, in such way as shall appear practicable, with a copy of your notes; and should you be of opinion that the return of your party by the way they went will be imminently dangerous, then ship the whole, and return by sea, by the way either of Cape Horn, or the Cape of Good Hope, as you shall be able. As you will be without money, clothes, or provisions, you must endeavour to use the credit of the United States to obtain them; for which purpose open letters of credit shall be furnished you, authorizing you to draw on the executive of the United States, or any of its officers, in any part of the world, on which draughts can be disposed of, and to apply with our recommendations to the consuls, agents, merchants, or citizens of any nation with which we have intercourse, assuring them, in our name, that any aids they may furnish you shall be honourably repaid, and on demand. Our consuls, Thomas Hewes, at Batavia, in Java, William Buchanan, in the Isles of France and Bourbon, and John Elmslie, at the Cape of Good Hope, will be able to supply your necessities, by draughts on us.

"Should you find it safe to return by the way you go, after sending two of your party round by sea, or with your whole party, if no conveyance by sea can be found, do so; making such observations on your return as may serve to supply, correct, or confirm those made on your outward journey.

"On re-entering the United States and reaching a place of safety, discharge any of your attendants who may desire and deserve it, procuring for them immediate payment of all arrears of pay and clothing which may have incurred since their departure, and assure them that they shall be recommended to the liberality of the legislature for the grant of a soldier's portion of land each, as proposed in my message to congress, and repair yourself, with your papers, to the seat of government.

"To provide, on the accident of your death, against anarchy, dispersion, and the consequent danger to your party, and total failure of the enterprise, you are hereby authorized, by any instrument signed and in your own hand, to name the person among them who shall succeed to the command on your decease, and by like instruments to change the nomination, from time to time, as further experience of the characters accompanying you shall point out superior fitness; and all the powers and authorities given to yourself are, in the event of your death, transferred to, and vested in the successor so named, with further power to him and his successors, in like manner to name each his successor, who, on the death of his predecessor, shall be invested with all the powers and authorities given to yourself.

Given under my hand, at the city of Washington, this twentieth day of June, 1803.

"THOMAS JEFFERSON,
*"President of the United States of America."*

While these things were going on here, the country of Louisiana, lately ceded by Spain to France, had been the subject of negociation at Paris between us and this last power; and had actually been transferred to us by treaties executed at Paris on the thirtieth of April. This information, received about the first day of July, increased infinitely the interest we felt in the expedition, and lessened the apprehensions of interruption from other powers. Everything in this quarter being now prepared, Captain Lewis left Washington on the fifth of July, 1803, and proceeded to Pittsburg, where other articles had been ordered to be provided for him. The men, too, were to be selected from the military stations on the Ohio. Delays of preparation, difficulties of navigation down the Ohio, and other untoward obstructions, retarded his arrival at Cahokia until the season was so far advanced as to render it prudent to suspend his entering the Missouri before the ice should break up in the succeeding spring.

From this time his journal, now published, will give the history of his journey to and from the Pacific ocean, until his return to St. Louis on the twenty-third of September, 1806. Never did a similar event excite more joy through the United States. The humblest of its citizens had taken a lively interest in the issue of this journey, and looked forward with impatience for the information it would furnish. Their anxieties, too, for the safety of the corps had been kept in a state of excitement by lugubrious rumours, circulated from time to time on uncertain authorities, and uncontradicted by letters, or other direct information, from the time they had left the Mandan towns, on their ascent up the river in April of the preceding year, 1805, until their actual return to St. Louis.

It was the middle of February, 1807, before Captain Lewis, with his companion Captain Clark, reached the city of Washington, where Congress was then in session. That body granted to the two chiefs and their followers the donation of lands which they had been encouraged to expect in reward of their toil and dangers. Captain Lewis was soon after appointed Governor of Louisiana, and Captain Clark a general of its militia, and agent of the United States for Indian affairs in that department.

A considerable time intervened before the Governor's arrival at St. Louis. He found the territory distracted by feuds and contentions among the officers of the government, and the people themselves divided by these into factions and parties. He determined at once to take no side with either; but to use every endeavour to conciliate and harmonize them. The even-handed justice he administered to all soon established a respect for his person and authority; and perseverance and time wore down animosities, and reunited the citizens again into one family.

Governor Lewis had, from early life, been subject to hypochondriac affections. It was a constitutional disposition in all the nearer branches of the family of his name, and was more immediately inherited by him from his father. They had not, however, been so strong as to give uneasiness to his family. While he lived with me in Washington I observed at times sensible depressions of mind; but knowing their constitutional source, I estimated their course by what I had seen in the family. During his western expedition, the constant exertion which that required of all the faculties of body

*Life of Captain Lewis*                              lv.

and mind, suspended these distressing affections; but after his establishment at St. Louis in sedentary occupations, they returned upon him with redoubled vigour, and began seriously to alarm his friends. He was in a paroxysm of one of these, when his affairs rendered it necessary for him to go to Washington. He proceeded to the Chickasaw Bluffs, where he arrived on the sixteenth of September, 1809, with a view of continuing his journey thence by water. Mr. Neely, agent of the United States with the Chickasaw Indians, arriving there two days after, found him extremely indisposed, and betraying at times some symptoms of a derangement of mind. The rumours of a war with England, and apprehensions that he might lose the papers he was bringing on, among which were the vouchers of his public accounts and the journals and papers of his western expedition, induced him here to change his mind, and to take his course by land through the Chickasaw country. Although he appeared somewhat relieved, Mr. Neely kindly determined to accompany and watch over him. Unfortunately, at their encampment, after having passed the Tennessee one day's journey, they lost two horses, which obliging Mr. Neely to halt for their recovery, the Governor proceeded, under a promise to wait for him at the house of the first white inhabitant on his road. He stopped at the house of a Mr. Grinder, who not being at home, his wife, alarmed at the symptoms of derangement she discovered, gave him up the house and retired to rest herself in an out-house, the Governor's and Neely's servants lodging in another. About three o'clock in the night he did the deed which plunged his friends into affliction, and deprived his country of one of her most valued citizens, whose valour and intelligence would have been now employed in avenging the wrongs of his country, and in emulating by land the splendid deeds which have honoured her arms on the ocean. It lost, too, to the nation the benefit of receiving from his own hand the narrative now offered them of his sufferings and successes, in endeavouring to extend for them the boundaries of science, and to present to their knowledge that vast and fertile country, which their sons are destined to fill with arts, with science, with freedom, and happiness.

To this melancholy close of the life of one whom posterity will declare not to have lived in vain, I have only to add that all the facts I have stated are either known to myself, or communicated by his family or others, for whose truth I have no hesitation to make myself responsible; and I conclude with tendering you the assurances of my respect and consideration.

TH. JEFFERSON.

Mr. PAUL ALLEN, Philadelphia.

# THE EXPEDITION

# THE EXPEDITION
OF
# CAPTAINS LEWIS AND CLARK
## 1804–5–6.

## CHAPTER I.

The party set out on the expedition and pass Wood river—Description of the town of St. Charles—Osage Woman river—Gasconade and Osage rivers described—Character of the Osage Indians—Curious traditionary account of their origin—The party proceed, and pass the Mine river—The two Charitons—The Kanzas, Nodawa, Newahaw, Neeshnabatona, Little Nemahar, each of which are particularly described—They encamp at the mouth of the river Platte—A particular description of the surrounding country—The various creeks, bays, islands, prairies, etc., given in the course of the route.

ON the acquisition of Louisiana, in the year 1803, the attention of the government of the United States was early directed towards exploring and improving the new territory. Accordingly, in the summer of the same year, an expedition was planned by the President for the purpose of discovering the courses and sources of the Missouri, and the most convenient water communication thence to the Pacific ocean. His private secretary, Captain Meriwether Lewis, and Captain William Clark, both officers of the army of the United States, were associated in the command of this enterprise. After receiving the requisite instructions, Captain Lewis left the seat of government, and being joined by Captain Clark at Louisville, in Kentucky, proceeded to St. Louis, where they arrived in the month of December. Their original intention was to pass the winter at La Charrette, the highest settlement on the Missouri. But the Spanish commandant of the province, not having received an official account of its transfer to the United States, was obliged by the general policy of his government to prevent strangers from passing through the Spanish territory. They therefore encamped at the mouth of Wood river, on the eastern side of the Mississippi, out of his jurisdiction, where they passed the winter in disciplining the men and making the necessary preparations for setting out early in the Spring, before which the cession was officially announced. The party consisted of nine young men from Kentucky, fourteen soldiers of the United States army who volunteered their services, two French watermen, an interpreter and hunter, and a black servant belonging to Captain Clark. All these, except the last, were enlisted to serve as privates during the expedition, and three sergeants appointed from amongst them by the captains. In addition to these were engaged a corporal and six soldiers, and nine watermen, to accompany the expedition as far as the Mandan nation, in order to assist in carrying the stores, or repelling an attack which was most to be apprehended between Wood river and that tribe. The necessary stores were subdivided into seven bales, and one box containing a small portion of each article in case of accident. They consisted of a great variety of clothing, working utensils, locks, flints, powder, ball, and articles of the greatest use. To these were added fourteen bales and one box of Indian presents, distributed in the same manner, and composed of richly-laced coats and other articles of dress, medals, flags, knives, and tomahawks for the chiefs, ornaments of different kinds, particularly beads, looking-glasses, handkerchiefs, paints, and generally such articles as were deemed best calculated for the taste of the Indians. The party was to embark on board of three boats;

the first was a keel boat fifty-five feet long, drawing three feet water, carrying one large square-sail and twenty-two oars. A deck of ten feet in the bow and stern formed a forecastle and cabin, while the middle was covered by lockers, which might be raised so as to form a breastwork in case of attack. This was accompanied by two periogues or open boats, one of six and the other of seven oars. Two horses were at the same time to be led along the banks of the river for the purpose of bringing home game, or hunting in case of scarcity.

Of the proceedings of this expedition, the following is a succinct and circumstantial narrative.

All the preparations being completed, we left our encampment on Monday, May 14, 1804. This spot is at the mouth of Wood river, a small stream which empties itself into the Mississippi, opposite to the entrance of the Missouri. It is situated in latitude 38° 55' $19\frac{6}{10}$" north, and longitude from Greenwich, 89° 57' 45". On both sides of the Mississippi the land for two or three miles is rich and level, but gradually swells into a high, pleasant country, with less timber on the western than on the eastern side, but all susceptible of cultivation. The point which separates the two rivers on the north extends for fifteen or twenty miles, the greater part of which is an open level plain, in which the people of the neighbourhood cultivate what little grain they raise. Not being able to set sail before four o'clock P. M., we did not make more than four miles, and encamped on the first island opposite a small creek called Cold Water.

May 15. The rain, which had continued yesterday and last night, ceased this morning. We then proceeded, and after passing two small islands about ten miles farther, stopped for the night at Piper's landing, opposite another island. The water is here very rapid and the banks falling in. We found that our boat was too heavily laden in the stern, in consequence of which she ran on logs three times to-day. It became necessary to throw the greatest weight on the bow of the boat, a precaution very necessary in ascending both the Missouri and Mississippi rivers, in the beds of which there lie great quantities of concealed timber.

The next morning we set sail at five o'clock. At the distance of a few miles we passed a remarkable large coal hill on the north side, called by the French La Charbonniere, and arrived at the town of St. Charles. Here we remained a few days.

St. Charles is a small town on the north bank of the Missouri, about twenty-one miles from its confluence with the Mississippi. It is situated in a narrow plain, sufficiently high to protect it from the annual risings of the river in the month of June, and at the foot of a range of small hills, which have occasioned its being called Petite Coté, a name by which it is more known to the French than by that of St. Charles. One principal street, about a mile in length and running parallel with the river, divides the town, which is composed of nearly one hundred small wooden houses, besides a chapel. The inhabitants, about four hundred and fifty in number, are chiefly descendants from the French of Canada; and, in their manners, they unite all the careless gayety and the amiable hospitality of the best times of France; yet, like most of their countrymen in America, they are but ill qualified for the rude life of a frontier; not that they are without talent, for they possess much natural genius and vivacity; nor that they are destitute of enterprise, for their hunting excursions are long, laborious, and hazardous; but their exertions are all desultory; their industry is without system, and without perseverance. The surrounding country, therefore, though rich, is not, in general, well cultivated; the inhabitants chiefly subsisting by hunting and trade with the Indians, and confine their culture to gardening, in which they excel.

Being joined by Captain Lewis, who had been detained by business at St. Louis, we again set sail on Monday, May 21, in the afternoon, but were prevented by wind and rain from going more than about three miles, when we encamped on the upper point of an island, nearly opposite a creek which falls in on the south side.

On the 22d we made about eighteen miles, passing several small farms on the bank of the river, a number of islands, and a large creek on the south side, called Bonhomme, or Goodman's river. A small number of emigrants from the United States have settled on the sides of this creek, which are very fertile. We also passed some high lands, and encamped on the north side, near a small creek. Here we met with a camp of Kickapoo Indians who had left us at St. Charles, with a promise of procuring us some provisions by the time we overtook them. They now made us a present of four deer, and we gave them in return two quarts of whiskey. This tribe reside on the heads of the Kaskaskia and Illinois river, on the other side of the Mississippi, but occasionally hunt on the Missouri.

May 23. Two miles from our camp of last night, we reached a river emptying itself on the north side, called Osage Woman river. It is about thirty yards wide, and has now a settlement of thirty or forty families from the United States. About a mile and a half beyond this is a large cave, on the south side, at the foot of cliffs nearly three hundred feet high, overhanging the water, which becomes very swift at this place. The cave is one hundred and twenty feet wide, forty feet deep, and twenty high; it is known by the name of the Tavern, among the traders, who have written their names on the rock, and painted some images which command the homage of the

Indians and French. About a mile farther we passed a small creek called Tavern creek, and encamped on the south side of the river, having gone nine miles.

Early the next morning we ascended a very difficult rapid, called the Devil's Race Ground, where the current sets for half a mile against some projecting rocks on the south side. We were less fortunate in attempting a second place of equal difficulty. Passing near the southern shore, the bank fell in so fast as to oblige us to cross the river instantly, between the northern side and a sandbar which is constantly moving and banking with the violence of the current. The boat struck on it, and would have upset immediately if the men had not jumped into the water and held her till the sand washed from under her. We encamped on the south side, having ascended ten miles, and the next day, May 25, passed on the south side the mouth of Wood river, on the north two small creeks and several islands, and stopped for the night at the entrance of a creek on the north side, called by the French La Charrette, ten miles from our last encampment, and a little above a small village of the same name. It consists of seven small houses, and as many poor families who have fixed themselves here for the convenience of trade, and form the last establishment of whites on the Missouri. It rained last night, yet we found this morning that the river had fallen several inches.

May 26. The wind being favourable we made eighteen miles to-day. We passed in the morning several islands, the largest of which is Buffaloe island, separated from the southern side by a small channel which receives the waters of Buffaloe creek. On the same side is Shepherd's creek, a little beyond which we encamped on the northern side. The next day we sailed along a large island called Otter island, on the northern side, extending nearly ten miles in length, nar-

# UP THE MISSOURI

row but high in its situation, and one of the most fertile in the whole river. Between it and the northern shore, three small creeks, one of which has the same name with the island, empty themselves. On the southern shore is a creek twenty yards wide, called Ash creek. In the course of the day we met two canoes loaded with furs, which had been two months on their route from the Mahar* nation, residing more than seven hundred miles up the river,—one large raft from the Pawnees on the river Platte, and three others from the Grand Osage river. At the distance of fifteen miles we encamped on a willow island, at the entrance of the river Gasconade. This river falls into the Missouri from the south, one hundred miles from the Mississippi. Its length is about one hundred and fifty miles in a course generally northeast through a hilly country. On its banks are a number of saltpetre caves, and it is believed some mines of lead are in the vicinity. Its width at the mouth is one hundred and fifty-seven yards, and its depth nineteen feet.

Here we halted for the purpose of hunting and drying our provisions, and making the necessary celestial observations. This being completed, we set sail on the 29th at four o'clock, and at four miles distance encamped on the south side, above a small creek, called Deer creek. The next day, 30th, we set out early, and at two miles distant reached a large cave, on the north, called Montbrun's tavern, after a French trader of that name, just above a creek called after the same person. Beyond this is a large island, and, at the distance of four miles, Rush creek coming in from the south; at eleven, Big-muddy river on the north, about fifty yards wide; three miles farther is Little-muddy river on the same side, opposite to which we encamped at the mouth of Grindstone creek. The rain which began last night continued through the day, accompanied with high wind and some hail. The river has been

*\* omaha*

rising fast for two days, and the country around appears full of water. Along the sides of the river to-day we observed much timber, the cottonwood, the sycamore, hickory, white walnut, some grapevines, and rushes. The high west wind and rain compelled us to remain all the next day, May 31. In the afternoon a boat came down from the Grand Osage river, bringing a letter from a person sent to the Osage nation on the Arkansaw river, which mentioned that the letter announcing the cession of Louisiana was committed to the flames —that the Indians would not believe that the Americans were owners of that country, and disregarded St. Louis and its supplies. The party was occupied in hunting, in the course of which they caught in the woods several very large rats. We set sail early the next morning, June 1, and at six miles distant passed Bear creek, a stream of about twenty-five yards width; but the wind being ahead and the current rapid, we were unable to make more than thirteen miles to the mouth of the Osage river, where we encamped and remained the following day, for the purpose of making celestial observations. The Osage river empties itself into the Missouri at one hundred and thirty-three miles distance from the mouth of the latter river. Its general course is west and west-southwest through a rich and level country. At the junction the Missouri is about eight hundred and seventy-five yards wide, and the Osage three hundred and ninety-seven. The low point of junction is in latitude 38° 31′ 16′′, and at a short distance from it is a high commanding position, whence we enjoyed a delightful prospect of the country.

The Osage river gives or owes its name to a nation inhabiting its banks at a considerable distance from this place. Their present name, however, seems to have originated from the French traders, for both among themselves and their neighbours they are called the Wasbashas. They number between twelve and thirteen hundred warriors, and consist of three tribes: the Great Osages, of about five hundred warriors, living in a village on the south bank of the river; the Little Osages, of nearly half that number, residing at the distance of six miles from them; and the Arkansaw band, a colony of Osages, of six hundred warriors, who left them some years ago, under the command of a chief called the Bigfoot, and settled on the Vermillion river, a branch of the Arkansaw. In person the Osages are among the largest and best-formed Indians, and are said to possess fine military capacities; but residing as they do in villages, and having made considerable advance in agriculture, they seem less addicted to war than their northern neighbours, to whom the use of rifles gives a great superiority. Among the peculiarities of this people, there is nothing more remarkable than the tradition relative to their origin. According to universal belief, the founder of the nation was a snail passing a quiet existence along the banks of the Osage, till a high flood swept him down to the Missouri, and left him exposed on the shore. The heat of the sun at length ripened him into a man, but with the change of his nature he had not forgotten his native seats on the Osage, towards which he immediately bent his way. He was, however, soon overtaken by hunger and fatigue, when happily the Great Spirit appeared, and giving him a bow and arrow, showed him how to kill and cook deer, and cover himself with the skin. He then proceeded to his original residence, but as he approached the river he was met by a beaver, who inquired haughtily who he was, and by what authority he came to disturb his possession. The Osage answered that the river was his own, for he had once lived on its borders. As they stood disputing, the daughter of the beaver came, and having by her entreaties reconciled her father to this young stranger, it was proposed that the Osage should marry the young beaver, and share with her family the enjoyment of the river. The Osage readily consented, and from this happy union there soon came the village and the nation of the Wasbasha, or Osages, who have ever since preserved a pious reverence for their ancestors, abstaining from the chace of the beaver, because in killing that animal they killed a brother of the Osage. Of late years, however, since the trade with the whites has rendered beaver skins more valuable, the sanctity of these maternal relatives has visibly reduced, and the poor animals have nearly lost all the privileges of kindred.

On the afternoon of June 3 we proceeded, and at three miles distant reached a creek called Cupboard creek, from a rock of that appearance near its entrance. Two miles farther we encamped at Moreau creek, a stream of twenty yards width, on the southern side. The next morning, we passed at an early hour Cedar island on the north, so called from the abundance of the tree of that name; near which is a small creek, named Nightingale creek, from a bird of that species, which sang for us during the night. Beyond Cedar island are some others of a smaller extent, and at seven miles distance a creek fifteen or twenty yards wide, entering from the north, and known by the name of Cedar creek. At seven and a half miles farther, we passed on the south side another creek, which we called Mast creek, from the circumstance of our mast being broken by running under a concealed tree; a little above is another creek on the left, one mile beyond which we encamped on the southern shore under high projecting cliffs. The French had reported that lead ore was to be found in this place, but on examining the hills we could discern no appearance of that mineral. Along the river on the south is a low land covered with rushes and high nettles, and, near the mouths of the creeks, supplied with oak, ash, and walnut timber. On the north the land is rich and well situated. We made seventeen and a half miles this day. The river is falling slowly. We continued our route the next morning early: a small creek called Lead creek, on the south; another on the north, known to the French by the name of Little Good Woman's creek; and again Big Rock creek on the south were the only streams we passed this morning. At eleven o'clock we met a raft made of two canoes joined together, in which two French traders were descending from eighty leagues up the river Kanzas, where they had wintered and caught great quantities of beaver, but had lost much of their game by fires from the prairies. They told us that the Kanzas nation is now hunting buffaloe in the plains, having passed the last winter on this river. Two miles farther, we reached on the south Little Manitou creek, which takes its name from a strange figure resembling the bust of a man, with the horns of a stag, painted on a projecting rock, which may represent some spirit or deity. Near this is a sandbar extending several miles, which renders the navigation difficult, and a small creek called Sand creek on the south, where we stopped for dinner, and gathered wild cresses and tongue grass from the sandbar. The rapidity of the currents, added to our having broken our mast, prevented our going more than twelve and a half miles. The scouts and hunters whom we always kept out report that they have seen fresh tracks of Indians. The next morning we left our camp, which was on the south side, opposite to a large island in the middle of the river, and at five miles reached a creek on the north side, of about twenty yards wide, called Split Rock creek, from a fissure in the point of a neighbouring rock. Three miles beyond this, on the south, is Saline river; it is about thirty yards wide, and has its name from the number of salt licks and springs, which render its water brackish; the river is very rapid and the banks falling in. After leaving Saline creek we passed one large island and several smaller ones, having made fourteen miles. The water rose a foot during the last night.

The next day, June 7, we passed at four and a half miles Big Manitou creek, near which is a limestone rock inlaid with flint of various colours, and embellished, or at least covered, with uncouth paintings of animals and inscriptions. We landed to examine it, but found the place occupied by a nest of rattlesnakes, of which we killed three. We also examined some licks and springs of salt water, two or three miles up this creek. We then proceeded by some small willow islands, and encamped at the mouth of Good Woman river on the north. It is about thirty-five yards wide, and said to be navigable for boats several leagues. The hunters, who had hitherto given us only deer, brought in this evening three bears, and had seen some indication of buffaloe. We had come fourteen miles.

June 8. We saw several small willow islands, and a creek on the south, near which are a number of deer licks; at nine miles distance we came to Mine river. This river, which falls into the Missouri from the south, is said to be navigable for boats eighty or ninety miles, and is about seventy yards wide at its mouth. It forks about five or six leagues from the Missouri, and at the point of junction are some very rich salt springs; the west branch in particular is so much impregnated that, for twenty miles, the water is not palatable; several branches of the Manitou and Good Woman are equally tinctured. The French report also that lead ore has been found on different parts of the river. We made several excursions near the river through the low rich country on its banks, and after dinner went on to the island of Mills, where we encamped. We met with a party of three hunters from the Sioux river; they had been out for twelve months, and collected about nine hundred dollars' worth of peltries and furs. We ascended this river twelve miles.

On the 9th we set out early, and reached a cliff of rocks, called the Arrow rock, near to which is a prairie called the Prairie of Arrows, and Arrow creek, a small stream about eight yards wide, whose source is in the adjoining prairies on the south. At this cliff the Missouri is confined within a bed of two hundred yards; and about four miles to the southeast is a large lick and salt spring of great strength. About three miles further is Blackbird creek on the north side, opposite to which is an island and a prairie inclosing a small lake. Five miles beyond this we encamped on the south side, after making, in the course of the day, thirteen miles. The land on the north is a high rich plain. On the south it is also even, of a good quality, and rising from fifty to one hundred feet.

The next morning, 10th, we passed Deer creek, and at the distance of five miles the two rivers called by the French the two Charatons, a corruption of Thieraton, the first of which is thirty, the second seventy yards wide, enter the Missouri together. They are both navigable for boats; the country through which they pass is broken, rich, and thickly covered with timber. The Ayauway nation, consisting of three hundred men, have a village near its head-waters on the river Des Moines. Farther on we passed a large island called *Chicot* or Stump island, and encamped on the south, after making ten miles. A head wind forced us to remain there all the next day, during which we dried the meat we had killed, and examined the surrounding country, which consists of good land, well watered and supplied with timber; the prairies also differ from those eastward of the Mississippi, inasmuch as the latter are generally without any covering except grass, whilst the former abound with hazel, grapes, and other fruits, among which is the Osage plum of a superior size and quality. On the morning of the 12th we passed through difficult places in the river, and reached Plum creek on the south side. At one o'clock we met two rafts loaded, the one with furs, the other with the tallow of buffaloe; they were from the Sioux nation, and on their way to St. Louis; but we were fortunate enough to engage one of them, a Mr. Durion, who had lived with that nation more than twenty years, and was high in their confidence, to accompany us thither. We made nine miles.

On the 13th, we passed at between four and five miles a bend of the river, and two creeks on the north called the Round Bend creeks. Between these two creeks is the prairie, in which once stood the ancient village of the Missouris. Of this village there remains no vestige, nor is there anything to recall this great and numerous nation except a feeble remnant of about thirty families. They were driven from their original seats by the invasions of the Sauks and other Indians from the Mississippi, who destroyed at this village two hundred of them in one contest, and sought refuge near the Little Osage, on the other side of the river. The encroachment of the same enemies forced, about thirty years since, both these nations from the banks of the Missouri. A few retired with the Osage, and the remainder found an asylum on the river Platte, among the Ottoes, who are themselves declining. Opposite the plain there was an island and a French fort, but there is now no appearance of either, the successive inundations having probably washed them away, as the willow island, which is in the situation described by Du Pratz, is small and of recent formation. Five miles from this place is the mouth of Grand river, where we encamped. This river follows a course nearly south, or southeast, and is between eighty and a hundred yards wide where it enters the Missouri, near a delightful and rich plain. A raccoon, a bear, and some deer were obtained to-day. We proceeded at six o'clock the next morning. The current was so rapid, and the banks on the north falling in so constantly, that we were obliged to approach the sandbars on the south. These were moving continually, and formed the worst passage we had seen, and which we surmounted with much difficulty. We met a trading raft from the Pawnee nation on the river Platte, and attempted unsuccessfully to engage one of their party to return with us. At the distance of eight miles we came to some high cliffs, called the Snake bluffs, from the number of that animal in the neighbourhood, and immediately above these bluffs, Snake creek, about eighteen yards wide, on which we encamped. One of our hunters, a half Indian, brought us an account of his having to-day passed a small lake, near which a number of deer were feeding, and in the pond he heard a snake making a guttural noise like a turkey. He fired his gun, but the noise became louder. He adds that he has heard the Indians mention this species of snake, and this story is confirmed by a Frenchman of our party. All the next day, the river being very high, the sandbars were so rolling and numerous, and the current so strong, that we were unable to stem it even with oars added to our sails; this obliged us to go nearer the banks, which were falling in, so that we could not make, though the boat was occasionally towed, more than fourteen miles. We passed several islands and one creek on the south side, and encamped on the north opposite a beautiful plain, which extends as far back as the Osage river, and some miles up the Missouri. In front of our encampment are the remains of an old village of the Little Osage, situated at some distance from the river, and at the foot of a small hill. About three miles above them, in view of our camp, is the situation of the old village of the Missouris after they fled from the Sauks.

The inroads of the same tribe compelled the Little Osage to retire from the Missouri a few years ago and establish themselves near the Great Osage. The river, which is here about one mile wide, had risen in the morning, but fell towards evening. Early this morning, June 16, we joined the camp of our hunters, who had provided two deer and two bear, and then passing an island and a prairie on the north covered with a species of timothy, made our way through bad sandbars and a swift current, to an encampment for the evening, on the north side, at ten miles distance. The timber which we examined to-day was not sufficiently strong for oars; the musquitoes and ticks are exceedingly troublesome. On the 17th, we set out early, and having come to a convenient place at one mile distance for procuring timber and making oars, we occupied ourselves in that way on this and the following day. The country on the north of the river is rich and covered with timber, among which we procured the ash for oars. At two miles it changes into extensive prairies, and at seven or eight miles distance becomes higher and waving. The prairie and high lands on the south commence more immediately on the river; the whole is well watered and provided with game, such as deer, elk, and bear. The hunters brought in a fat horse which was probably lost by some war party—this being the crossing place for the Sauks, Ayauways, and Sioux, in their excursions against the Osage.

June 19. The oars being finished, we proceeded under a gentle breeze by two large and some smaller islands. The sandbars are numerous, and so bad that at one place we were forced to clear away the driftwood in order to pass; the water, too, was so rapid that we were under the necessity of towing the boat for half a mile round a point of rocks on the south side. We passed two creeks, one called Tiger creek on the north, twenty-five yards wide, at the extremity of a large island called Panther island; the other, Tabo creek on the south, fifteen yards wide. Along the shores are gooseberries and raspberries in abundance. At the distance of seventeen and a half miles we encamped on the south, near a lake about two miles from the river and several in circumference, and much frequented by deer and all kinds of fowls. On the north the land is higher and better calculated for farms than that on the south, which ascends more gradually, but is still rich and pleasant. The musquitoes and other animals are so troublesome that musquitoe biers or nets were distributed to the party. The next morning we passed a large island, opposite to which on the north is a large and beautiful prairie called Sauk prairie, the land being fine and well timbered on both sides of the river. Pelicans were seen to-day. We made six and three-quarter miles, and encamped at the lower point of a small island, along the north side of which we proceeded the next day, June 21, but not without danger in consequence of the sands and the rapidity of the water, which rose three inches last night. Behind another island come in from the south two creeks, called Eau Beau, or Clear Water creeks; on the north is a very remarkable bend, where the high lands approach the river, and form an acute angle at the head of a large island produced by a narrow channel through the point of the bend. We passed several other islands, and encamped at seven and a half miles on the south.

22d. The river rose during the night four inches. The water is very rapid and crowded with concealed timber. We passed two large islands and an extensive prairie on the south, beginning with a rich low land, and rising to the distance of seventy or eighty feet of rolling clear country. The thermometer at three o'clock P.M. was at 87°. After coming ten and a half miles we encamped on the south, opposite a large creek called Fire Prairie river.

23d. The wind was against us this morning, and became so violent that we made only three and a half miles, and were obliged to lie to during the day at a small island. This is separated from the northern side by a narrow channel which cannot be passed by boats, being choked by trees and drifted wood. Directly opposite on the south is a high commanding position, more than seventy feet above high water mark, and overlooking the river which is here of but little width; this spot has many advantages for a fort and trading house with the Indians.* The river fell eight inches last night.

The next day, 24th, we passed, at eight miles distance, Hay Cabin creek coming in from the south, about twenty yards wide, and so called from camps of straw built on it; to the north are some rocks projecting into the river, and a little beyond them a creek on the same side, called Charaton Scarty,—that is, Charaton like the Otter. We halted, after making eleven and a half miles, the country on both sides being fine and interspersed with prairies, in which we now see numerous herds of deer, pasturing in the plains or feeding on the young willows of the river.

25th. A thick fog detained us till eight o'clock, when we set sail, and at three miles reached a bank of stone coal on the north, which appeared to be very abundant; just below it is a creek called after the bank La Charbonniere. Four miles farther, and on the southern side, comes in a small creek, called La Benite. The prairies here approach the river and contain many fruits, such as plums, raspberries, wild apples, and nearer the river vast quantities of mulberries. Our encampment was at thirteen miles distance on an island to the north, opposite some hills higher than usual and almost one hundred and sixty or one hundred and eighty feet.

---

* The United States built in September, 1808, a factory and fort at this spot, which is very convenient for trading with the Osages, Ayauways, and Kanzas.

## UP THE MISSOURI

26th. At one mile we passed at the end of a small island Blue Water creek, which is about thirty yards wide at its entrance from the south.* Here the Missouri is confined within a narrow bed, and the current still more so by counter currents or whirls on one side and a high bank on the other. We passed a small island and a sandbar, where our tow-rope broke twice, and we rowed round with great exertions. We saw a number of parroquets, and killed some deer; after nine and three-quarter miles we encamped at the upper point of the mouth of the river Kanzas; here we remained two days, during which we made the necessary observations, recruited the party, and repaired the boat. The river Kanzas takes its rise in the plains between the Arkansaw and Platte rivers, and pursues a course generally east till its junction with the Missouri, which is in latitude 38° 31' 13"; here it is three hundred and forty and a quarter yards wide, though it is wider a short distance above the mouth. The Missouri itself is about five hundred yards in width; the point of union is low and subject to inundations for two hundred and fifty yards; it then rises a little above high water mark, and continues so as far back as the hills. On the south of the Kanzas the hills or highlands come within one mile and a half of the river; on the north of the Missouri they do not approach nearer than several miles; but on all sides the country is fine. The comparative specific gravities of the two rivers is, for the Missouri seventy-eight, the Kanzas seventy-two degrees; the waters of the latter have a very disagreeable taste; the former has risen during yesterday and to-day about two feet. On the banks of the Kanzas reside the Indians of the same name, consisting of two villages, one at about twenty, the other forty leagues from its mouth, and amounting to about three hun-

---

* A few miles up the Blue Water creek are quarries of plaster of paris, since worked and brought down to St. Louis.

dred men. They once lived twenty-four leagues higher than the Kanzas, on the south bank of the Missouri, and were then more numerous, but they have been reduced and banished by the Sauks and Ayauways, who, being better supplied with arms, have an advantage over the Kanzas, though the latter are not less fierce or warlike than themselves. This nation is now hunting in the plains for the buffaloe, which our hunters have seen for the first time.

On the 29th, we set out late in the afternoon, and having passed a sandbar, near which the boat was almost lost, and a large island on the north, we encamped at seven and a quarter miles on the same side in the low lands, where the rushes are so thick that it is troublesome to walk through them. Early the next morning, 30th, we reached, at five miles distance, the mouth of a river coming in from the north, and called by the French, Petite Riviere Platte, or Little Shallow river; it is about sixty yards wide at its mouth. A few of the party who ascended informed us that the lands on both sides are good, and that there are several falls well calculated for mills; the wind was from the southwest, and the weather oppressively warm, the thermometer standing at 96° above zero at three o'clock P. M. One mile beyond this is a small creek on the south, at five miles from which we encamped on the same side, opposite the lower point of an island called Diamond island. The land on the north between the Little Shallow river and the Missouri is not good and subject to overflow, —on the south it is higher and better timbered.

July 1. We proceeded along the north side of Diamond island, where a small creek called Biscuit creek empties itself. One and a half miles above the island is a large sandbar in the middle of the river, beyond which we stopped to refresh the men, who suffered very much from the heat. Here we observed great quantities of grapes and raspberries. Between one and two miles farther are three islands and a creek on the south known by the French name of Remore. The main current, which is now on the south side of the largest of the three islands, ran three years, as we were told, on the north, and there was then no appearance of the two smaller islands. At the distance of four and a half miles we reached the lower point of a cluster of small islands, two large and two small, called Isles des Parcs or Field islands. Paccaun trees were this day seen, and large quantities of deer and turkies on the banks. We had advanced twelve miles.

July 2. We left our encampment, opposite to which is a high and beautiful prairie on the southern side, and passed up the south of the islands, which are high meadows, and a creek on the north called Parc creek. Here for half an hour the river became covered with driftwood, which rendered the navigation dangerous, and was probably caused by the giving way of some sandbar, which had detained the wood. After making five miles we passed a stream on the south called Turky creek, near a sandbar, where we could scarcely stem the current with twenty oars and all the poles we had. On the north at about two miles farther is a large island called by the Indians, Wau-car-da-war-card-da, or the Bear Medicine island. Here we landed and replaced our mast, which had been broken three days ago by running against a tree overhanging the river. Thence we proceeded, and after night stopped on the north side, above the island, having come eleven and a half miles. Opposite our camp is a valley, in which was situated an old village of the Kanzas, between two high points of land, and on the bank of the river. About a mile in the rear of the village was a small fort, built by the French, on an elevation. There are now no traces of the village, but the situation of the fort may be recognized by some remains of chimnies and the general outline of the fortification, as well as by the fine spring which supplied it with water. The party who were stationed here were probably cut off by the Indians, as there are no accounts of them.

July 3. A gentle breeze from the south carried us eleven and a quarter miles this day, past two islands, one a small willow island, the other large, and called by the French Isle des Vaches, or Cow island. At the head of this island, on the northern shore, is a large pond containing beaver, and fowls of different kinds. After passing a bad sandbar, we stopped on the south side at an old trading house, which is now deserted, and a half mile beyond it encamped on the south. The land is fine along the rivers and some distance back. We observed the black walnut and oak, among the timber; and the honeysuckle and the buck's-eye, with the nuts on them.

The morning of the 4th July was announced by the discharge of our gun. At one mile we reached the mouth of a bayeau or creek, coming from a large lake on the north side, which appears as if it had once been the bed of the river, to which it runs parallel for several miles. The water of it is clear and supplied by a small creek and several springs, and the number of goslings which we saw on it induced us to call it the Gosling lake. It is about three-quarters of a mile wide, and seven or eight miles long. One of our men was bitten by a snake, but a poultice of bark and gunpowder was sufficient to cure the wound. At ten and a quarter miles we reached a creek on the south about twelve yards wide and coming from an extensive prairie, which approached the borders of the river. To this creek, which had no name, we gave that of Fourth of July creek; above it is a high mound, where three Indian paths centre, and from which is a very extensive prospect. After fifteen miles' sail we came to on the north, a little above a creek on the southern side, about thirty yards wide, which we called Independence creek, in honour of the day, which we could celebrate only by an evening gun and an additional gill of whiskey to the men.

The next day, 5th, we crossed over to the south and came along the bank of an extensive and beautiful prairie, interspersed with copses of timber, and watered by Independence creek. On this bank formerly stood the second village of the Kanzas; from the remains it must have been once a large town. We passed several bad sandbars, and a small creek to the south, which we called Yellow Ochre creek, from a bank of that mineral a little above it. The river continues to fall. On the shores are great quantities of summer and fall grapes, berries, and wild roses. Deer is not so abundant as usual, but there are numerous tracks of elk around us. We encamped at ten miles distance on the south side under a high bank, opposite to which was a low land covered with tall rushes and some timber.

July 6. We set sail, and at one mile passed a sandbar, three miles farther an island, a prairie to the north, at the distance of four miles, called Reevey's prairie, after a man who was killed there; at which place the river is confined to a very narrow channel, and by a sandbar from the south. Four miles beyond is another sandbar terminated by a small willow island, and forming a very considerable bend in the river towards the north. The sand of the bar is light, intermixed with small pebbles and some pit coal. The river falls slowly, and, owing either to the muddiness of its water or the extreme heat of the weather, the men perspire profusely. We encamped on the south, having made twelve miles. The bird called whip-poor-will sat on the boat for some time.

In the morning, July 7, the rapidity of the water obliged us to draw the boat along with ropes. At six and three-quarter miles, we came to a sandbar, at a point opposite a fine rich prairie on the north, called St. Michael's. The prairies of this neighbourhood have the appearance of distinct farms, divided by narrow strips of woodland, which follow the borders of the small runs leading to the river. Above this, about a mile, is a cliff of yellow clay on the north. At four o'clock we passed a narrow part of the channel, where the water is confined within a bed of two hundred yards wide, the current running directly against the southern bank, with no sand on the north to confine it or break its force. We made fourteen miles, and halted on the north, after which we had a violent gust about seven o'clock. One of the hunters saw in a pond to the north which we passed yesterday a number of young swans. We saw a large rat, and killed a wolf. Another of our men had a stroke of the sun; he was bled, and took a preparation of nitre which relieved him considerably.

July 8. We set out early, and soon passed a small creek on the north which we called Ordway's creek, from our sergeant of that name, who had been sent on shore with the horses, and went up it. On the same side are three small islands, one of which is the Little Nodawa, and a large island called the Great Nowada, extending more than five miles, and containing seven or eight thousand acres of high good land, rarely overflowed, and one of the largest islands of the Missouri. It is separated from the northern shore by a small channel of from forty-five to eighty yards wide, up which we passed, and found near the western extremity of the island the mouth of the river Nodawa. This river pursues nearly a southern course, is navigable for boats to some distance, and about seventy yards wide above the mouth, though not so wide immediately there, as the mud from the Missouri contracts its channel. At twelve and a quarter miles we encamped on the north side, near the head of Nodawa island,

and opposite a smaller one in the middle of the river. Five of the men were this day sick with violent headaches. The river continues to fall.

July 9. We passed the island opposite to which we last night encamped, and saw near the head of it a creek falling in from a pond on the north, to which we gave the name of Pike pond, from the numbers of that animal which some of our party saw from the shore. The wind changed at eight from N. E. to S. W. and brought rain. At six miles we passed the mouth of Monter's creek on the south, and two miles above a few cabins, where one of our party had encamped with some Frenchmen about two years ago. Farther on we passed an island on the north, opposite some cliffs on the south side, near which Loup or Wolf river falls into the Missouri. This river is about sixty yards wide; it heads near the same sources as the Kanzas, and is navigable for boats at some distance up. At fourteen miles we encamped on the south side.

Tuesday, 10th. We proceeded on by a prairie on the upper side of Wolf river, and at four miles passed a creek fifteen yards wide on the south, called Pape's creek after a Spaniard of that name who killed himself there. At six miles we dined on an island called by the French Isle de Salomon, or Solomon's island, opposite to which on the south is a beautiful plain covered with grass, intermixed with wild rye and a kind of wild potatoe. After making ten miles we stopped for the night on the northern side, opposite a cliff of yellow clay. The river has neither risen nor fallen to-day. On the north the lowland is very extensive, and covered with vines; on the south the hills approach nearer the river, and back of them commence the plains. There are a great many goslings along the banks.

Wednesday, 11th. After three miles' sailing we came to a

willow island on the north side, behind which enters a creek called by the Indians Tarkio. Above this creek on the north the lowlands are subject to overflow, and farther back the undergrowth, of vines particularly, is so abundant that they can scarcely be passed. Three miles from the Tarkio we encamped on a large sand island on the north, immediately opposite the river Nemahaw.

Thursday, 12th. We remained here to-day for the purpose of refreshing the party and making lunar observations. The Nemahaw empties itself into the Missouri from the south, and is eighty yards wide at the confluence, which is in latitude 39° 55' 56". Captain Clark ascended it in the periogue about two miles to the mouth of a small creek on the lower side. On going ashore he found in the level plain several artificial mounds or graves, and on the adjoining hills others of a larger size. This appearance indicates sufficiently the former population of this country, the mounds being certainly intended as tombs; the Indians of the Missouri still preserving the custom of interring the dead on high ground. From the top of the highest mound a delightful prospect presented itself,—the level and extensive meadows watered by the Nemahaw, and enlivened by the few trees and shrubs skirting the borders of the river and its tributary streams; the lowland of the Missouri covered with undulating grass, nearly five feet high, gradually rising into a second plain, where rich weeds and flowers are interspersed with copses of the Osage plum. Farther back are seen small groves of trees; an abundance of grapes; the wild cherry of the Missouri, resembling our own, but larger, and growing on a small bush; and the chokecherry, which we observed for the first time. Some of the grapes gathered to-day are nearly ripe. On the south of the Nemahaw, and about a quarter of a mile from its mouth, is a cliff of freestone, in which are various inscriptions and marks made by the Indians. The sand island where we are encamped is covered with the two species of willow, broad and narrow leaf.

July 13. We proceeded at sunrise with a fair wind from the south, and at two miles passed the mouth of a small river on the north, called Big Tarkio. A channel from the bed of the Missouri once ran into this river, and formed an island called St. Joseph's, but the channel is now filled up, and the island is added to the northern shore. Farther on to the south is situated an extensive plain, covered with a grass resembling timothy in its general appearance, except the seed which is like flaxseed, and also a number of grapevines. At twelve miles, we passed an island on the north, above which is a large sandbar covered with willows; and at twenty and a half miles, stopped on a large sandbar, in the middle of the river opposite a high handsome prairie, which extends to the hills four or five miles distant, though near the bank the land is low, and subject to be overflowed. This day was exceedingly fine and pleasant, a storm of wind and rain from north-northeast last night having cooled the air.

July 14. We had some hard showers of rain before seven o'clock, when we set out. We had just reached the end of the sand island, and seen the opposite banks falling in, and so lined with timber that we could not approach it without danger, when a sudden squall from the northeast struck the boat on the starboard quarter, and would have certainly dashed her to pieces on the sand island, if the party had not leaped into the river, and with the aid of the anchor and cable kept her off, the waves dashing over her for the space of forty minutes; after which, the river became almost instantaneously calm and smooth. The two periogues were ahead, in a situation nearly similar, but fortunately no damage was done to the boats or the loading. The wind having shifted to the southeast, we came, at the distance of two miles, to an island on the north, where we dined. One mile above, on the same side of the river, is a small factory, where a merchant of St. Louis traded with the Ottoes and Pawnees two years ago. Near this is an extensive lowland, part of which is overflowed occasionally; the rest is rich and well timbered. The wind again changed to northwest by north. At seven and a half miles, we reached the lower point of a large island, on the north side. A small distance above this point is a river, called by the Maha Indians Nishnahbatona. This is a considerable creek, nearly as large as the Mine river, and runs parallel to the Missouri the greater part of its course, being fifty yards wide at the mouth. In the prairies or glades we saw wild timothy, lambsquater, cuckleberries, and on the edges of the river, summer-grapes, plums, and gosseberries. We also saw to-day, for the first time, some elk, at which some of the party shot, but at too great a distance. We encamped on the north side of the island, a little above Nishnahbatona, having made nine miles. The river fell a little.

July 15. A thick fog prevented our leaving the encampment before seven. At about four miles, we reached the extremity of the large island, and crossing to the south, at the distance of seven miles, arrived at the little Nemaha, a small river from the south, forty yards wide a little above its mouth, but contracting, as do almost all the waters emptying into the Missouri, at its confluence. At nine and three-quarter miles, we encamped on a woody point, on the south. Along the southern bank is a rich lowland covered with peavine and rich weeds, and watered by small streams rising in the adjoining prairies. They, too, are rich, and though with abundance of grass, have no timber except what grows near the water; interspersed through both are grapevines, plums of two kinds, two species of wild cherries, hazlenuts, and gosseberries. On the south there is one unbroken plain; on the north the river is skirted with some timber, behind which the plain extends four or five miles to the hills, which seem to have little wood.

July 16. We continued our route between a large island opposite to our last night's encampment and an extensive prairie on the south. About six miles, we came to another large island, called Fairsun island, on the same side; above which is a spot where about twenty acres of the hill have fallen into the river. Near this is a cliff of sandstone for two miles, which is much frequented by birds. At this place the river is about one mile wide, but not deep, as the timber, or sawyers, may be seen, scattered across the whole of its bottom. At twenty miles distance we saw on the south an island called by the French l'Isle Chauve, or Bald island, opposite to a large prairie, which we called Baldpated prairie, from a ridge of naked hills which bound it, running parallel with the river as far as we could see, and from three to six miles distance. To the south the hills touch the river. We encamped a quarter of a mile beyond this, in a point of woods on the north side. The river continues to fall.

Tuesday, July 17. We remained here this day, in order to make observations and correct the chronometer, which ran down on Sunday. The latitude we found to be 40° 27' $5\frac{4}{10}$". The observations of the time proved our chronometer too slow by 6' $51\frac{6}{10}$". The highlands bear from our camp north 25° west, up the river. Captain Lewis rode up the country, and saw the Nishnahbatona, about ten or twelve miles from its mouth, at a place not more than three hundred yards from the Missouri, and a little above our camp. It then passes near the foot of the Baldhills, and is at least six feet below the level of the Missouri. On its banks are the oak, walnut, and mulberry. The common current of the Missouri, taken with the log, is 50 fathoms in 40" at some places, and even 20".

Wednesday, July 18. The morning was fair, and a gentle wind from southeast by south carried us along between the prairie on the north and Bald island to the south, opposite the middle of which the Nishnahbatona approaches the nearest to the Missouri. The current here ran fifty fathoms in 41". At thirteen and a half miles, we reached an island on the north, near to which the banks overflow; while on the south the hills project over the river and form high cliffs. At one point a part of the cliff, nearly three-quarters of a mile in length, and about two hundred feet in height, has fallen into the river. It is composed chiefly of sandstone intermixed with an iron ore of bad quality; near the bottom is a soft slatestone with pebbles. We passed several bad sandbars in the course of the day, and made eighteen miles, and encamped on the south, opposite to the lower point of the Oven islands. The country around is generally divided into prairies, with little timber, except on low points, islands, and near creeks, and that consisting of cottonwood, mulberry, elm, and sycamore. The river falls fast. An Indian dog came to the bank; he appeared to have been lost, and was nearly starved; we gave him some food, but he would not follow us.

Thursday, July 19. The Oven islands are small, and two in number; one near the south shore, the other in the middle of the river. Opposite to them is the prairie, called Terrien's Oven, from a trader of that name. At four and a half miles, we reached some high cliffs of a yellow earth, on the south, near which are two beautiful runs of water, rising in the adjacent prairies, and one of them with a deer-lick about two hundred yards from its mouth. In this neighbourhood we observed some iron ore in the bank. At two and a half miles above the runs, a large portion of the hill, for nearly three-quarters of a mile, has fallen into the river. We encamped on the western extremity of an island in the middle of the river, having made ten and three-quarter miles. The river falls a little. The sandbars which we passed to-day are more numerous, and the rolling sands more frequent and dangerous than any we have seen, these obstacles increasing as we approach the river Platte. The Missouri here is wider also than below, where the timber on the banks resists the current; while here the prairies which approach are more easily washed and undermined. The hunters have brought for the last few days no quadruped but deer; great quantities of young geese are seen to-day; one of them brought calamus, which he had gathered opposite our encampment, and a large quantity of sweetflag.

Friday, July 20. There was a heavy dew last night, and this morning was foggy and cool. We passed at about three miles distance a small willow island to the north, and a creek on the south, about twenty-five yards wide, called by the French L'eau qui Pleure, or the Weeping Water, and emptying itself just above a cliff of brown clay. Thence we made two and a half miles to another island; three miles farther to a third; six miles beyond which is a fourth island, at the head of which we encamped on the southern shore,—in all eighteen miles. The party, who walked on the shore to-day, found the plains to the south rich, but much parched with frequent fires, and with no timber except the scattering trees about the sources of the runs, which are numerous and fine. On the north is a similar prairie country. The river continues to fall. A large yellow wolf was this day killed. For a month past the party have been troubled with biles, and occasionally with the dysentery. These biles were large tumors which broke out under the arms, on the legs, and generally in the parts most exposed to action, which sometimes became too painful to permit the men to work. After remaining some days, they disappeared without any assistance, except a poultice of the bark of the elm, or of Indian meal. This disorder, which we ascribe to the muddiness of the river water, has not affected the general health of the party, which is quite as good, if not better, than that of the same number of men in any other situation.

Saturday, July 21. We had a breeze from the southeast, by the aid of which we passed, at about ten miles, a willow island on the south, near highlands covered with timber at the bank, and formed of limestone with cemented shells; on the opposite side is a bad sandbar, and the land near it is cut through at high water by small channels, forming a number of islands. The wind lulled at seven o'clock, and we reached, in the rain, the mouth of the great river Platte, at the distance of fourteen miles. The highlands which had accompanied us on the south, for the last eight or ten miles, stopped at about three-quarters of a mile from the entrance of the Platte. Captains Lewis and Clark ascended the river in a periogue for about one mile, and found the current very rapid, rolling over sands and divided into a number of channels, none of which are deeper than five or six feet. One of our Frenchmen, who spent two winters on it, says that it spreads much more at some distance from the mouth; that its depth is generally not more than five or six feet; that there are many small islands scattered through it, and that from its rapidity and the quantity of its sand it cannot be navigated by boats or periogues, though the Indians pass it in small flat canoes made of hides; that the Saline or Salt river, which in some seasons is too brackish to be drank, falls into it from the south about thirty miles up, and a little above it Elkhorn river from the north, running nearly parallel with the Missouri. The river is, in fact, much more rapid than the Missouri, the bed of which it fills with moving sands, and drives the current on the northern shore, on which it is constantly encroaching. At its junction the Platte is about six hundred yards wide, and the same number of miles from the Mississippi. With much difficulty we worked round the sandbars near the mouth, and came to above the point, having made fifteen miles. A number of wolves were seen and heard around us in the evening.

July 22. The next morning we set sail, and having found at the distance of ten miles from the Platte a high and shaded situation on the north, we encamped there, intending to make the requisite observations, and to send for the neighbouring tribes, for the purpose of making known the recent change in the government, and the wish of the United States to cultivate their friendship.

# CHAPTER II.

Some account of the Pawnee Indians—Council held with the Ottoe and Missouri Indians—Council held with another party of the Ottoes—Death of Sergeant Floyd—The party encamp near the mouth of Whitestone river—The character of the Missouri, with the rivers that enter it—The surrounding country—The various islands, bays, creeks, etc., given in the course of the expedition.

OUR camp is by observation in latitude 41° 3' 11". Immediately behind it is a plain about five miles wide, one half covered with wood, the other dry and elevated. The low grounds on the south, near the junction of the two rivers, are rich, but subject to be overflowed. Farther up, the banks are higher, and opposite our camp the first hills approach the river, and are covered with timber, such as oak, walnut, and elm. The intermediate country is watered by the Papillon, or Butterfly creek, of about eighteen yards wide, and three miles from the Platte; on the north are high open plains and prairies, and at nine miles from the Platte the Musquitoe creek and two or three small willow islands. We stayed here several days, during which we dried our provisions, made new oars, and prepared our despatches and maps of the country we had passed, for the President of the United States, to whom we intend to send them by a periogue from this place. The hunters have found game scarce in this neighbourhood; they have seen deer, turkies, and grouse; we have also an abundance of ripe grapes; and one of our men caught a white catfish, the eyes of which were small, and its tail resembling that of a dolphin. The present season is that in which the Indians go out into the prairies to hunt the buffaloe; but as we discovered some hunters' tracks, and observed the plains on fire in the direction of their villages, we hoped that they might have returned to gather the green Indian corn, and therefore despatched two men to the Ottoes or Pawnee villages with a present of tobacco, and an invitation to the chiefs to visit us. They returned after two days absence. Their first course was through an open prairie to the south, in which they crossed Butterfly creek. They then reached a small beautiful river, called Corne de Cerf, or Elkhorn river, about one hundred yards wide, with clear water and a gravelly channel. It empties a little below the Ottoe village into the Platte, which they crossed, and arrived at the town about forty-five miles from our camp. They found no Indians there, though they saw some fresh tracks of a small party. The Ottoes were once a powerful nation, and lived about twenty miles above the Platte, on the southern bank of the Missouri. Being reduced, they migrated to the neighbourhood of the Pawnees, under whose protection they now live. Their village is on the south side of the Platte, about thirty miles from its mouth; and their number is two hundred men, including about thirty families of Missouri Indians, who are incorporated with them. Five leagues above them, on the same side of the river, resides the nation of Pawnees. This people were among the most numerous of the Missouri Indians, but have gradually been dispersed and broken, and even since the year 1797 have undergone some sensible changes. They now consist of four bands; the first is the one just mentioned, of about five hundred men, to whom of late years have been added the second band, who are called Republican Pawnees, from their having lived on the Republican branch of the river Kanzas, whence they emigrated to join the principal band of Pawnees; the Republican Pawnees amount to nearly two hundred and fifty men. The third are the Pawnees Loups, or Wolf Pawnees, who reside on the Wolf fork of the Platte, about ninety miles from the principal Pawnees, and number two hundred and eighty men. The fourth band originally resided on the Kanzas and Arkansaw, but in their wars with the Osages they were so often defeated that they at last retired to their present position on the Red river, where they form a tribe of four hundred men. All these tribes live in villages, and raise corn; but during the intervals of culture rove in the plains in quest of buffaloe.

Beyond them on the river, and westward of the Black mountains, are the Kaninaviesch, consisting of about four hundred men. They are supposed to have emigrated originally from the Pawnees nation; but they have degenerated from the improvements of the parent tribe, and no longer live in villages, but rove through the plains.

Still farther to the westward are several tribes, who wander and hunt on the sources of the river Platte, and thence to Rock Mountain. These tribes, of which little more is known than the names and the population, are, first, the Staitan, or Kite Indians, a small tribe of one hundred men. They have acquired the name of Kites, from their flying: that is, their being always on horseback; and the smallness of their numbers is to be attributed to their extreme ferocity; they are the most warlike of all the western Indians; they never yield in battle, they never spare their enemies, and the retaliation of this barbarity has almost extinguished the nation. Then come the Wetapahato and Kiawa tribes, associated together, and amounting to two hundred men; the Castahana, of three hundred men, to which are to be added the Cataka of seventy-five men, and the Dotami. These wandering tribes are conjectured to be the remnants of the Great Padouca nation, who occupied the country between the upper parts of the river Platte and the river Kanzas. They were visited by Bourgemont, in 1724, and then lived on the Kanzas river. The seats, which he describes as their residence, are now occupied by the Kanzas nation; and of the Padoucas, there does not now exist even the name.

July 27. Having completed the object of our stay, we set sail, with a pleasant breeze from the N. W. The two horses swam over to the southern shore, along which we went, passing by an island, at three and a half miles, formed by a pond, fed by springs; three miles farther is a large sand island, in the middle of the river; the land on the south being high, and covered with timber; that on the north, a high prairie. At ten and a half miles from our encampment we saw and examined a curious collection of graves or mounds, on the south side of the river. Not far from a low piece of land and a pond is a tract of about two hundred acres in circumference, which is covered with mounds of different heights, shapes, and sizes; some of sand, and some of both earth and sand, the largest being nearest the river. These mounds indicate the position of the ancient village of the Ottoes, before they retired to the protection of the Pawnees. After making fifteen miles, we encamped on the south, on the bank of a high handsome prairie, with lofty cottonwood in groves, near the river.

July 28. At one mile, this morning, we reached a bluff, on the north, being the first highlands which approach the river on that side since we left the Nadawa. Above this is an island, and a creek about fifteen yards wide, which, as it has no name, we called Indian Knob creek, from a number of round knobs bare of timber, on the highlands to the north. A little below the bluff, on the north, is the spot where the Ayauway Indians formerly lived. They were a branch of the Ottoes, and emigrated from this place to the river Desmoines. At ten and three-quarter miles, we encamped on the north, opposite an island in the middle of the river. The land, generally, on the north, consists of high prairie and hills, with timber; on the south, low and covered with cottonwood. Our hunter brought to us in the evening a Missouri Indian, whom he had found, with two others, dressing an elk; they were perfectly friendly, gave him some of the meat, and one of them agreed to accompany him to the boat. He is one of the few remaining Missouris, who live with the Ottoes; he belongs to a small party, whose camp is four miles from the river, and he says that the body of the nation is now hunting buffaloe in the plains. He appeared quite sprightly, and his language resembled that of the Osage, particularly in his calling a chief, inca. We sent him back with one of our party next morning,

Sunday, July 29, with an invitation to the Indians to meet us above on the river, and then proceeded. We soon came to a northern bend in the river, which runs within twenty yards of Indian Knob creek, the water of which is five feet higher than that of the Missouri. In less than two miles we passed Boyer's creek on the north, of twenty-five yards width. We stopped to dine under a shade, near the highland on the south, and caught several large catfish, one of them nearly white, and all very fat. Above this highland we observed the traces of a great hurricane, which passed the river obliquely from N. W. to S. E. and tore up large trees, some of which, perfectly sound and four feet in diameter, were snapped off near the ground. We made ten miles to a wood on the north, where we encamped. The Missouri is much more crooked since we passed the river Platte, though, generally speaking, not so rapid; more of prairie, with less timber, and cottonwood in the low grounds, and oak, black walnut, hickory, and elm.

July 30. We went early in the morning three and a quarter miles, and encamped on the south, in order to wait for the Ottoes. The land here consists of a plain, above the high-water level, the soil of which is fertile, and covered with a grass from five to eight feet high, interspersed with copses of large plums, and a currant, like those of the United States. It also furnishes two species of honeysuckle, one growing to a kind of shrub, common about Harrodsburgh (Kentucky), the other is not so high; the flowers grow in clusters, are short, and of a light pink colour; the leaves, too, are distinct, and do not surround the stalk, as do those of the common honeysuckle of the United States. Back of this plain is a woody ridge about seventy feet above it, at the end of which we formed our camp. This ridge separates the lower from a higher prairie, of a good quality, with grass of ten or twelve inches in height, and extending back about a mile to another elevation of eighty or ninety feet, beyond which is one continued plain. Near our camp, we enjoy from the bluffs a most beautiful view of the river and the adjoining country. At a distance, varying from four to ten miles, and of a height between seventy and three hundred feet, two parallel ranges of highland afford a passage to the Missouri, which enriches the low grounds between them. In its winding course it nourishes the willow islands, the scattered cottonwood, elm, sycamore, lynn, and ash, and the groves are interspersed with hickory, walnut, coffeenut, and oak.

July 31. The meridian altitude of this day made the latitude of our camp $41° 18' 1\frac{4}{10}''$. The hunters supplied us with deer, turkies, geese, and beaver; one of the last was caught alive, and in a very short time was perfectly tamed. Catfish are very abundant in the river, and we have also seen a buffaloe-fish. One of our men brought in yesterday an animal called by the Pawnees chocartoosh, and by the French, blaireau, or badger. The evening is cool, yet the musquitoes are still very troublesome.

We waited with much anxiety the return of our messenger to the Ottoes. The men whom we despatched to our last encampment returned without having seen any appearance of its having been visited. Our horses, too, had strayed; but we were so fortunate as to recover them at the distance of twelve miles. Our apprehensions were at length relieved by the arrival of a party of about fourteen Ottoe and Missouri Indians, who came at sunset on the second of August, accompanied by a Frenchman who resided among them and interpreted for us. Captains Lewis and Clark went out to meet them, and told them that we would hold a council in the morning. In the mean time we sent them some roasted meat, pork, flour, and meal, in return for which they made us a present of watermelons. We learnt that our man Liberte had set out from their camp a day before them; we were in hopes that he had fatigued his horse, or lost himself in the woods, and would soon return; but we never saw him again.

August 3. The next morning, the Indians, with their six chiefs, were all assembled under an awning, formed with the mainsail, in presence of all our party, paraded for the occasion. A speech was then made, announcing to them the change in the government, our promises of protection, and advice as to their future conduct. All the six chiefs replied to our speech, each in his turn, according to rank: they expressed their joy at the change in the government; their hopes that we would recommend them to their great father (the President), that they might obtain trade and necessaries; they wanted arms as well for hunting as for defence, and asked our mediation between them and the Mahas, with whom they are now at war. We promised to do so, and wished some of them to accompany us to that nation, which they declined, for fear of being killed by them. We then proceeded to distribute our presents. The grand chief of the nation not being of the party, we sent him a flag, a medal, and some ornaments for clothing. To the six chiefs who were present, we gave a medal of the second grade to one Ottoe chief and one Missouri chief; a medal of the third grade to two inferior chiefs of each nation,—the customary mode of recognizing a chief being to place a medal round his neck, which is considered among his tribe as a proof of his consideration abroad. Each of these medals was accompanied by a present of paint, garters, and cloth ornaments of dress; and to this we added a cannister of powder, a bottle of whiskey, and a few presents to the whole, which appeared to make them perfectly satisfied. The airgun, too, was fired, and astonished them greatly. The absent grand chief was an Ottoe named Weahrushhah, which in English degenerates into Little Thief. The two principal chieftains present were Shongotongo, or Big Horse; and Wethea, or Hospitality; also Shosguscan, or White Horse, an Ottoe; the first an Ottoe, the second a Missouri. The incidents just related induced us to give to this place the name of the Council-bluff; the situation of it is exceedingly favourable for a fort and trading factory, as the soil is well calculated for bricks, and there is an abundance of wood in the neighbourhood, and the air being pure and healthy. It is also central to the chief resorts of the Indians: one day's journey to the Ottoes; one and a half to the great Pawnees; two days from the Mahas; two and a quarter from the Pawnees Loups village; convenient to the hunting grounds of the Sioux; and twenty-five days journey to Santa Fee.

The ceremonies of the council being concluded, we set sail in the afternoon, and encamped at the distance of five miles, on the south side, where we found the musquitoes very troublesome.

August 4. A violent wind, accompanied by rain, purified

and cooled the atmosphere last night; we proceeded early, and reached a very narrow part of the river, where the channel is confined within a space of two hundred yards by a sand point on the north and a bend on the south; the banks in the neighbourhood washing away, the trees falling in, and the channel filled with buried logs. Above this is a trading house, on the south, where one of our party passed two years, trading with the Mahas. At nearly four miles is a creek on the south, emptying opposite a large island of sand; between this creek and our last night's encampment the river has changed its bed and encroached on the southern shore. About two miles farther is another creek on the south, which, like the former, is the outlet of three ponds, communicating with each other, and forming a small lake, which is fed by streams from the highlands. At fifteen miles, we encamped on the south. The hills on both sides of the river are nearly twelve or fifteen miles from each other, those of the north containing some timber, while the hills of the south are without any covering except some scattering wood in the ravines and near where the creeks pass into the hills, rich plains and prairies occupying the intermediate space, and partially covered, near the water, with cottonwood. There has been a great deal of pumice stone on the shore to-day.

August 5. We set out early, and by means of our oars made twenty and a half miles, though the river was crowded with sandbars. On both sides the prairies extend along the river, the banks being covered with great quantities of grapes, of which three different species are now ripe; one large, and resembling the purple grape. We had some rain this morning, attended by high wind; but generally speaking, have remarked that thunder storms are less frequent than in the Atlantic states, at this season. Snakes, too, are less frequent, though we killed one to-day of the shape and size of the rattlesnake, but of a lighter colour. We fixed our camp on the north side. In the evening, Captain Clark, in pursuing some game in an eastern direction, found himself at the distance of three hundred and seventy yards from the camp, at a point of the river whence we had come twelve miles. When the water is high this peninsula is overflowed, and judging from the customary and notorious changes in the river, a few years will be sufficient to force the main current of the river across, and leave the great bend dry. The whole lowland between the parallel range of hills seems formed of mud or ooze of the river, at some former period, mixed with sand and clay. The sand of the neighbouring banks accumulates with the aid of that brought down the stream, and forms sandbars, projecting into the river; these drive the channel to the opposite banks, the loose texture of which it undermines, and at length deserts its ancient bed for a new and shorter passage; it is thus that the banks of the Missouri are constantly falling, and the river changing its bed.

August 6. In the morning, after a violent storm of wind and rain from N. W., we passed a large island to the north. In the channel separating it from the shore, a creek called Soldier's river enters; the island kept it from our view, but one of our men who had seen it represents it as about forty yards wide at its mouth. At five miles, we came to a bend of the river towards the north; a sandbar, running in from the south, had turned its course so as to leave the old channel quite dry. We again saw the same appearance at our encampment twenty and a half miles distant on the north side. Here the channel of the river had encroached south, and the old bed was without water, except a few ponds. The sandbars are still very numerous.

August 7. We had another storm from the N. W. in the course of the last evening; in the morning we proceeded, having the wind from the north, and encamped on the northern shore, having rowed seventeen miles. The river is here encumbered with sandbars, but no islands except two small ones called Detachment islands, and formed on the south side by a small stream.

We despatched four men back to the Ottoes village in quest of our man, Liberte, and to apprehend one of the soldiers, who left us on the 4th, under pretense of recovering a knife which he had dropped a short distance behind, and who we fear has deserted. We also sent small presents to the Ottoes and Missouris, and requested that they would join us at the Maha village, where a peace might be concluded between them.

August 8. At two miles distance, this morning, we came to a part of the river where there was concealed timber difficult to pass. The wind was from the N. W., and we proceeded in safety. At six miles a river empties on the northern side called by the Sioux Indians Eanchwadepon, or Stone river; and by the French, Petite Riviere des Sioux, or Little Sioux river. At its confluence it is eighty yards wide. Our interpreter, Mr. Durion, who has been to the sources of it and knows the adjoining country, says that it rises within about nine miles of the river Desmoines; that within fifteen leagues of that river it passes through a large lake nearly sixty miles in circumference, and divided into two parts by rocks which approach each other very closely; its width is various, it contains many islands, and is known by the name of the Lac d'Esprit; it is near the Dogplains, and within four days march of the Mahas. The country watered by it is open and undulating, and may be visited in boats up the river for some distance. The Desmoines, he adds, is about eighty yards wide where the Little Sioux river approaches it; it is shoaly, and one of its principal branches is called the Cat river. Two miles beyond this river is a long island which we called Pelican island, from the numbers of that animal which were feeding on it; one of these being killed, we poured into his bag five gallons of water. An elk, too, was shot, and we had again to remark that snakes are rare in this part of the Missouri. A meridian altitude near the Little Sioux river made the latitude 41° 42' 34". We encamped on the north, having come sixteen miles.

August 9. A thick fog detained us until past seven o'clock, after which we proceeded with a gentle breeze from the southeast. After passing two sandbars, we reached, at seven and a half miles, a point of highland on the left, near which the river has forced itself a channel across a peninsula, leaving on the right a circuit of twelve or eighteen miles, which is now recognised by the ponds and islands it contains. At seventeen and a half miles we reached a point on the north, where we encamped. The hills are at a great distance from the river for the last several days; the land on both sides low, and covered with cottonwood and abundance of grapevines. An elk was seen to-day, a turkey also shot, and near our camp is a beaver den; the musquitoes have been more troublesome than ever for the last two days.

August 10. At two and a half miles, we came to a place called Coupée á Jacques, where the river has found a new bed, and abridged a circuit of several miles; at twelve and a half miles, a cliff of yellow stone on the left. This is the first highland near the river above the Council-bluff. After passing a number of sandbars we reached a willow island at the distance of twenty-two and a half miles, which we were enabled to do with our oars and a wind from the S. W., and encamped on the north side.

August 11. After a violent wind from the N. W. attended with rain, we sailed along the right of the island. At nearly five miles, we halted on the south side for the purpose of examining a spot where one of the great chiefs of the Mahas named Blackbird, who died about four years ago of the smallpox, was buried. A hill of yellow soft sandstone rises from the river in bluffs of various heights, till it ends in a knoll about three hundred feet above the water; on the top of this a mound, of twelve feet diameter at the base and six feet high, is raised over the body of the deceased king; a pole of about eight feet high is fixed in the centre, on which we placed a white flag, bordered with red, blue, and white. The Blackbird seems to have been a personage of great consideration; for ever since his death he is supplied with provisions, from time to time, by the superstitious regard of the Mahas. We descended to the river and passed a small creek on the south called by the Mahas, Waucandipeeche (Great Spirit is bad). Near this creek and the adjoining hills the Mahas had a village, and lost four hundred of their nation by the dreadful malady which destroyed the Blackbird. The meridian altitude made the latitude 42° 1' 3$\frac{8}{10}$" north. We encamped, at seventeen miles distance, on the north side in a bend of the river. During our day's course it has been crooked; we observed a number of places in it where the old channel is filled up, or gradually becoming covered with willow and cottonwood; great numbers of *herrons* are observed to-day, and the musquitoes annoy us very much.

August 12. A gentle breeze from the south carried us along about ten miles, when we stopped to take a meridian altitude, and sent a man across to our place of observation; yesterday he stepped nine hundred and seventy-four yards, and the distance we had come round was eighteen miles and three-quarters. The river is wider and shallower than usual. Four miles beyond this bend a bluff begins, and continues several miles; on the south it rises from the water at different heights, from twenty to one hundred and fifty feet, and higher as it recedes on the river; it consists of yellow and brown clay, with soft sandstone imbedded in it, and is covered with timber, among which may be observed some red cedar; the lands on the opposite side are low and subject to inundation, but contain willows, cottonwood, and many grapes. A prairie wolf came near the bank and barked at us; we attempted unsuccessfully to take him. This part of the river abounds in beaver. We encamped on a sand-island in a bend to the north, having made twenty miles and a quarter.

August 13. Set out at daylight with a breeze from the southeast, and passed several sandbars. Between ten and eleven miles we came to a spot on the south where a Mr. Mackay had a trading establishment in the year 1795 and 1796, which he called Fort Charles. At fourteen miles, we reached a creek on the south on which the Mahas reside; and at seventeen miles and a quarter, formed a camp on a sand-bar, to the south side of the river, opposite the lower point of a large island. From this place Sergeant Ordway and four men were detached to the Maha village with a flag and a present, in order to induce them to come and hold a council with us. They returned at twelve o'clock the next day, August 14. After crossing a prairie covered with high grass, they reached the Maha creek, along which they proceeded to its three forks, which join near the village; they crossed the north branch and went along the south; the walk was very fatiguing, as they were forced to break their way through grass, sunflowers, and thistles, all above ten feet high and interspersed with wild pea. Five miles from our camp they reached the position of the ancient Maha village; it had once consisted of three hundred cabins, but was burnt about four years ago, soon after the smallpox had destroyed four hundred men and a proportion of women and children. On a hill in the rear of the village are the graves of the nation, to the south of which runs the fork of the Maha creek; this they crossed where it was about ten yards wide, and followed its course to the Missouri, passing along a ridge of hill for one and a half mile, and a long pond between that and the Missouri; they then recrossed the Maha creek and arrived at the camp, having seen no tracks of Indians nor any sign of recent cultivation.

In the morning, 15th, some men were sent to examine the cause of a large smoke from the northeast, and which seemed to indicate that some Indians were near; but they found that a small party who had lately passed that way had left some trees burning, and that the wind from that quarter blew the smoke directly towards us. Our camp lies about three miles northeast from the old Maha village, and is in latitude 42° 13' 41". The accounts we have had of the effects of the smallpox on that nation are most distressing; it is not known in what way it was first communicated to them, though probably by some war party. They had been a military and powerful people; but when these warriors saw their strength wasting before a malady which they could not resist, their phrenzy was extreme; they burnt their village, and many of them put to death their wives and children, to save them from so cruel an affliction, and that all might go together to some better country.

On the 16th, we still waited for the Indians; a party had gone out yesterday to the Maha creek, which was dammed up by the beaver between the camp and the village; a second went to-day. They made a kind of drag with small willows and bark, and swept the creek; the first company brought three hundred and eighteen, the second upwards of eight hundred, consisting of pike, bass, fish resesmbling salmon, trout, redhorse, buffaloe, one rockfish, one flatback, perch, catfish,

## UP THE MISSOURI

a small species of perch called on the Ohio silverfish, a shrimp of the same size, shape, and flavour of those about Neworleans and the lower part of the Mississippi. We also found very fat muscles, and on the river, as well as the creek, are different kinds of ducks and plover. The wind, which in the morning had been from the northwest, shifted round in the evening to the southeast, and as usual we had a breeze, which cooled the air and relieved us from the musquitoes, who generally give us great trouble.

Friday, 17. The wind continued from the southeast, and the morning was fair. We observed about us a grass resembling wheat, except that the grain is like rye, also some similar to both rye and barley, and a kind of timothy, the seed of which branches from the main stock, and is more like a flaxseed than a timothy. In the evening, one of the party sent to the Ottoes returned with the information that the rest were coming on with the deserter; they had also caught Liberte, but by a trick he made his escape; they were bringing three of the chiefs in order to engage our assistance in making peace with the Mahas. This nation having left their village, that desirable purpose cannot be effected; but in order to bring in any neighbouring tribes, we set the surrounding prairies on fire. This is the customary signal made by traders to apprize the Indians of their arrival; it is also used between different nations as an indication of any event which they have previously agreed to announce in that way, and as soon as it is seen collects the neighbouring tribes, unless they apprehend that it is made by their enemies.

August 18. In the afternoon the party arrived with the Indians, consisting of the Little Thief and the Big Horse, whom we had seen on the third, together with six other chiefs and a French interpreter. We met them under a shade, and after they had finished a repast with which we supplied them, we inquired into the origin of the war between them and the Mahas, which they related with great frankness. It seems that two of the Missouris went to the Mahas to steal horses, but were detected and killed; the Ottoes and Missouris thought themselves bound to avenge their companions, and the whole nations were at last obliged to share in the dispute; they are also in fear of a war from the Pawnees, whose village they entered this summer, while the inhabitants were hunting, and stole their corn. This ingenuous confession did not make us the less desirous of negociating a peace for them; but no Indians have as yet been attracted by our fire. The evening was closed by a dance; and the next day,

August 19, the chiefs and warriors being assembled at ten o'clock, we explained the speech we had already sent from the Council-bluffs, and renewed our advice. They all replied in turn, and the presents were then distributed; we exchanged the small medal we had formerly given to the Big Horse for one of the same size with that of Little Thief; we also gave a small medal to a third chief, and a kind of certificate or letter of acknowledgment to five of the warriors expressive of our favour and their good intentions; one of them, dissatisfied, returned us the certificate; but the chief, fearful of our being offended, begged that it might be restored to him; this we declined, and rebuked them severely for having in view mere traffic instead of peace with their neighbours. This displeased them at first; but they at length all petitioned that it should be given to the warrior, who then came forward and made an apology to us; we then delivered it to the chief to be given to the most worthy, and he bestowed it on the same warrior, whose name was Great Blue Eyes. After a more substantial present of small articles and tobacco, the council was ended with a dram to the Indians. In the evening we exhibited different objects of curiosity, and particularly the airgun, which gave them great surprise. Those people are almost naked, having no covering, except a sort of breechcloth round the middle, with a loose blanket or buffaloe robe painted, thrown over them. The names of these warriors, besides those already mentioned, were Karkapaha (or Crow's head), and Nenasawa (or Black Cat), Missouris; and Sananona (or Iron Eyes), Neswaunja (or Big Ox), Stageaunja (or Big Blue Eyes), and Wasashaco (or Brave Man), all Ottoes. These two tribes speak very nearly the same language; they all begged us to give them whiskey.

The next morning, August 20, the Indians mounted their horses and left us, having received a cannister of whiskey at parting. We then set sail, and after passing two islands on the north, came to on that side under some bluffs, the first near the river since we left the Ayauwa village. Here we had the misfortune to lose one of our sergeants, Charles Floyd. He was yesterday seized with a bilious colic, and all our care and attention were ineffectual to relieve him. A little before his death he said to Captain Clark, "I am going to leave you"; his strength failed him as he added, "I want you to write me a letter"; but he died with a composure which justified the high opinion we had formed of his firmness and good conduct. He was buried on the top of the bluff with the honours due to a brave soldier; and the place of his interment marked by a cedar post, on which his name and the day of his death were inscribed. About a mile beyond this place, to which we gave his name, is a small river about thirty yards wide, on the north, which we called Floyd's river, where we encamped. We had a breeze from the southeast, and made thirteen miles.

August 21. The same breeze from the southeast carried us by a small willow creek on the north, about one mile and a half above Floyd's river. Here began a range of bluffs which continued till near the mouth of the great Sioux river, three miles beyond Floyd's. This river comes in from the north, and is about one hundred and ten yards wide. Mr. Durion, our Sioux interpreter, who is well acquainted with it, says that it is navigable upwards of two hundred miles to the falls, and even beyond them; that its sources are near those of the St. Peters. He also says that below the falls a creek falls in from the eastward, after passing through cliffs of red rock; of this the Indians make their pipes, and the necessity of procuring that article has introduced a sort of law of nations, by which the banks of the creek are sacred, and even tribes at war meet without hostility at these quarries, which possess a right of asylum. Thus we find even among savages certain principles deemed sacred, by which the rigours of their merciless system of warfare are mitigated. A sense of common danger, where stronger ties are wanting, gives all the binding force of more solemn obligations. The importance of preserving the known and settled rules of warfare among civilized nations, in all their integrity, becomes strikingly evident; since even savages, with their few precarious wants, cannot exist in a state of peace or war where this faith is once violated. The wind became southerly, and blew with such violence that we took a reef in our sail; it also blew the sand from the bars in such quantities that we could not see the channel at any distance ahead. At four and a quarter miles we came to two willow islands, beyond which are several sandbars; and at twelve miles, a spot where the Mahas once had a village, now no longer existing. We again passed a number of sandbars, and encamped on the south, having come twenty-four and three-quarter miles. The country through which we passed has the same uniform appearance ever since we left the river Platte: rich low grounds near the river, succeeded by undulating prairies, with timber near the waters. Some wolves were seen to-day on the sandbeaches to the south; we also procured an excellent fruit, resembling a red currant, growing on a shrub like the privy, and about the height of a wild plum.

August 22. About three miles distance, we joined the men who had been sent from the Maha village with our horses, and who brought us two deer. The bluffs or hills which reach the river at this place, on the south, contain allum, copperas, cobalt, which had the appearance of soft isinglass, pyrites, and sandstone, the two first very pure. Above this bluff comes in a small creek on the south, which we call Rologe creek. Seven miles above is another cliff, on the same side, of allum rock, of a dark brown colour, containing in its crevices great quantities of cobalt, cemented shells, and red earth. From this the river bends to the eastward, and approaches the Sioux river within three or four miles. We sailed the greater part of the day, and made nineteen miles to our camp on the north side. The sandbars are as usual numerous; there are also considerable traces of elk, but none are yet seen. Captain Lewis, in proving the quality of some of the substances in the first cliff, was considerably injured by the fumes and taste of the cobalt, and took some strong medicine to relieve him from its effects. The appearance of these mineral substances enable us to account for disorders of the stomach with which the party had been affected since they left the river Sioux. We had been in the habit of dipping up the water of the river inadvertently and making use of it, till, on examination, the sickness was thought to proceed from a scum covering the surface of the water along the southern shore, and which, as we now discovered, proceeded from these bluffs. The men had been ordered, before we reached the bluffs, to agitate the water, so as to disperse the scum, and take the water, not at the surface, but at some depth. The consequence was that these disorders ceased; the biles, too, which had afflicted the men, were not observed beyond the Sioux river. In order to supply the place of Sergeant Floyd, we permitted the men to name three persons, and Patrick Gass, having the greatest number of votes, was made a sergeant.

August 23. We set out early, and at four miles came to a small run between cliffs of yellow and blue earth; the wind, however, soon changed, and blew so hard from the west that we proceeded very slowly, the fine sand from the bar being driven in such clouds that we could scarcely see. Three and a quarter miles beyond this run, we came to a willow island, and a sand island opposite, and encamped on the south side, at ten and a quarter miles. On the north side is an extensive and delightful prairie, which we called Buffaloe prairie, from our having here killed the first buffaloe. Two elk swam the river to-day and were fired at, but escaped; a deer was killed from the boat; one beaver was killed, and several prairie wolves were seen.

August 24. It began to rain last night, and continued this morning; we proceeded, however, two and a quarter miles, to the commencement of a bluff of blue clay, about one hundred and eighty or one hundred and ninety feet on the south side; it seems to have been lately on fire, and even now the ground is so warm that we cannot keep our hands in it at any depth; there are strong appearances of coal, and also great quantities of cobalt, or a crystalized substance resembling it. There is a fruit now ripe which looks like a currant, except that it is double the size, and grows on a bush like a privy, the size of a damson, and of a delicious flavour;

its Indian name means rabbit-berries. We then passed, at the distance of about seven miles, the mouth of a creek on the north side, called by an Indian name, meaning Whitestone river. The beautiful prairie of yesterday has changed into one of greater height, and very smooth and extensive. We encamped on the south side, at ten and a quarter miles, and found ourselves much annoyed by the musquitoes.

# CHAPTER III.

Whimsical instance of superstition of the Sioux Indians—Council held with the Sioux—Character of that tribe, their manners, etc.—A ridiculous instance of their heroism—Ancient fortifications—Quicurre river described—Vast herds of Buffaloe—Account of the Petit Chien or Little Dog—Narrow escape of George Shannon—Description of White river—Surprising fleetness of the Antelope—Pass the river of the Sioux—Description of the Grand Le Tour, or Great Bend—Encamp on the Teton river.

AUGUST 25. Captains Lewis and Clark, with ten men, went to see an object deemed very extraordinary among all the neighbouring Indians. They dropped down to the mouth of Whitestone river, about thirty yards wide, where they left the boat, and at the distance of two hundred yards ascended a rising ground, from which a plain extended itself as far as the eye could discern. After walking four miles, they crossed the creek where it is twenty-three yards wide, and waters an extensive valley. The heat was so oppressive that we were obliged to send back our dog to the creek, as he was unable to bear the fatigue; and it was not till after four hours march that we reached the object of our visit. This was a large mound in the midst of the plain about N. 20° W. from the mouth of Whitestone river, from which it is nine miles distant. The base of the mound is a regular parallelogram, the longest side being about three hundred yards, the shorter sixty or seventy; from the longest side it rises with a steep ascent from the north and south to the height of sixty-five or seventy feet, leaving on the top a level plain of twelve feet in breadth and ninety in length. The north and south extremities are connected by two oval borders which serve as new bases, and divide the whole side into three steep but regular gradations from the plain. The only thing characteristic in this hill is its extreme symmetry, and this, together with its being totally detached from the other hills, which are at the distance of eight or nine miles, would induce a belief that it was artificial; but as the earth and the loose pebbles which compose it are arranged exactly like the steep grounds on the borders of the creek, we concluded from this similarity of texture that it might be natural. But the Indians have made it a great article of their superstition: it is called the mountain of Little People, or Little Spirits, and they believe that it is the abode of little devils, in the human form, of about eighteen inches high and with remarkably large heads; they are armed with sharp arrows, with which they are very skilful, and are always on the watch to kill those who should have the hardihood to approach their residence. The tradition is that many have suffered from these little evil spirits, and among others three Maha Indians fell a sacrifice to them a few years since. This has inspired all the neighbouring nations, Sioux, Mahas, and Ottoes, with such terror that no consideration could tempt them to visit the hill. We saw none of these wicked little spirits, nor any place for them, except some small holes scattered over the top; we were happy enough to escape their vengeance, though we remained some time on the mound to enjoy the delightful prospect of the plain, which spreads itself out till the eye rests upon the N. W. hills at a great distance, and those of the N. E. still farther off, enlivened by large herds of buffaloe feeding at a distance. The soil of these plains is exceedingly fine; there is, however, no timber except on the Missouri, all the wood of the Whitestone river not being sufficient to cover thickly one hundred acres. The plain country which surrounds this mound has contributed not a little to its bad reputation; the wind driving from every direction over the level ground obliges the insects to seek shelter on its leeward side, or be driven against us by the wind. The small birds, whose food they are, resort of course in great numbers in quest of subsistence; and the Indians always seem to discover an unusual assemblage of birds as produced by some supernatural cause; among them we observed the brown martin employed in looking for insects, and so gentle that they did not fly until we got within a few feet of them. We have also distinguished among the numerous birds of the plain the blackbird, the wren or prairie bird, and a species of lark about the size of a partridge, with a short tail. The excessive heat and thirst forced us from the hill, about one o'clock, to the nearest water, which we found in the creek at three miles distance, and remained an hour and a half. We then went down the creek, through a lowland about one mile in width, and crossed it three times, to the spot where we first reached it in the morning. Here we gathered some delicious plums, grapes, and blue currants, and afterwards arrived at the mouth of the river about sunset. To this place the course from the mound is S. twenty miles, E. nine miles; we there resumed our periogue, and on reaching our encampment of last night set the prairies on fire, to warn the Sioux of our approach. In the mean time the boat, under Sergeant Pryor, had proceeded in the afternoon one mile, to a bluff of blue clay on the south, and after passing a sandbar and two sand islands fixed their camp at the distance of six miles on the south. In the evening some rain fell. We had killed a duck and several birds; in the boat they had caught some large catfish.

Sunday, August 26. We rejoined the boat at nine o'clock before she set out, and then passing by an island, and under a cliff on the south, nearly two miles in extent and composed of white and blue earth, encamped at nine miles distance on a sandbar towards the north. Opposite to this, on the south,

## UP THE MISSOURI

is a small creek called Petit Arc or Little Bow, and a short distance above it an old village of the same name. This village, of which nothing remains but the mound of earth about four feet high surrounding it, was built by a Maha chief named Little Bow, who, being displeased with Blackbird, the late king, seceded with two hundred followers and settled at this spot, which is now abandoned, as the two villages have reunited since the death of Blackbird. We have great quantities of grapes, and plums of three kinds—two of a yellow colour and distinguished by one of the species being longer than the other, and a third round and red; all have an excellent flavour, particularly those of the yellow kind.

August 27. The morning star appeared much larger than usual. A gentle breeze from the southeast carried us by some large sandbars, on both sides and in the middle of the river, to a bluff, on the south side, at seven and a half miles distant; this bluff is of white clay or chalk, under which is much stone, like lime, incrusted with a clear substance, supposed to be cobalt, and some dark ore. Above this bluff we set the prairie on fire, to invite the Sioux. After twelve and a half-miles, we had passed several other sandbars, and now reached the mouth of a river called by the French Jacques (James river) or Yankton, from the tribe which inhabits its banks. It is about ninety yards wide at the confluence; the country which it waters is rich prairie, with little timber; it becomes deeper and wider above its mouth, and may be navigated a great distance, as its sources rise near those of St. Peter's, of the Mississippi, and the Red river of Lake Winnipeg. As we came to the mouth of the river, an Indian swam to the boat; and on our landing we were met by two others, who informed us that a large body of Sioux were encamped near us; they accompanied three of our men, with an invitation to meet us at a spot above the river; the third

Indian remained with us; he is a Maha boy, and says that his nation have gone to the Pawnees to make peace with them. At fourteen miles, we encamped on a sandbar to the north. The air was cool, the evening pleasant, the wind from the southeast, and light. The river has fallen gradually, and is now low.

Tuesday, 28th. We passed, with a stiff breeze from the south, several sandbars. On the south is a prairie which rises gradually from the water to the height of a bluff, which is, at four miles distance, of a whitish colour, and about seventy or eighty feet high. Farther on is another bluff, of a brownish colour, on the north side; and at the distance of eight and a half miles is the beginning of Calumet bluff, on the south side, under which we formed our camp, in a beautiful plain, to wait the arrival of the Sioux. At the first bluff the young Indian left us and joined their camp. Before reaching Calumet bluff one of the periogues ran upon a log in the river and was rendered unfit for service, so that all our loading was put into the second periogue. On both sides of the river are fine prairies, with cottonwood; and near the bluff there is more timber in the points and valleys than we have been accustomed to see.

Wednesday, 29th. We had a violent storm of wind and rain last evening, and were engaged during the day in repairing the periogue and other necessary occupations, when at four o'clock in the afternoon, Sergeant Pryor and his party arrived on the opposite side, attended by five chiefs and about seventy men and boys. We sent a boat for them and they joined us, as did also Mr. Durion, the son of our interpreter, who happened to be trading with the Sioux at this time. He returned with Sergeant Pryor to the Indians, with a present of tobacco, corn, and a few kettles, and told them that we would speak to their chiefs in the morning. Sergeant

Pryor reported that on reaching their village, which is at twelve miles distance from our camp, he was met by a party with a buffaloe robe, on which they desired to carry their visitors, an honour which they declined, informing the Indians that they were not the commanders of the boats. As a great mark of respect they were then presented with a fat dog, already cooked, of which they partook heartily, and found it well flavoured. The camps of the Sioux are of a conical form, covered with buffaloe robes, painted with various figures and colours, with an aperture in the top for the smoke to pass through. The lodges contain from ten to fifteen persons, and the interior arrangement is compact and handsome, each lodge having a place for cooking detached from it.

August 30, Thursday. The fog was so thick that we could not see the Indian camp on the opposite side, but it cleared off about eight o'clock. We prepared a speech, and some presents, and then sent for the chiefs and warriors, whom we received at twelve o'clock under a large oak tree, near to which the flag of the United States was flying. Captain Lewis delivered a speech, with the usual advice and counsel for their future conduct. We then acknowledged their chiefs, by giving to the grand chief a flag, a medal, a certificate, with a string of wampum; to which we added a chief's coat, that is, a richly laced uniform of the United States artillery corps, and a cocked hat and red feather. One second chief and three inferior ones were made or recognised by medals, and a suitable present of tobacco and articles of clothing. We then smoked the pipe of peace, and the chiefs retired to a bower, formed of bushes by their young men, where they divided among each other the presents, and smoked and eat, and held a council on the answer which they were to make us to-morrow. The young people exercised their bows and arrows in shooting at marks for beads, which we distributed to the best marksmen; and in the evening the whole party danced until a late hour, and in the course of their amusement we threw among them some knives, tobacco, bells, tape, and binding, with which they were much pleased. Their musical instruments were the drum, and a sort of little bag made of buffaloe hide, dressed white, with small shot or pebbles in it, and a bunch of hair tied to it. This produces a sort of rattling music, with which the party was annoyed by four musicians during the council this morning.

August 31. In the morning, after breakfast, the chiefs met and sat down in a row, with pipes of peace, highly ornamented, and all pointed towards the seats intended for Captains Lewis and Clark. When they arrived and were seated, the grand chief, whose Indian name, Weucha, is in English Shake Hand, and in French is called Le Liberateur (the deliverer), rose and spoke at some length, approving what we had said, and promising to follow our advice.

"I see before me," said he, "my great father's two sons. You see me, and the rest of our chiefs and warriors. We are very poor; we have neither powder nor ball nor knives, and our women and children at the village have no clothes. I wish that as my brothers have given me a flag and a medal, they would give something to those poor people, or let them stop and trade with the first boat which comes up the river. I will bring chiefs of the Pawnees and Mahas together, and make peace between them; but it is better that I should do it than my great father's sons, for they will listen to me more readily. I will also take some chiefs to your country in the spring; but before that time I cannot leave home. I went formerly to the English, and they gave me a medal and some clothes; when I went to the Spanish they gave me a medal, but nothing to keep it from my skin; but now you give me a medal and clothes. But still we are poor; and I wish, brothers, you would give us something for our squaws."

When he sat down, Mahtoree, or White Crane, rose.

"I have listened," said he, "to what our father's words were yesterday; and I am to-day glad to see how you have dressed our old chief. I am a young man, and do not wish to take much; my fathers have made me a chief; I had much sense before, but now I think I have more than ever. What the old chief has declared I will confirm, and do whatever he and you please; but I wish that you would take pity on us, for we are very poor."

Another chief, called Pawnawneahpahbe, then said:

"I am a young man, and know but little; I cannot speak well; but I have listened to what you have told the old chief, and will do whatever you agree."

The same sentiments were then repeated by Aweawechache.

We were surprised at finding that the first of these titles means, "Struck by the Pawnee," and was occasioned by some blow which the chief had received in battle, from one of the Pawnee tribe. The second is, in English, "Half Man," which seems a singular name for a warrior, till it was explained to have its origin, probably, in the modesty of the chief, who, on being told of his exploits, would say, "I am no warrior: I am only half a man." The other chiefs spoke very little; but after they had finished, one of the warriors delivered a speech, in which he declared he would support them. They promised to make peace with the Ottoes and Missouris, the only nations with whom they are at war. All these harangues concluded by describing the distress of the nation; they begged us to have pity on them; to send them traders; that they wanted powder and ball; and seemed anxious that we should supply them with some of their great father's milk, the name by which they distinguish ardent spirits. We then gave some tobacco to each of the chiefs, and a certificate to two of the warriors who attended the chief. We prevailed on Mr. Durion to remain here, and accompany as many of the Sioux chiefs as he could collect down to the seat of government. We also gave his son a flag, some clothes, and provisions, with directions to bring about a peace between the surrounding tribes, and to convey some of their chiefs to see the President. In the evening they left us, and encamped on the opposite bank, accompanied by the two Durions. During the evening and night we had much rain, and observed that the river rises a little. The Indians who have just left us are the Yanktons, a tribe of the great nation of Sioux. These Yanktons are about two hundred men in number, and inhabit the Jacques, Desmoines, and Sioux rivers. In person they are stout, well proportioned, and have a certain air of dignity and boldness. In their dress they differ nothing from the other bands of the nation whom we saw and will describe afterwards; they are fond of decorations, and use paint and porcupine quills and feathers. Some of them wore a kind of necklace of white bear's claws, three inches long, and closely strung together round their necks. They have only a few fowling pieces, being generally armed with bows and arrows, in which, however, they do not appear as expert as the more northern Indians. What struck us most was an institution, peculiar to them, and to the Kite Indians farther to the westward, from whom it is said to have been copied. It is an association of the most active and brave young men, who are bound to each other by attachment, secured by a vow never to retreat before any danger, or give way to their enemies. In war they go forward without sheltering themselves behind trees, or aiding their natural valour by any artifice. This punctilious determination not to be turned from their course became heroic, or ridiculous, a short time since, when the Yanktons were crossing the Missouri on the ice. A hole lay immediately in their course, which might easily have been avoided by going round. This the foremost of the band disdained to do, but went straight forward, and was lost. The others would have followed his example, but were forcibly prevented by the rest of the tribe. These young men sit, and encamp, and dance together, distinct from the rest of the nation; they are generally about thirty or thirty-five years old; and such is the deference paid to courage that their seats in council are superior to those of the chiefs, and their persons more respected. But, as may be supposed, such indiscreet bravery will soon diminish the numbers of those who practice it; so that the band is now reduced to four warriors, who were among our visitors. These were the remains of twenty-two, who composed the society not long ago; but in a battle with the Kite Indians, of the Black mountains, eighteen of them were killed, and these four were dragged from the field by their companions.

Whilst these Indians remained with us we made very minute inquiries relative to their situation and numbers and trade and manners. This we did very satisfactorily, by means of two different interpreters; and from their accounts, joined to our interviews with other bands of the same nation, and much intelligence acquired since, we were enabled to understand, with some accuracy, the condition of the Sioux, hitherto so little known.

The Sioux, or Dacorta Indians, originally settled on the Mississippi, and called by Carver, Madowesians, are now subdivided into tribes, as follows:

First, the Yanktons. This tribe inhabits the Sioux, Desmoines, and Jacque rivers, and numbers about two hundred warriors.

Second, the Tetons of the burnt woods. This tribe numbers about three hundred men, who rove on both sides of the Missouri, the White, and the Teton rivers.

Third, the Tetons Okandandas, a tribe consisting of about one hundred and fifty men, who inhabit both sides of the Missouri below the Chayenne river.

Fourth, Tetons Minnakenozzo, a nation inhabiting both sides of the Missouri above the Chayenne river, and containing about two hundred and fifty men.

Fifth, Tetons Saone. These inhabit both sides of the Missouri below the Warreconne river, and consist of about three hundred men.

Sixth, Yanktons of the Plains, or Big Devils, who rove on the heads of the Sioux, Jacques, and Red rivers; the most numerous of all the tribes, and number about five hundred men.

Seventh, Wahpatone, a nation residing on the St. Peter's just above the mouth of that river, and numbering two hundred men.

Eighth, Mindawarcarton, or proper Dacorta or Sioux Indians. These possess the original seat of the Sioux, and are properly so denominated. They rove on both sides of the Mississippi about the falls of St. Anthony, and consist of three hundred men.

Ninth, The Wahpatoota, or Leaf Beds. This nation inhabits both sides of the river St. Peter's below Yellowwood river, amounting to about one hundred and fifty men.

Tenth, Sistasoone. This nation numbers two hundred men, and reside at the head of the St. Peter's Of these several tribes, more particular notice will be taken hereafter.

Saturday, September 1, 1804. We proceeded this morning under a light southern breeze, and passed the Calumet bluffs; these are composed of a yellowish red and brownish clay as hard as chalk, which it much resembles, and are one hundred and seventy or one hundred and eighty feet high. At this place the hills on each side come to the verge of the river, those on the south being higher than on the north. Opposite the bluffs is a large island covered with timber, above which the highlands form a cliff over the river on the north side called White Bear cliff, an animal of that kind being killed in one of the holes in it, which are numerous and apparently deep. At six miles we came to a large sand island covered with cottonwood; the wind was high, and the weather rainy and cloudy during the day. We made fifteen miles to a place on the north side, at the lower point of a large island called Bonhomme, or Goodman's island. The country on both sides has the same character of prairies, with no timber; with occasional lowlands covered with cottonwood, elm, and oak; our hunters had killed an elk and a beaver; the catfish, too, are in great abundance.

September 2. It rained last night, and this morning we had a high wind from the N. W. We went three miles to the lower part of an ancient fortification on the south side, and passed the head of Bonhomme island, which is large and well timbered; after this the wind became so violent, attended by a cold rain, that we were compelled to land at four miles on the northern side, under a high bluff of yellow clay about one hundred and ten feet in height. Our hunters supplied us with four elk, and we had grapes and plums on the banks; we also saw the beargrass and rue, on the side of the bluffs. At this place there are highlands on both sides of the river, which become more level at some distance back, and contain but few streams of water. On the southern bank, during this day, the grounds have not been so elevated. Captain Clark crossed the river to examine the remains of the fortification we had just passed.

This interesting object is on the south side of the Missouri, opposite the upper extremity of Bonhomme island, and in a low level plain, the hills being three miles from the river. It begins by a wall composed of earth, rising immediately from the bank of the river and running in a direct course S. 76°, W. ninety-six yards; the base of this wall or mound is seventy-five feet, and its height about eight. It then diverges in a course S. 84° W., and continues at the same height and depth to the distance of fifty-three yards, the angle being formed by a sloping descent; at the junction of these two is an appearance of a hornwork of the same height with the first angle; the same wall then pursues a course N. 69° W. for three hundred yards; near its western extremity is an opening or gateway at right angles to the wall, and projecting inwards; this gateway is defended by two nearly semi-circular walls placed before it, lower than the large walls, and from the gateway there seems to have been a covered way communicating with the interval between these two walls; westward of the gate the wall becomes much larger, being about one hundred and five feet at its base, and twelve feet high; at the end of this high ground the wall extends for fifty-six yards on a course N. 32° W.; it then turns N. 23° W. for seventy-three yards; these two walls seem to have had a double or covered way; they are from ten to fifteen feet eight inches in height, and from seventy-five to one hundred and five feet in width at the base, the descent inwards being steep, whilst outwards it forms a sort of glacis. At the distance of seventy-three yards, the wall ends abruptly at a large hollow place much lower than the general level of the plain, and from which is some indication of a covered way to the water. The space between them is occupied by several mounds scattered promiscuously through the gorge, in the centre of which is a deep round hole. From the extremity of the last wall, in a course N. 32°

W., is a distance of ninety-six yards over the low ground, where the wall recommences and crosses the plain in a course N. 81° W. for eighteen hundred and thirty yards to the bank of the Missouri. In this course its height is about eight feet, till it enters, at the distance of five hundred and thirty-three yards, a deep circular pond of seventy-three yards diameter; after which it is gradually lower towards the river; it touches the river at a muddy bar, which bears every mark of being an encroachment of the water, for a considerable distance, and a little above the junction is a small circular redoubt. Along the bank of the river, and at eleven hundred yards distance, in a straight line from this wall, is a second, about six feet high and of considerable width; it rises abruptly from the bank of the Missouri, at a point where the river bends and goes straight forward, forming an acute angle with the last wall till it enters the river again, not far from the mounds just described, towards which it is obviously tending. At the bend the Missouri is five hundred yards wide; the ground on the opposite side highlands, or low hills on the bank; and where the river passes between this fort and Bonhomme island, all the distance from the bend, it is constantly washing the banks into the stream, a large sandbank being already taken from the shore near the wall. During the whole course of this wall, or glacis, it is covered with trees, among which are many large cotton trees two or three feet in diameter. Immediately opposite the citadel, or the part most strongly fortified, on Bonhomme island, is a small work in a circular form, with a wall surrounding it about six feet in height. The young willows along the water, joined to the general appearance of the two shores, induce a belief that the bank of the island is encroaching, and the Missouri indemnifies itself by washing away the base of the fortification. The citadel contains about twenty acres, but the parts between the long walls must embrace nearly five hundred acres.

These are the first remains of the kind which we have had an opportunity of examining; but our French interpreters assure us that there are great numbers of them on the Platte, the Kanzas, the Jacques, etc., and some of our party say that they observed two of those fortresses on the upper side of the Petit Arc creek not far from its mouth; that the wall was about six feet high, and the sides of the angles one hundred yards in length.

September 3. The morning was cold, and the wind from the northwest. We passed at sunrise three large sandbars, and at the distance of ten miles reached a small creek about twelve yards wide coming in from the north, above a white bluff; this creek has obtained the name of Plum creek, from the number of that fruit which are in the neighbourhood, and of a delightful quality. Five miles farther, we encamped on the south near the edge of a plain; the river is wide, and covered with sandbars to-day; the banks are high and of a whitish colour; the timber scarce, but an abundance of grapes. Beaver houses, too, have been observed in great numbers on the river, but none of the animals themselves.

September 4. We set out early, with a very cold wind from S. S. E., and at one mile and a half reached a small creek called Whitelime creek, on the south side. Just above this is a cliff covered with cedar trees, and at three miles a creek called Whitepaint creek, of about thirty yards wide; on the same side, and at four and a half miles distance from the Whitepaint creek, is the Rapid river, or, as it is called by the French, la Riviere qui Court; this river empties into the Missouri in a course S. W. by W., and is one hundred and fifty-two yards wide and four feet deep at the confluence. It rises in the Black mountains, and passes through a hilly country with a poor soil. Captain Clark ascended three miles to a beautiful plain on the upper side, where the Pawnees once had a village; he found that the river widened above its mouth, and much divided by sands and islands, which, joined to the great rapidity of the current, makes the navigation very difficult even for small boats. Like the Platte its waters are of a light colour; like that river, too, it throws out into the Missouri great quantities of sand, coarser even than that of the Platte, which form sandbars and shoals near its mouth.

We encamped just above it on the south, having made only eight miles, as the wind shifted to the south, and blew so hard that in the course of the day we broke our mast. We saw some deer, a number of geese, and shot a turkey and a duck. The place in which we halted is a fine low ground, with much timber, such as red cedar, honey-locust, oak, arrow-wood, elm, and coffeenut.

September 5, Wednesday. The wind was again high from the south. At five miles we came to a large island called Pawnee island, in the middle of the river, and stopped to breakfast at a small creek on the north, which has the name of Goat creek, at eight and a half miles. Near the mouth of this creek the beaver had made a dam across so as to form a large pond, in which they built their houses. Above this island the river Poncara falls into the Missouri from the south, and is thirty yards wide at the entrance, Two men whom we despatched to the village of the same name returned with information that they had found it on the lower side of the creek; but as this is the hunting season the town was so completely deserted that they had killed a buffaloe in the village itself. This tribe of Poncaras, who are said to have once numbered four hundred men, are now reduced to about fifty, and have associated for mutual protection with the

Mahas, who are about two hundred in number. These two nations are allied by a similarity of misfortune; they were once both numerous, both resided in villages, and cultivated Indian corn; their common enemies, the Sioux and small-pox, drove them from their towns, which they visit only occasionally for the purposes of trade; and they now wander over the plains on the sources of the Wolf and Quicourt rivers. Between the Pawnee island and Goat creek on the north is a cliff of blue earth, under which are several mineral springs impregnated with salts; near this we observed a number of goats, from which the creek derives its name. At three and a half miles from the creek we came to a large island on the south, along which we passed to the head of it, and encamped about four o'clock. Here we replaced the mast we had lost with a new one of cedar; some bucks and an elk were procured to-day, and a black-tailed deer was seen near the Poncara's village.

Thursday, September 6. There was a storm this morning from the N. W., and though it moderated, the wind was still high and the weather very cold; the number of sandbars, too, added to the rapidity of the current, obliged us to have recourse to the towline; with all our exertions we did not make more than eight and a half miles, and encamped on the north, after passing high cliffs of soft, blue, and red-coloured stone, on the southern shore. We saw some goats and great numbers of buffaloe, in addition to which the hunters furnished us with elk, deer, turkies, geese, and one beaver; a large catfish, too, was caught in the evening. The ground near the camp was a low prairie, without timber, though just below is a grove of cottonwood.

Friday, September 7. The morning was very cold and the wind southeast. At five and a half miles we reached and encamped at the foot of a round mountain, on the south, having passed two small islands. This mountain, which is about three hundred feet at the base, forms a cone at the top, resembling a dome at a distance, and seventy feet or more above the surrounding highlands. As we descended from this dome we arrived at a spot, on the gradual descent of the hill, nearly four acres in extent, and covered with small holes; these are the residence of a little animal called by the French petit chien (little dog), who sit erect near the mouth and make a whistling noise, but when alarmed take refuge in their holes. In order to bring them out we poured into one of the holes five barrels of water without filling it, but we dislodged and caught the owner. After digging down another of the holes for six feet, we found, on running a pole into it, that we had not yet dug half way to the bottom; we discovered, however, two frogs in the hole, and near it we killed a dark rattlesnake, which had swallowed a small prairie dog; we were also informed, though we never witnessed the fact, that a sort of lizard and a snake live habitually with these animals. The petit chien are justly named, as they resemble a small dog in some particulars, though they have also some points of similarity to the squirrel. The head resembles the squirrel in every respect, except that the ear is shorter, the tail like that of the ground-squirrel, the toe-nails are long, the fur is fine, and the long hair is gray.

Saturday, September 8. The wind still continued from the southeast, but moderately. At seven miles we reached a house on the north side, called the Pawnee house, where a trader named Trudeau wintered in the year 1796-7; behind this, hills much higher than usual appear to the north, about eight miles off. Before reaching this house we came by three small islands on the north side, and a small creek on the south; and after leaving it reached another, at the end of seventeen miles, on which we encamped, and called it Boat island; we here saw herds of buffaloe, and some elk, deer, turkies, beaver, a squirrel, and a prairie dog. The party on the north represent the country through which they passed as poor, rugged, and hilly, with the appearance of having been lately burnt by the Indians; the broken hills, indeed, approach the river on both sides, though each is bordered by a strip of woodland near the water.

Sunday, September 9. We coasted along the island on which we had encamped, and then passed three sand and willow islands and a number of smaller sandbars. The river is shallow, and joined by two small creeks from the north and one from the south. In the plains to the south are great numbers of buffaloe, in herds of nearly five hundred; all the copses of timber appear to contain elk or deer. We encamped on a sandbar on the southern shore at the distance of fourteen and a quarter miles.

September 10, Monday. The next day we made twenty miles. The morning was cloudy and dark, but a light breeze from the southeast carried us past two small islands on the south and one on the north, till, at the distance of ten and a half miles, we reached an island extending for two miles in the middle of the river, covered with red cedar, from which it derives its name of Cedar island. Just below this island, on a hill to the south, is the backbone of a fish, forty-five feet long, tapering towards the tail, and in a perfect state of petrifaction, fragments of which were collected and sent to Washington. On both sides of the river are high darkcoloured bluffs. About a mile and a half from the island, on the southern shore, the party on that side discovered a large and very strong impregnated spring of water; and another, not so strongly impregnated, half a mile up the hill. Three miles beyond Cedar island is a large island on the north, and a number of sandbars; after which is another, about a mile in length, lying in the middle of the river, and separated by a small channel at its extremity from another above it, on which we encamped. These two islands are called Mud islands. The river is shallow during this day's course, and is falling a little. The elk and buffaloe are in great abundance, but the deer have become scarce.

September 11, Tuesday. At six and a half miles we passed the upper extremity of an island on the south, four miles beyond which is another on the same side of the river; and about a quarter of a mile distant we visited a large village of the barking squirrel. It was situated on a gentle declivity, and covered a space of nine hundred and seventy yards long and eight hundred yards wide; we killed four of them. We then resumed our course, and during five and a half miles passed two islands on the north, and then encamped at the distance of sixteen miles, on the south side of the river and just above a small run. The morning had been cloudy, but in the afternoon it began raining, with a high northwest wind, which continued during the greater part of the night. The country seen to-day consists of narrow strips of lowland, rising into uneven grounds, which are succeeded, at the distance of three miles, by rich and level plains, but without any timber. The river itself is wide, and crowded with sandbars. Elk, deer, squirrels, a pelican, and a very large porcupine were our game this day; some foxes, too, were seen, but not caught.

In the morning we observed a man riding on horseback down towards the boat, and we were much pleased to find that it was George Shannon, one of our party, for whose safety we had been very uneasy. Our two horses having strayed from us on the 26th of August, he was sent to search for them. After he had found them he attempted to rejoin us, but seeing some other tracks, which must have been those of Indians, and which he mistook for our own, he concluded that we were ahead, and had been for sixteen days following the bank of the river above us. During the first four days he exhausted his bullets, and was then nearly starved, being obliged to subsist for twelve days on a few grapes and a rabbit which he killed by making use of a hard piece of stick for a ball. One of his horses gave out and was left behind, the other he kept as a last resource for food. Despairing of overtaking us, he was returning down the river in hopes of meeting some other boat, and was on the point of killing his horse when he was so fortunate as to join us.

Wednesday, September 12. The day was dark and cloudy, the wind from the northwest. At a short distance we reached an island in the middle of the river which is covered with timber, a rare object now. We with great difficulty were enabled to struggle through the sandbars, the water being very rapid and shallow, so that we were several hours in making a mile. Several times the boat wheeled on the bar, and the men were obliged to jump out and prevent her from upsetting; at others, after making a way up one channel the shoalness of the water forced us back to seek the deep channel. We advanced only four miles in the whole day, and encamped on the south. Along both sides of the river are high grounds; on the southern side, particularly, they form dark bluffs, in which may be observed slate and coal intermixed. We saw also several villages of barking squirrels, great numbers of growse, and three foxes.

September 13, Thursday. We made twelve miles to-day through a number of sandbars, which make it difficult to find the proper channel. The hills on each side are high, and separated from the river by a narrow plain on its borders. On the north these lowlands are covered in part with timber, and great quantities of grapes, which are now ripe; on the south we found plenty of plums, but they are not yet ripe, and near the dark bluffs a run tainted with allum and copperas, the southern side being more strongly impregnated with minerals than the northern. Last night four beaver were caught in the traps; a porcupine was shot as it was upon a cottontree, feeding on its leaves and branches. We encamped on the north side, opposite to a small willow island. At night the musquitoes were very troublesome, though the weather was cold and rainy and the wind from the northwest.

Friday, September 14. At two miles we reached a round island on the northern side; at about five, a run on the south; two and a half miles farther, a small creek; and at nine miles encamped near the mouth of a creek, on the same side. The sandbars are very numerous and render the river wide and shallow, and obliged the crew to get into the water and drag the boat over the bars several times. During the whole day we searched along the southern shore and at some distance into the interior, to find an ancient volcano which we heard at St. Charles was somewhere in this neighbourhood; but we could not discern the slightest appearance of anything volcanic. In the course of their search the party shot a buckgoat and a hare. The hills, particularly on the south, continue high, but the timber is confined to the islands and banks of the river. We had occasion here to observe the rapid undermining of these hills by the Missouri. The first attacks seem to be on the hills which overhang the river; as soon as the violence of the current destroys the grass at the foot of them the whole texture appears loosened, and the ground dissolves and mixes with the water; the muddy mixture is then forced over the low grounds, which it covers sometimes to the depth of three inches, and gradually destroys the herbage, after which it can offer no resistance to the water and becomes at last covered with sand.

Saturday, September 15. We passed at an early hour the creek near our last night's encampment, and at two miles distance reached the mouth of White river, coming in from the south. We ascended a short distance, and sent a sergeant and another man to examine it higher up. This river has a bed of about three hundred yards, though the water is confined to one hundred and fifty; in the mouth is a sand island and several sandbars. The current is regular and swift, with sandbars projecting from the points. It differs very much from the Platte and Quicurre in throwing out comparatively little sand, but its general character is like that of the Missouri. This resemblance was confirmed by the sergeant, who ascended about twelve miles, at which distance it was about the same width as near the mouth, and the course, which was generally west, had been interrupted by islands and sandbars. The timber consisted chiefly of elm; they saw pine burrs, and sticks of birch were seen floating down the river; they had also met with goats such as we have heretofore seen, great quantities of buffaloe, near to which were wolves, some deer, and villages of barking squirrels. At the confluence of White river with the Missouri is an excellent position for a town, the land rising by three gradual ascents, and the neighbourhood furnishing more timber than is usual in this country. After passing high dark bluffs on both sides, we reached the lower point of an island towards the south, at the distance of six miles. The island bears an abundance of grapes, and is covered with red cedar; it also contains a number of rabbits. At the end of this island, which is small, a narrow channel separates it from a large sand island, which we passed and encamped eight miles on the north, under a high point of land opposite a large creek to the south, on which we observe an unusual quantity of timber. The wind was from the northwest this afternoon and high, the weather cold, and its dreariness increased by the howling of a number of wolves around us.

September 16, Sunday. Early this morning, having reached a convenient spot on the south side and at one mile and a quarter distance, we encamped just above a small creek, which we called Corvus, having killed an animal of that genus near it. Finding that we could not proceed over the sandbars as fast as we desired while the boat was so heavily loaded, we concluded not to send back, as we originally intended, our third periogue, but to detain the soldiers until spring, and in the mean time lighten the boat by loading the periogue; this operation, added to that of drying all our wet articles, detained us during the day. Our camp is in a beautiful plain, with timber thinly scattered for three-quarters of a mile, and consisting chiefly of elm, cottonwood, some ash of an indifferent quality, and a considerable quantity of a small species of white oak; this tree seldom rises higher than thirty feet, and branches very much; the bark is rough, thick, and of a light colour; the leaves small, deeply indented, and of a pale green; the cup which contains the acorn is fringed on the edges, and embraces it about one-half; the acorn itself, which grows in great profusion, is of an excellent flavour and has none of the roughness which most other acorns possess; they are now falling, and have probably attracted the number of deer which we saw on this place, as all the animals we have seen are fond of that food. The ground, having been recently burnt by the Indians, is covered with young green grass, and in the neighbourhood are great quantities of fine plums. We killed a few deer for the sake of their skins, which we wanted to cover the periogues, the meat being too poor for food. The cold season coming on, a flannel shirt was given to each man, and fresh powder to those who had exhausted their supply.

Monday, September 17. Whilst some of the party were engaged in the same way as yesterday, others were employed in examining the surrounding country. About a quarter of a mile behind our camp, and at an elevation of twenty feet above it, a plain extends nearly three miles parallel to the river and about a mile back to the hills, towards which it gradually ascends. Here we saw a grove of plum trees loaded with fruit, now ripe, and differing in nothing from those of the Atlantic states except that the tree is smaller and more thickly set. The ground of the plain is occupied by the burrows of multitudes of barking squirrels, who entice hither the wolves of a small kind, hawks, and polecats, all of which animals we saw, and presumed that they fed on the squirrel. This plain is intersected nearly in its whole extent by deep ravines and steep irregular rising grounds from one to two hundred feet. On ascending the range of hills which border the plain, we saw a second high level plain stretching to the south as far as the eye could reach. To the westward a high range of hills, about twenty miles distant, runs nearly north and south, but not to any great extent, as their rise and termination is embraced by one view, and they seemed covered with a verdure similar to that of the plains. The same view extended over the irregular hills which border the northern side of the Missouri. All around the country had been recently burnt, and a young green grass about four inches high covered the ground, which was enlivened by herds of antelopes and buffaloe, the last of which were in such multitudes that we cannot exaggerate in saying that at a single glance we saw three thousand of them before us. Of all the animals we had seen, the antelope seems to possess the most wonderful fleetness; shy and timorous, they generally repose only on the ridges, which command a view of all the approaches of an enemy; the acuteness of their sight distinguishes the most distant danger, the delicate sensibility of their smell defeats the precautions of concealment, and when alarmed their rapid career seems more like the flight of birds than the movements of an earthly being. After many unsuccessful attempts, Captain Lewis at last, by winding around the ridges, approached a party of seven which were on an eminence towards which the wind was unfortunately blowing. The only male of the party frequently encircled the summit of the hill as if to announce any danger to the females, who formed a group at the top. Although they did not see Captain Lewis the smell alarmed them, and they fled when he was at the distance of two hundred yards; he immediately ran to the spot where they had been; a ravine concealed them from him, but the next moment they appeared on a second ridge at the distance of three miles. He doubted whether it could be the same, but their number and the extreme rapidity with which they continued their course convinced him that they must have gone with a speed equal to that of the most distinguished racehorse. Among our acquisitions to-day was a mule-deer, a magpie, the common deer, and buffaloe; Captain Lewis also saw a hare, and killed a rattlesnake near the burrows of the barking squirrels.

Tuesday, September 18. Having everything in readiness we proceeded, with the boat much lightened, but the wind being from the N. W. we made but little way. At one mile we reached an island in the middle of the river nearly a mile in length and covered with red cedar; at its extremity a small creek comes in from the north; we then met some sandbars, and the wind being very high and ahead we encamped on the south, having made only seven miles. In addition to the common deer, which were in great abundance, we saw goats, elk, buffaloe, the black-tailed deer; the large wolves, too, are very numerous, and have long hair with coarse fur, and arc of a light colour. A small species of wolf about the size of a gray fox was also killed, and proved to be the animal which we had hitherto mistaken for a fox; there are also many porcupines, rabbits, and barking squirrels in the neighbourhood.

September 19. We this day enjoyed a cool clear morning, and a wind from the southeast. We reached at three miles a bluff on the south, and four miles farther the lower point of Prospect island, about two and a half miles in length; opposite to this are high bluffs about eighty feet above the water, beyond which are beautiful plains gradually rising as they recede from the river; these are watered by three streams which empty near each other; the first is about thirty-five yards wide, the ground on its sides high and rich, with some timber; the second about twelve yards wide, but with less timber; the third is nearly of the same size and contains more water, but it scatters its waters over the large timbered plain, and empties itself into the river at three places. These rivers are called by the French Les trois rivières des Sioux, the three Sioux rivers; and as the Sioux generally cross the Missouri at this place, it is called the Sioux pass of the three rivers. These streams have the same right of asylum, though in a less degree than Pipestone creek, already mentioned.

Two miles from the island we passed a creek fifteen yards wide; eight miles farther another, twenty yards wide; three miles beyond which is a third of eighteen yards width, all on the south side; the second, which passes through a high plain, we called Elm creek; to the third we gave the name of Night creek, having reached it late at night. About a mile beyond this is a small island on the north side of the river, and is called Lower island, as it is situated at the commencement of what is known by the name of the Grand Detour, or Great Bend of the Missouri. Opposite is a creek on the south about ten yards wide, which waters a plain where there are great numbers of the prickly pear, which name we gave to the creek.

## UP THE MISSOURI

We encamped on the south opposite the upper extremity of the island, having made an excellent day's sail of twenty-six and a quarter miles. Our game this day consisted chiefly of deer; of these, four were black-tails, one a buck with two main prongs of horns on each side and forked equally. Large herds of buffaloe, elk, and goats were also seen.

Thursday, September 20. Finding we had reached the Big Bend, we despatched two men with our only horse across the neck to hunt there and wait our arrival at the first creek beyond it. We then set out, with fair weather and the wind from S. E., to make the circuit of the bend. Near the lower island the sandbars are numerous and the river shallow. At nine and a half miles is a sand island, on the southern side. About ten miles beyond it is a small island on the south, opposite to a small creek on the north. This island, which is near the N. W. extremity of the bend, is called Solitary island. At about eleven miles farther we encamped on a sandbar, having made twenty-seven and a half miles. Captain Clark, who early this morning had crossed the neck of the bend, joined us in the evening. At the narrowest part the gorge is composed of high and irregular hills of about one hundred and eighty or one hundred and ninety feet in elevation; from this descends an unbroken plain over the whole of the bend, and the country is separated from it by this ridge. Great numbers of buffaloe, elk, and goats are wandering over these plains, accompanied by grouse and larks. Captain Clark saw a hare also, on the Great Bend. Of the goats killed to-day one is a female, differing from the male in being smaller in size; its horns, too, are smaller and straighter, having one short prong and no black about the neck; none of these goats have any beard, but are delicately formed and very beautiful.

Friday, September 21. Between one and two o'clock the serjeant on guard alarmed us by crying that the sandbar on which we lay was sinking. We jumped up, and found that both above and below our camp the sand was undermined and falling in very fast; we had scarcely got into the boats and pushed off when the bank under which they had been lying fell in, and would certainly have sunk the two periogues if they had remained there. By the time we reached the opposite shore the ground of our encampment sunk also. We formed a second camp for the rest of the night, and at daylight proceeded on to the gorge or throat of the Great Bend, where we breakfasted. A man whom we had despatched to step off the distance across the bend made it two thousand yards; the circuit is thirty miles. During the whole course the land of the bend is low, with occasional bluffs; that on the opposite side, high prairie ground and long ridges of dark bluffs. After breakfast, we passed through a high prairie on the north side and a rich cedar lowland and cedar bluff on the south, till we reached a willow island below the mouth of a small creek. This creek, called Tyler's river, is about thirty-five yards wide, comes in on the south, and is at the distance of six miles from the neck of the Great Bend. Here we found a deer and the skin of a white wolf, left us by our hunters ahead; large quantities of different kinds of plover and brants are in this neighbourhood, and seen collecting and moving towards the south; the catfish are small, and not in such plenty as we had found them below this place. We passed several sandbars, which make the river very shallow and about a mile in width, and encamped on the south at the distance of eleven and a half miles. On each side the shore is lined with hard rough gulley-stones, rolled from the hills and small brooks. The most common timber is the cedar, though in the prairies there are great quantities of the prickly pear. From this place we passed several sandbars, which make the river shallow and about a mile in width. At the distance of eleven and a half miles we encamped on the north at the lower point of an ancient island, which has since been connected with the main land by the filling up of the northern channel, and is now covered with cottonwood. We here saw some tracks of Indians, but they appeared three or four weeks old. This day was warm.

September 22. A thick fog detained us until seven o'clock; our course was through inclined prairies on each side of the river, crowded with buffaloe. We halted at a point on the north side, near a high bluff on the south, and took a meridian altitude, which gave us the latitude of 44° 11' $33\frac{3}{10}$". On renewing our course, we reached first a small island on the south at the distance of four and a half miles, immediately above which is another island opposite to a creek fifteen yards wide. This creek and the two islands, one of which is half a mile long and the second three miles, are called the Three Sisters, a beautiful plain extending on both sides of the river. This is followed by an island on the north called Cedar island, about one mile and a half in length and the same distance in breadth, and deriving its name from the quality of the timber. On the south side of this island is a fort and a large trading house built by a Mr. Loisel, who wintered here during the last year in order to trade with the Sioux, the remains of whose camps are in great numbers about this place. The establishment is sixty or seventy feet square, built with red cedar and picketted in with the same materials. The hunters who had been sent ahead joined us here. They mention that the hills are washed in gullies, in passing over which some mineral substances had rotted and destroyed their moccasins; they had killed two deer and a beaver. At sixteen miles distance we came to on the north side at the mouth of a small creek. The large stones which we saw yesterday on the shores are now some distance in the river and render the navigation dangerous. The musquitoes are still numerous in the low grounds.

Sunday, September 23. We passed, with a light breeze from the southeast, a small island on the north called Goat island, above which is a small creek called by the party Smoke creek, as we observed a great smoke to the southwest on approaching it. At ten miles we came to the lower point of a large island, having passed two small willow islands with sandbars projecting from them. This island, which we called Elk island, is about two and a half miles long and three-quarters of a mile wide, situated near the south, and covered with cottonwood, the red currant, and grapes. The river is here almost straight for a considerable distance, wide and shallow, with many sandbars. A small creek on the north, about sixteen yards wide, we called Reuben's creek, as Reuben Fields, one of our men, was the first of the party who reached it. At a short distance above this we encamped for the night, having made twenty miles. The country generally consists of low, rich, timbered ground on the north and high barren lands on the south; on both sides great numbers of buffaloe are feeding. In the evening three boys of the Sioux nation swam across the river and informed us that two parties of Sioux were encamped on the next river, one consisting of eighty and the second of sixty lodges, at some distance above. After treating them kindly we sent them back, with a present of two carrots of tobacco to their chiefs, whom we invited to a conference in the morning.

Monday, September 24. The wind was from the east and the day fair; we soon passed a handsome prairie on the north side covered with ripe plums, and the mouth of a creek on the south called Highwater creek, a little above our encampment. At about five miles we reached an island two and a half miles in length, and situated near the south. Here we were joined by one of our hunters who procured four elk, but whilst he was in pursuit of the game the Indians had stolen his horse. We left the island and soon overtook five Indians on the shore; we anchored, and told them from the boat we were friends and wished to continue so, but were not afraid of any Indians; that some of their young men had stolen the horse which their great father had sent for their great chief, and that we could not treat with them until he was restored. They said that they knew nothing of the horse, but if he had been taken he should be given up. We went on, and at eleven and a half miles passed an island on the north which we called Goodhumored island; it is about one and a half miles long, and abounds in elk. At thirteen and a half miles we anchored one hundred yards off the mouth of a river on the south side, where we were joined by both the periogues, and encamped; two-thirds of the party remained on board, and the rest went as a guard on shore with the cooks and one periogue; we have seen along the sides of the hills on the north a great deal of stone; besides the elk, we also observed a hare; the five Indians whom we had seen followed us, and slept with the guard on shore. Finding one of them was a chief we smoked with him, and made him a present of tobacco. This river is about seventy yards wide, and has a considerable current. As the tribe of the Sioux which inhabit it are called Teton, we gave it the name of Teton river.

# CHAPTER IV.

Council held with the Tetons—Their manners, dances, etc.—Chayenne river—Council held with the Ricara Indians—Their manners and habits—Strange instance of Ricara idolatry—Another instance—Cannonball river—Arrival among the Mandans—Character of the surrounding country, and of the creeks, islands, etc.

SEPTEMBER 25. The morning was fine, and the wind continued from the southeast. We raised a flagstaff and an awning, under which we assembled at twelve o'clock with all the party parading under arms. The chiefs and warriors from the camp two miles up the river met us, about fifty or sixty in number, and after smoking delivered them a speech; but as our Sioux interpreter, Mr. Durion, had been left with the Yanktons, we were obliged to make use of a Frenchman who could not speak fluently, and therefore we curtailed our harangue. After this we went through the ceremony of acknowledging the chiefs, by giving to the grand chief a medal, a flag of the United States, a laced uniform coat, a cocked hat and feather; to the two other chiefs, a medal and some small presents; and to two warriors of consideration, certificates. The name of the great chief is Untongasabaw, or Black Buffaloe; the second, Tortohonga, or the Partisan; the third, Tartongawaka, or Buffaloe Medicine; the name of one of the warriors was Wawzinggo, that of the second, Matocoquepa, or Second Bear. We then invited the chiefs on board, and showed them the boat, the airgun, and such curiosities as we thought might amuse them; in this we succeeded too well, for after giving them a quarter of a glass of whiskey, which they seemed to like very much, and sucked the bottle, it was with much difficulty that we could get rid of them. They at last accompanied Captain Clark on shore in a periogue with five men; but it seems they had formed a design to stop us, for no sooner had the party landed than three of the Indians seized the cable of the periogue, and one of the soldiers of the chief put his arms round the mast; the second chief, who affected intoxication, then said that we should not go on, that they had not received presents enough from us. Captain Clark told him that he would not be prevented from going on; that we were not squaws, but warriors; that we were sent by our great father, who could in a moment exterminate them. The chief replied that he, too, had warriors, and was proceeding to offer personal violence to Captain Clark, who immediately drew his sword and made a signal to the boat to prepare for action. The Indians, who surrounded him, drew their arrows from their quivers and were bending their bows, when the swivel in the boat was instantly pointed towards them, and twelve of our most determined men jumped into the periogue and joined Captain Clark. This movement made an impression on them, for the grand chief ordered the young men away from the periogue, and they withdrew and held a short council with the warriors. Being unwilling to irritate them, Captain Clark then went forward and offered his hand to the first and second chiefs, who refused to take it. He then turned from them and got into the periogue, but had not gone more than ten paces when both the chiefs and two of the warriors waded in after him, and he brought them on board. We then proceeded on for a mile and anchored off a willow island, which, from the circumstances which had just occurred, we called Badhumoured island.

Wednesday, September 26. Our conduct yesterday seemed to have inspired the Indians with fear of us, and as we were desirous of cultivating their acquaintance we complied with their wish that we should give them an opportunity of treating us well, and also suffer their squaws and children to see us and our boat, which would be perfectly new to them. Accordingly, after passing at one and a half mile a small willow island and several sandbars, we came to on the south side, where a crowd of men, women, and children were waiting to receive us. Captain Lewis went on shore and remained several hours, and observing that their disposition was friendly we resolved to remain during the night to a dance which they were preparing for us. Captains Lewis and Clark, who went on shore one after the other, were met on landing by ten well-dressed young men, who took them up in a robe highly decorated and carried them to a large council house, where they were placed on a dressed buffaloe skin by the side of the grand chief. The hall or council-room was in the shape of three-quarters of a circle, covered at the top and sides with skins well dressed and sewed together. Under this shelter sat about seventy men, forming a circle round the chief, before whom were placed a Spanish flag and the one we had given them yesterday. This left a vacant circle of about six feet diameter, in which the pipe of peace was raised on two forked sticks about six or eight inches from the ground, and under it the down of the swan was scattered; a large fire in which there were cooking provisions stood near, and in the centre about four hundred pounds of excellent buffaloe meat as a present for us. As soon as we were seated an old man got up, and after approving what we had done begged us to take pity on their unfortunate situation. To this we replied with assurances of protection. After he had ceased, the great chief rose and delivered an harangue to the same effect; then with great solemnity he took some of the most delicate parts of the dog, which was cooked for the festival, and held it to the flag by way of sacrifice; this done, he held up the pipe of peace, and first pointed it towards the heavens, then to the four quarters of the globe, and then to the earth, made a short speech, lighted the pipe, and presented it to us. We smoked, and he again harangued his people, after which the repast was served up to us. It consisted of the dog which they had just been cooking, this being a great dish among the Sioux and used on all festivals; to this were added pemitigon, a dish made of buffaloe meat, dried or jerked, and then pounded and mixed raw with grease, and a kind of ground potatoe, dressed like the preparation of Indian corn called hominy, to which it is little inferior. Of all these luxuries, which were placed before us in platters with horn spoons, we took the pemitigon and the potatoe, which we found good, but we could as yet partake but sparingly of the dog. We eat and smoked for an hour, when it became dark; everything was then cleared away for the dance, a large fire being made in the centre of the house, giving at once light and warmth to the ballroom. The orchestra was composed of about ten men, who played on a sort of tambourin formed of skin stretched across a hoop, and made a jingling noise with a long stick to which the hoofs of deer and goats were hung; the third instrument was a small skin bag with pebbles in it; these, with five or six young men for the vocal part, made up the band. The women then came forward highly decorated, some with poles in their hands on which were hung the scalps of their enemies; others with guns, spears, or different trophies, taken in war by their husbands, brothers, or connexions. Having arranged themselves in two columns, one on each side of the fire, as soon as the music began they danced towards each other till they met in the centre, when the rattles were shaken, and they all shouted and returned back to their places. They have no step, but shuffle along the ground; nor does the music appear to be anything more than a confusion of noises, distinguished only by hard or gentle blows upon the buffaloe skin; the song is perfectly extemporaneous. In the pauses of the dance any man of the company comes forward and recites, in a sort of low guttural tone, some little story or incident, which is either martial or ludicrous, or, as was the case this evening, voluptuous and indecent; this is taken up by the orchestra and the dancers, who repeat it in a higher strain and dance to it. Sometimes they alternate, the orchestra first performing, and when it ceases the women raise their voices and make a music more agreeable, that is, less intolerable than that of the musicians. The dances of the men, which are always separate from those of the women, are conducted very nearly in the same way, except that the men jump up and down instead of shuffling, and in the war dances the recitations are all of a military cast. The harmony of the entertainment had nearly been disturbed by one of the musicians, who, thinking he had not received a due share of the tobacco we had distributed during the evening, put himself into a passion, broke one of the drums, threw two of them into the fire, and left the band. They were taken out of the fire; a buffaloe robe held in one hand and beaten with the other, by several of the company, supplied the place of the lost drum or tambourin, and no notice was taken of the offensive conduct of the man. We staid till twelve o'clock at night, when we informed the chiefs that they must be fatigued with all these attempts to amuse us, and retired, accompanied by four chiefs, two of whom spent the night with us on board.

While on shore we saw twenty-five squaws and about the same number of children who had been taken prisoners two weeks ago in a battle with their countrymen the Mahas. In this engagement the Sioux destroyed forty lodges, killed seventy-five men, of which we saw many of the scalps, and took these prisoners; their appearance is wretched and dejected; the women, too, seem low in stature, coarse, and ugly, though their present condition may diminish their beauty. We gave them a variety of small articles, such as awls and needles, and interceded for them with the chiefs, to whom we recommended to follow the advice of their great father, to restore the prisoners and live in peace with the Mahas, which they promised to do.

The tribe which we this day saw are a part of the great Sioux nation, and are known by the name of the Teton Okandandas; they are about two hundred men in number, and their chief residence is on both sides of the Missouri, between the Chayenne and Teton rivers. In their persons they are rather ugly and ill made, their legs and arms being too small, their cheekbones high, and their eyes projecting. The females, with the same character of form, are more handsome, and both sexes appear cheerful and sprightly, but in our intercourse with them we discovered that they were cunning and vicious.

The men shave the hair off their heads except a small tuft on the top, which they suffer to grow and wear in plaits over the shoulders; to this they seem much attached, as the loss of it is the usual sacrifice at the death of near relations. In full dress, the men of consideration wear a hawk's feather, or calumet feather worked with porcupine quills, and fastened to the top of the head, from which it falls back. The face and body are generally painted with a mixture of grease and coal. Over the shoulders is a loose robe or mantle of buffaloe skin dressed white, adorned with porcupine quills, loosely fixed so as to make a gingling noise when in motion, and painted with various uncouth figures, unintelligible to us, but to them emblematic of military exploits or any other incident; the hair of the robe is worn next the skin in fair weather, but when it rains the hair is put outside, and the robe is either thrown over the arm or wrapped round the body, all of which it may cover. Under this in the winter season they wear a kind of shirt resembling ours and made either of skin or cloth, and covering the arms and body. Round the middle is fixed a girdle of cloth or procured dressed elk-skin, about an inch in width and closely tied to the body; to this is attached a piece of cloth or blanket or skin about a foot wide, which passes between the legs and is tucked under the girdle both before and behind; from the hip to the ancle he is covered by leggings of dressed antelope skins, with seams at the sides two inches in width, and ornamented by little tufts of hair, the produce of the scalps they have made in war, which are scattered down the leg. The winter moccasins are of dressed buffaloe-skin, the hair being worn inwards, and soaled with thick elk-skin parchment; those for summer are of deer or elk-skin dressed without the hair, and with soals of elk-skin. On great occasions, or wherever they are in full dress, the young men drag after them the entire skin of a polecat fixed to the heel of the moccasin. Another skin of the same animal is either tucked into the girdle or carried in the hand, and serves as a pouch for their tobacco, or what the French traders call the bois roule; this is the inner bark of a species of red willow, which, being dried in the sun or over the fire, is rubbed between the hands and broken into small pieces, and is used alone or mixed with tobacco. The pipe is generally of red earth, the stem made of ash, about three or four feet long, and highly decorated with feathers, hair, and porcupine quills.

The hair of the women is suffered to grow long and is parted from the forehead across the head, at the back of which it is either collected into a kind of bag or hangs down over the shoulders. Their moccasins are like those of the men, as are also the leggings, which do not, however, reach beyond the knee, where it is met by a long loose shift of skin which reaches nearly to the ancles; this is fastened over the shoulders by a string and has no sleeves, but a few pieces of the skin hang a short distance down the arm. Sometimes a girdle fastens this skin round the waist, and over all is thrown a robe like that worn by the men. They seem fond of dress. Their lodges are very neatly constructed, in the same form as those of the Yanktons; they consist of about one hundred cabins, made of white buffaloe hide dressed, with a larger one in the centre for holding councils and dances. They are built round, with poles about fifteen or twenty feet high covered with white skins; these lodges may be taken to pieces, packed up, and carried with the nation wherever they go by dogs which bear great burdens. The women are chiefly employed in dressing buffaloe skins; they seem perfectly well disposed, but are addicted to stealing anything which they can take without being observed. This nation, although it makes so many ravages among its neighbours, is badly supplied with guns. The water which they carry with them is contained chiefly in the paunches of deer and other animals, and they make use of wooden bowls. Some had their heads shaved, which we found was a species of mourning for relations. Another usage on these occasions is to run arrows through the flesh both above and below the elbow.

While on shore to-day we witnessed a quarrel between two squaws, which appeared to be growing every moment more boisterous, when a man came forward, at whose approach everyone seemed terrified and ran. He took the squaws and without any ceremony whipped them severely. On inquiring into the nature of such summary justice we learnt that this man was an officer well known to this and many other tribes. His duty is to keep the peace, and the whole interior police of the village is confided to two or three of these officers, who are named by the chief and remain in power some days, at least till the chief appoints a successor; they seem to be a sort of constable or sentinel, since they are always on the watch to keep tranquility during the day and guarding the camp in the night. The short duration of their office is compensated by its authority; his power is supreme, and in the suppression of any riot or disturbance no resistance to him is suffered; his person is sacred, and if in the execution of his duty he strikes even a chief of the second class he cannot be punished for this salutary insolence. In general they accompany the person of the chief, and when ordered to any duty, however dangerous, it is a point of honour rather to die than to refuse obedience. Thus, when they attempted to stop us yesterday the chief ordered one of these men to take possession of the boat; he immediately put his arms round the mast, and, as we understood, no force except the command of the chief would have induced him to release his hold. Like the other men their bodies are blackened, but their distinguishing mark is a collection of two or three raven skins fixed to the girdle behind the back in such a way that the tails stick out horizontally from the body. On his head, too, is a raven skin split into two parts and tied so as to let the beak project from the forehead.

Thursday, September 27. We rose early, and the two chiefs took off, as a matter of course and according to their custom, the blanket on which they had slept. To this we added a peck of corn as a present to each. Captain Lewis and the chiefs went on shore to see a part of the nation that was expected, but did not come. He returned at two o'clock with four of the chiefs and a warrior of distinction, called Wadrapa (or on his guard); they examined the boat and admired whatever was strange during half an hour, when they left it with great reluctance. Captain Clark accompanied them to the lodge of the grand chief, who invited them to a dance, where, being joined by Captain Lewis, they remained till a late hour. The dance was very similar to that of yesterday. About twelve we left them, taking the second chief and one principal warrior on board. As we came near the boat the man who steered the periogue by mistake brought her broadside against the boat's cable and broke it. We called up all hands to their oars, but our noise alarmed the two Indians; they called out to their companions, and immediately the whole camp crowded to the shore; but after half an hour they returned, leaving about sixty men near us. The alarm given by the chiefs was said to be that the Mahas had attacked us, and that they were desirous of assisting us to repel it; but we suspected that they were afraid we meant to set sail and intended to prevent us from doing so, for in the night the Maha prisoners had told one of our men who understood the language that we were to be stopped. We therefore, without giving any indication of our suspicion, prepared everything for an attack, as the loss of our anchor obliged us to come to near a falling bank very unfavourable for defence. We were not mistaken in these opinions, for when in the morning,

Friday, September 28, after dragging unsuccessfully for the anchor, we wished to set sail, it was with great difficulty that we could make the chiefs leave the boat. At length we got rid of all except the great chief, when, just as we were setting out, several of the chief's soldiers sat on the rope which held the boat to the shore. Irritated at this, we got everything ready to fire on them if they persisted, but the great chief said that these were his soldiers and only wanted some tobacco. We had already refused a flag and some tobacco to the second chief, who had demanded it with great importunity; but willing to leave them without going to extremities, we threw him a carrot of tobacco, saying to him,

"You have told us that you were a great man and have influence; now show your influence by taking the rope from those men, and we will then go without any further trouble." This appeal to his pride had the desired effect; he went out of the boat, gave the soldiers the tobacco, and pulling the rope out of their hands delivered it on board, and we then set sail under a breeze from the S. E. After sailing about two miles we observed the third chief beckoning to us; we took him on board, and he informed us that the rope had been held by the order of the second chief, who was a double-faced man. A little farther on we were joined by the son of the chief, who came on board to see his father. On his return we sent a speech to the nation, explaining what we had done and advising them to peace, but if they persisted in their attempts to stop us, we were willing and able to defend ourselves. After making six miles, during which we passed a willow island on the south and one sandbar, we encamped on another in the middle of the river. The country on the south side was a low prairie, that on the north highland.

September 29. We set out early, but were again impeded by sandbars, which made the river shallow; the weather was, however, fair; the land on the north side, low and covered with timber, contrasted with the bluffs to the south. At nine o'clock we saw the second chief and two women and three men on shore, who wished us to take two women offered by the second chief to make friends, which was refused; he then requested us to take them to the other band of their nation, who were on the river not far from us; this we declined, but in spite of our wishes they followed us along shore. The chief asked us to give them some tobacco; this we did, and gave more as a present for that part of the nation which we did not see. At seven and a half miles we came to a small creek on the southern side, where we saw great numbers of

## UP THE MISSOURI

elk, and which we called Notimber creek from its bare appearance. Above the mouth of this stream a Ricara band of Pawnees had a village five years ago, but there are no remains of it except the mound which encircled the town. Here the second chief went on shore. We then proceeded, and at the distance of eleven miles encamped on the lower part of a willow island in the middle of the river, being obliged to substitute large stones in the place of the anchor which we lost.

September 30. The wind was this morning very high from the southeast, so that we were obliged to proceed under a double-reefed mainsail through the rain. The country presented a large low prairie covered with timber on the north side; on the south, we first had high barren hills, but after some miles it became of the same character as that on the opposite side. We had not gone far when an Indian ran after us and begged to be carried on board as far as the Ricaras, which we refused; soon after, we discovered on the hills at a distance a great number of Indians, who came towards the river and encamped ahead of us. We stopped at a sandbar at about eleven miles, and after breakfasting proceeded on a short distance to their camp, which consisted of about four hundred souls. We anchored one hundred yards from the shore, and discovering that they were Tetons belonging to the band which we had just left, we told them that we took them by the hand, and would make each chief a present of tobacco; that we had been badly treated by some of their band, and that having waited for them two days below we could not stop here, but referred them to Mr. Durion for our talk and an explanation of our views. They then apologized for what had past, assured us that they were friendly and very desirous that we should land and eat with them; this we refused, but sent the periogue on shore with the tobacco, which was delivered to one of the soldiers of the chief whom we had on board. Several of them now ran along the shore after us, but the chief threw them a twist of tobacco and told them to go back and open their ears to our counsels, on which they immediately returned to their lodges. We then proceeded past a continuation of the low prairie on the north, where we had large quantities of grapes, and on the south saw a small creek and an island. Six miles above this two Indians came to the bank, looked at us about half an hour, and then went without speaking over the hills to the southwest. After some time the wind rose still higher and the boat struck a log, turned, and was very near taking in water. The chief became so much terrified at the danger that he hid himself in the boat, and as soon as we landed got his gun and told us that he wanted to return, that we would now see no more Tetons, and that we might proceed unmolested; we repeated the advice we had already given, presented him with a blanket, a knife, some tobacco, and after smoking with him he set out. We then continued to a sandbar on the north side, where we encamped, having come twenty and a half miles. In the course of the day we saw a number of sandbars which impede the navigation. The only animal which we observed was the white gull, then in great abundance.

October 1, 1804. The weather was very cold and the wind high from the southeast during the night, and continued so this morning. At three miles distance we had passed a large island in the middle of the river, opposite to the lower end of which the Ricaras once had a village on the south side of the river; there are, however, no remnants of it now except a circular wall three or four feet in height which encompassed the town. Two miles beyond this island is a river coming in from the southwest about four hundred yards wide, the current gentle and discharging not much water and very little sand; it takes its rise in the second range of the Cote Noire or Black mountains, and its general course is nearly east. This river has been occasionally called Dog river, under a mistaken opinion that its French name was Chien, but its true appellation is Chayenne, and it derives this title from the Chayenne Indians. Their history is the short and melancholy relation of the calamities of almost all the Indians. They were a numerous people and lived on the Chayenne, a branch of the Red river of Lake Winnipeg. The invasion of the Sioux drove them westward; in their progress they halted on the southern side of the Missouri below the Warreconne, where their ancient fortifications still exist; but the same impulse again drove them to the heads of the Chayenne, where they now rove, and occasionally visit the Ricaras. They are now reduced, but still number three hundred men.

Although the river did not seem to throw out much sand, yet near and above its mouth we find a great many sandbars difficult to pass. On both sides of the Missouri near the Chayenne are rich thinly timbered lowlands, behind which are bare hills. As we proceeded we found that the sandbars made the river so shallow and the wind was so high that we could scarcely find the channel, and at one place were forced to drag the boat over a sandbar, the Missouri being very wide and falling a little. At seven and a half miles we came to at a point and remained three hours, during which time the wind abated; we then passed within four miles two creeks on the south, one of which we called Centinel creek, and the other Lookout creek. This part of the river has but little timber; the hills are not so high as we have hitherto seen, and the number of sandbars extends the river to more than a mile in breadth. We continued about four and a half miles farther to a sandbar in the middle of the river, where we spent the night, our progress being sixteen miles. On the opposite shore we saw a house among the willows, and a boy to whom we called, and brought him on board. He proved to be a young Frenchman in the employ of a Mr. Valle, a trader, who is now here pursuing his commerce with the Sioux.

Tuesday, October 2. There had been a violent wind from the S. E. during the night, which having moderated we set sail with Mr. Valle, who visited us this morning and accompanied us for two miles. He is one of three French traders who have halted here, expecting the Sioux who are coming down from the Ricaras, where they now are, for the purposes of traffic. Mr. Valle tells us that he passed the last winter three hundred leagues up the Chayenne under the Black mountains. That river he represents as very rapid, liable to sudden swells, the bed and shores formed of coarse gravel, and difficult of ascent even for canoes. One hundred leagues from its mouth it divides into two branches, one coming from the south, the other at forty leagues from the junction enters the Black mountains. The land which it waters from the Missouri to the Black mountains resembles the country on the Missouri, except that the former has even less timber, and of that the greater proportion is cedar. The Chayennes reside chiefly on the heads of the river, and steal horses from the Spanish settlement, a plundering excursion which they perform in a month's time. The Black mountains he observes are very high, covered with great quantities of pine, and in some parts the snow remains during the summer. There are also great quantities of goats, white bear, prairie cocks, and a species of animal which from his description must resemble a small elk, with large circular horns.

At two and a half miles we had passed a willow island on the south; on the north side of the river were dark bluffs, and on the south low rich prairies. We took a meridian altitude on our arrival at the upper end of the isthmus of the bend, which we called the Lookout bend, and found the latitude to be 44° 19' 36". This bend is nearly twenty miles round, and not more than two miles across.

In the afternoon we heard a shot fired, and not long after observed some Indians on a hill; one of them came to the shore and wished us to land, as there were twenty lodges of Yanktons or Boisbrule there. We declined doing so, telling him that we had already seen his chiefs, and that they might learn from Mr. Durion the nature of the talk we had delivered to them. At nine miles we came to the lower point of a long island on the north, the banks of the south side of the river being high, those of the north forming a low rich prairie. We coasted along this island, which we called Caution island, and after passing a small creek on the south encamped on a sandbar in the middle of the river, having made twelve miles. The wind changed to the northwest and became very high and cold. The current of the river is less rapid, and the water, though of the same colour, contains less sediment than below the Chayenne, but its width continues the same. We were not able to hunt to-day, for as there are so many Indians in the neighbourhood we were in constant expectation of being attacked, and were therefore forced to keep the party together and be on our guard.

Wednesday, October 3. The wind continued so high from the northwest that we could not set out till after seven; we then proceeded till twelve o'clock, and landed on a bar towards the south, where we examined the periogues and the forecastle of the boat, and found that the mice had cut several bags of corn and spoiled some of our clothes. About one o'clock an Indian came running to the shore with a turkey on his back; several others soon joined him, but we had no intercourse with them. We then went on for three miles, but the ascent soon became so obstructed by sandbars and shoal water that, after attempting in vain several channels, we determined to rest for the night under some high bluffs on the south and send out to examine the best channel. We had made eight miles along high bluffs on each side. The birds we saw were the white gulls and the brant, which were flying to the southward in large flocks.

Thursday, 4th. On examination we found that there was no outlet practicable for us in this channel, and that we must retread our steps. We therefore returned three miles and attempted another channel, in which we were more fortunate. The Indians were in small numbers on the shore, and seemed willing, had they been more numerous, to molest us. They called to desire that we would land, and one of them gave three yells and fired a ball ahead of the boat; we, however, took no notice of it, but landed on the south to breakfast. One of these Indians swam across and begged for some powder; we gave him a piece of tobacco only. At eight and a half miles we had passed an island in the middle of the river, which we called Goodhope island. At one and a half mile we reached a creek on the south side about twelve yards wide, to which we gave the name of Teal creek. A little above this is an island, on the north side of the current, about one and a half mile in length and three-quarters of a mile in breadth. In the centre of this island is an old village of the Ricaras, called Lahoocat; it was surrounded by a circular wall containing seventeen lodges. The Ricaras are known to have lived there in 1797, and the village seems to have been deserted about five years since; it does not contain much timber. We encamped on a sandbar making out from the upper end of this island, our journey to-day being twelve miles.

Friday, October 5. The weather was very cold; yesterday evening and this morning there was a white frost. We sailed along the highlands on the north side, passing a small creek on the south, between three and four miles. At seven o'clock we heard some yells and saw three Indians of the Teton band, who asked us to come on shore and begged for some tobacco, to all which we gave the same answer as hitherto. At eight miles we reached a small creek on the north. At fourteen we passed an island on the south, covered with wild rye, and at the head a large creek comes in from the south, which we named Whitebrant creek, from seeing several white brants among flocks of dark-coloured ones. At the distance of twenty miles we came to on a sandbar towards the north side of the river, with a willow island opposite; the hills or bluffs come to the banks of the river on both sides, but are not so high as they are below; the river itself, however, continues of the same width, and the sandbars are quite as numerous. The soil of tile banks is dark-coloured, and many of the bluffs have the appearance of being on fire. Our game this day was a deer, a prairie wolf, and some goats out of a flock that was swimming across the river.

Saturday, October 6. The morning was still cold, the wind being from the north. At eight miles we came to a willow island on the north opposite a point of timber, where there are many large stones near the middle of the river which seem to have been washed from the hills and high plains on both sides, or driven from a distance down the stream. At twelve miles we halted for dinner at a village which we suppose to have belonged to the Ricaras; it is situated in a low plain on the river, and consists of about eighty lodges, of an octagon form, neatly covered with earth, placed as close to each other as possible and picketted round. The skin canoes, mats, buckets, and articles of furniture found in the lodges induce us to suppose that it had been left in the spring. We found three different sorts of squashes growing in the village; we also killed an elk near it and saw two wolves. On leaving the village the river became shallow, and after searching a long time for the main channel, which was concealed among sandbars, we at last dragged the boat over one of them rather than go back three miles for the deepest channel. At fourteen and a half miles we stopped for the night on a sandbar, opposite a creek on the north called Otter creek, twenty-two yards in width, and containing more water than is common for creeks of that size. The sides of the river during the day are variegated with high bluffs and low timbered grounds on the banks; the river is very much obstructed by sandbars. We saw geese, swan, brants, and ducks of different kinds on the sandbars, and on shore numbers of the prairie hen; the magpie, too, is very common, but the gulls and plover, which we saw in such numbers below, are now quite rare.

Sunday, October 7. There was frost again last evening, and this morning was cloudy and attended with rain. At two miles we came to the mouth of a river, called by the Ricaras Sawawkawna, or Pork river; the party who examined it for about three miles up say that its current is gentle and that it does not seem to throw out much sand. Its sources are in the first range of the Black mountains, and though it has now only water of twenty yards width, yet when full it occupies ninety. Just below the mouth is another village or wintering camp of the Ricaras composed of about sixty lodges, built in the same form as those passed yesterday, with willow and straw mats, baskets, and buffaloe-skin canoes remaining entire in the camp. We proceeded under a gentle breeze from the southwest; at ten o'clock we saw two Indians on the north side, who told us they were a part of the lodge of Tartongawaka, or Buffaloe Medicine, the Teton chief whom we had seen on the twenty-fifth; that they were on the way to the Ricaras, and begged us for something to eat, which we of course gave them. At seven and a half miles is a willow island on the north, and another on the same side five miles beyond it, in the middle of the river, between highlands on both sides. At eighteen and a half miles is an island called Grouse island, on which are the walls of an old village; the island has no timber, but is covered with grass and wild rye, and owes its name to the number of grouse that frequent it. We then went on till our journey for the day was twenty-two miles; the country presented the same appearance as usual. In the low timbered ground near the mouth of the Sawawkawna we saw the tracks of large white bear, and on Grouse island killed a female blaireau, and a deer of the black-tailed species, the largest we have ever seen.

Monday, October 8. We proceeded early with a cool northwest wind, and at two and a half miles above Grouse island reached the mouth of a creek on the south, then a small willow island, which divides the current equally; and at four and a half miles came to a river on the southern side, where we halted. This river, which our meridian altitude fixes at 45° 39' 5" north latitude, is called by the Ricaras Wetawhoo; it rises in the Black mountains, and its bed, which flows at the mouth over a low soft slate stone, is one hundred and twenty yards wide; but the water is now confined within twenty yards, and is not very rapid, discharging mud with a small proportion of sand. Here, as in every bend of the river, we again observe the red berries resembling currants, which we mentioned before. Two miles above the Wetawhoo, and on the same side, is a small river called Maropa by the Indians; it is twenty yards in width, but so dammed up by mud that the stream creeps through a channel of not more than an inch in diameter, and discharges no sand. One mile farther we reached an island close to the southern shore, from which it is separated by a deep channel of sixty yards. About half way a number of Ricara Indians came out to see us. We stopped and took a Frenchman on board, who accompanied us past the island to our camp on the north side of the river, which is at the distance of twelve miles from that of yesterday. Captain Lewis then returned with four of the party to see the village; it is situated in the centre of the island, near the southern shore, under the foot of some high, bald, uneven hills, and contains about sixty lodges. The island itself is three miles long, and covered with fields in which the Indians raise corn, beans, and potatoes. Several Frenchmen living among these Indians, as interpreters or traders, came back with Captain Lewis, and particularly a Mr. Gravelines, a man who has acquired the language. On setting out we had a low prairie covered with timber on the north, and on the south highlands, but at the mouth of the Wetawhoo the southern country changes, and a low timbered plain extends along the south, while the north has a ridge of barren hills during the rest of the day's course.

Tuesday, 9th. The wind was so cold and high last night, and during all the day, that we could not assemble the Indians in council; but some of the party went to the village. We received the visits of the three principal chiefs with many others, to whom we gave some tobacco and told them that we would speak to them to-morrow. The names of these chiefs were, first, Kakawissassa, or Lighting Crow; second chief, Pocasse, or Hay; third chief, Piaheto, or Eagle's Feather. Notwithstanding the high waves, two or three squaws rowed to us in little canoes made of a single buffaloe skin stretched over a frame of boughs interwoven like a basket, and with the most perfect composure. The object which appeared to astonish the Indians most was Captain Clark's servant York, a remarkable stout strong negro. They had never seen a being of that colour, and therefore flocked round him to examine the extraordinary monster. By way of amusement he told them that he had once been a wild animal, and caught and tamed by his master, and to convince them showed them feats of strength which, added to his looks, made him more terrible than we wished him to be. Opposite our camp is a small creek on the south, which we distinguished by the name of the chief, Kakawissassa.

Wednesday, 10th. The weather was this day fine, and as we were desirous of assembling the whole nation at once, we despatched Mr. Gravelines, who, with Mr. Tabeau, another French trader, had breakfasted with us, to invite the chiefs of the two upper villages to a conference. They all assembled at one o'clock, and after the usual ceremonies we addressed them in the same way in which we had already spoken to the Ottoes and Sioux; we then made or acknowledged three chiefs, one for each of the three villages, giving to each a flag, a medal, a red coat, a cocked hat and feather, also some goods, paint and tobacco, which they divided among themselves; and after this the airgun was exhibited, very much to their astonishment; nor were they less surprised at the colour and manner of York. On our side we were equally gratified at discovering that these Ricaras made use of no spirituous liquors of any kind, the example of the traders who bring it to them so far from tempting having in fact disgusted them. Supposing that it was as agreeable to them as to the other Indians, we had at first offered them whiskey; but they refused it, with this sensible remark that they were surprised that their father should present to them a liquor which would make them fools. On another occasion they observed to Mr. Tabeau that no man could be their friend who tried to lead them into such follies. The council being over, they retired to consult on their answer, and the next morning,

Thursday, 11th, at eleven o'clock, we again met in council at our camp. The grand chief made a short speech of thanks for the advice we had given, and promised to follow it, adding that the door was now open and no one dare shut it, and that we might depart whenever we pleased, alluding to the treatment we had received from the Sioux; they also brought us some corn, beans, and dried squashes, and in return we gave them a steel mill with which they were much pleased. At one o'clock we left our camp with the grand chief and his nephew on board, and at about two miles anchored below a creek on the south separating the second and third village of the Ricaras, which are about a half mile distant from each other. We visited both the villages and sat conversing with the chiefs for some time, during which they presented us with a bread made of corn and beans, also corn and beans boiled, and a large rich bean which they take from the mice of the prairie, who discover and collect it. These two villages are placed near each other in a high smooth prairie, a fine situation, except that having no wood the inhabitants are obliged to go for it across the river to a timbered lowland opposite to them. We told them that we would speak to them in the morning at their villages, separately.

Friday, 12th. Accordingly, after breakfast we went on shore to the house of the chief of the second village named Lassel, where we found his chiefs and warriors. They made us a present of about seven bushels of corn, a pair of leggings, a twist of their tobacco, and the seeds of two different species of tobacco. The chief then delivered a speech expressive of his gratitude for the presents and the good counsels which we had given him; his intention of visiting his great father but for fear of the Sioux; and requested us to take one of the Ricara chiefs up to the Mandans and negotiate a peace between the two nations. To this we replied in a suitable way, and then repaired to the third village. Here we were addressed by the chief in nearly the same terms as before, and entertained with a present of ten bushels of corn,

## UP THE MISSOURI

some beans, dried pumpkins, and squashes. After we had answered and explained the magnitude and power of the United States, the three chiefs came with us to the boat. We gave them some sugar, a little salt, and a sun-glass. Two of them then left us, and the chief of the third, by name Ahketahnasha or Chief of the Town, accompanied us to the Mandans. At two o'clock we left the Indians, who crowded to the shore to take leave of us, and after making seven and a half miles landed on the north side, and had a clear, cool, pleasant evening.

The three villages which we have just left are the residence of a nation called the Ricaras. They were originally colonies of Pawnees, who established themselves on the Missouri, below the Chayenne, where the traders still remember that twenty years ago they occupied a number of villages. From that situation a part of the Ricaras emigrated to the neighbourhood of the Mandans, with whom they were then in alliance. The rest of the nation continued near the Chayenne till the year 1797, in the course of which, distressed by their wars with the Sioux, they joined their countrymen near the Mandans. Soon after a new war arose between the Ricaras and the Mandans, in consequence of which the former came down the river to their present position. In this migration those who had first gone to the Mandans kept together and now live in the two lower villages, which may thence be considered as the Ricaras proper. The third village was composed of such remnants of the villages as had survived the wars, and as these were nine in number, a difference of pronunciation and some difference of language may be observed between them and the Ricaras proper, who do not understand all the words of these wanderers. The villages are within the distance of four miles of each other, the two lower ones consisting of between one hundred and fifty and two hundred men each, the third of three hundred. The Ricaras are tall and well proportioned, the women handsome and lively, and, as among other savages, to them falls all the drudgery of the field and the labours of procuring subsistence, except that of hunting; both sexes are poor, but kind and generous, and although they receive with thankfulness what is given to them, do not beg as the Sioux did, though this praise should be qualified by mentioning that an axe was stolen last night from our cooks. The dress of the men is a simple pair of moccasins, legings, and a cloth round the middle, over which a buffaloe robe is occasionally thrown, with their hair, arms, and ears decorated with different ornaments. The women wear moccasins, legings, a long shirt made of goats' skins, generally white and fringed, which is tied round the waist; to these they add, like the men, a buffaloe robe without the hair in summer. These women are handsomer than the Sioux; both of them are, however, disposed to be amorous, and our men found no difficulty in procuring companions for the night by means of the interpreters. These interviews were chiefly clandestine, and were of course to be kept a secret from the husband or relations. The point of honour, indeed, is completely reversed among the Ricaras; that the wife or the sister should submit to a stranger's embraces without the consent of her husband or brother is a cause of great disgrace and offence, especially as for many purposes of civility or gratitude the husband and brother will themselves present to a stranger these females, and be gratified by attentions to them. The Sioux had offered us squaws, but while we remained there having declined, they followed us with offers of females for two days. The Ricaras had been equally accomodating; we had equally withstood their temptation; but such was their desire to oblige that two very handsome young squaws were sent on board this evening, and persecuted us with civilities.

## UP THE MISSOURI

The black man York participated largely in these favours, for instead of inspiring any prejudice, his colour seemed to procure him additional advantages from the Indians, who desired to preserve among them some memorial of this wonderful stranger. Among other instances of attention, a Ricara invited him into his house, and presenting his wife to him, retired to the outside of the door; while there one of York's comrades who was looking for him came to the door, but the gallant husband would permit no interruption before a reasonable time had elapsed.

The Ricara lodges are in a circular or octagonal form, and generally about thirty or forty feet in diameter; they are made by placing forked posts about six feet high round the circumference of the circle; these are joined by poles from one fork to another, which are supported also by other forked poles slanting from the ground; in the centre of the lodge are placed four higher forks about fifteen feet in length, connected together by beams; from these to the lower poles the rafters of the roof are extended so as to leave a vacancy in the middle for the smoke; the frame of the building is then covered with willow branches, with which is interwoven grass, and over this mud or clay; the aperture for the door is about four feet wide, and before it is a sort of entry about ten feet from the lodge. They are very warm and compact.

They cultivate maize or Indian corn, beans, pumpkins, watermelons, squashes, and a species of tobacco peculiar to themselves.

Their commerce is chiefly with the traders, who supply them with goods in return for peltries, which they procure not only by their own hunting but in exchange for corn from their less civilized neighbours. The object chiefly in demand seemed to be red paint, but they would give anything they had to spare for the most trifling article. One of the men to-day gave an Indian a hook made out of a pin, and he gave him in return a pair of moccasins.

They express a disposition to keep at peace with all nations, but they are well armed with fusils, and being much under the influence of the Sioux, who exchanged the goods which they got from the British for Ricara corn, their minds are sometimes poisoned and they cannot be always depended on. At the present moment they are at war with the Mandans. We are informed by Mr. Gravelines, who had passed through that country, that the Yankton or Jacques river rises about forty miles to the east or northeast of this place, the Chayenne branch of the Red river about twenty miles farther, passing the Sioux, and the St. Peter's about eighty.

Saturday, 13th. In the morning our visitors left us, except the brother of the chief who accompanies us and one of the squaws. We passed at an early hour a camp of Sioux on the north bank, who merely looked at us without saying a word, and from the character of the tribe we did not solicit a conversation. At ten and a half miles we reached the mouth of a creek on the north, which takes its rise from some ponds a short distance to the northeast; to this stream we gave the name of Stoneidol creek, for after passing a willow and sand island just above its mouth we discovered that a few miles back from the Missouri there are two stones resembling human figures, and a third like a dog,—all which are objects of great veneration among the Ricaras. Their history would adorn the metamorphoses of Ovid. A young man was deeply enamoured with a girl whose parents refused their consent to the marriage. The youth went out into the fields to mourn his misfortunes; a sympathy of feeling led the lady to the same spot, and the faithful dog would not cease to follow his master. After wandering together and having nothing but grapes to subsist on, they were at last converted into stone, which, beginning at the feet, gradually invaded the nobler parts, leaving nothing unchanged but a bunch of grapes which the female holds in her hands to this day. Whenever the Ricaras pass these sacred stones they stop to make some offering of dress to propitiate these deities. Such is the account given by the Ricara chief, which we had no mode of examining, except that we found one part of the story very agreeably confirmed, for on the river near where the event is said to have occurred we found a greater abundance of fine grapes than we had yet seen. Above this is a small creek four and a half miles from Stoneidol creek, which is fifteen yards wide, comes in from the south, and received from us the name of Pocasse, or Hay creek, in honour of the chief of the second village. Above the Ricara island the Missouri becomes narrow and deeper, the sandbars being generally confined to the points; the current, too, is much more gentle; the timber on the lowlands is also in much greater quantities, though the high grounds are still naked. We proceeded on under a fine breeze from the southeast, and after making eighteen miles encamped on the north near a timbered low plain, after which we had some rain and the evening was cold. The hunters killed one deer only.

Sunday, 14th. We set out in the rain, which continued during the day. At five miles we came to a creek on the south about fifteen yards wide, and named by us Piaheto, or Eagle's Feather, in honour of the third chief of the Ricaras. After dinner we stopped on a sandbar and executed the sentence of a court martial, which inflicted corporal punishment on one of the soldiers. This operation affected the Indian chief very sensibly, for he cried aloud during the punishment; we explained the offence and the reasons of it. He acknowledged that examples were necessary, and that he himself had given them by punishing with death; but his nation never whipped even children from their birth. After this we continued, with the wind from the northeast, and at the distance of twelve miles encamped in a cove of the southern bank. Immediately opposite our camp, on the north side, are the ruins of an ancient fortification, the greater part of which is washed into the river; nor could we distinguish more than that the walls were eight or ten feet high. The evening is wet and disagreeable, and the river, which is somewhat wider than yesterday, continues to have an unusual quantity of timber. The country was level on both sides in the morning, but afterwards we passed some black bluffs on the south.

Monday, 15th. We stopped at three miles on the north, a little above a camp of Ricaras who are hunting, where we were visited by about thirty Indians. They came over in their skin canoes, bringing us meat, for which we returned them beads and fishhooks. About a mile higher we found another encampment of Ricaras on the south, consisting of eight lodges; here we again ate and exchanged a few presents. As we went we discerned numbers of other Indians on both sides of the river; and at about nine miles we came to a creek on the south, where we saw many high hills resembling a house with a slanting roof, and a little below the creek an old village of the Sharha or Chayenne Indians. The morning had been cloudy, but the evening became pleasant, the wind from the northeast, and at sunset we halted, after coming ten miles over several sandbars and points, above a camp of ten Ricara lodges on the north side. We visited their camp and smoked and eat with several of them; they all appeared kind and pleased with our attentions, and the fair sex received our men with more than hospitality. York was here again an object of astonishment; the children would follow him constantly, and if he chanced to turn towards them run

# UP THE MISSOURI 117

with great terror. The country of to-day is generally low and covered with timber on both sides, though in the morning we passed some barren hills on the south.

Tuesday, 16th. At this camp the squaw who accompanied the chief left us; two others were very anxious to go on with us. Just above our camp we passed a circular work or fort where the Sharha or Chayennes formerly lived; and a short distance beyond, a creek which we called Chayenne creek. At two miles is a willow island with a large sandbar on both sides above it, and a creek, both on the south, which we called Sohawch, the Ricara name for girl; and two miles above, a second creek, to which we gave the name of Chapawt, which means woman in the same language. Three miles farther is an island situated in a bend to the north, about a mile and a half long and covered with cottonwood. At the lower end of this island comes in a small creek from the north called Keetooshsahawna, or Place of Beaver. At the upper extremity of the island a river empties itself from the north; it is called Warreconne, or Elk Shed their Horns, and is about thirty-five yards wide; the island itself is named Carp island by Evans, a former traveller. As we proceeded there were great numbers of goats on the banks of the river, and we soon after saw large flocks of them in the water; they had been gradually driven into the river by the Indians, who now lined the shore so as to prevent their escape, and were firing on them, while sometimes boys went into the river and killed them with sticks; they seemed to be very successful, for we counted fifty-eight which they had killed. We ourselves killed some, and then passing the lodges to which these Indians belonged, encamped at the distance of half a mile on the south, having made fourteen and a half miles. We were soon visited by numbers of these Ricaras, who crossed the river hallooing and singing; two of them then returned for some goats' flesh and buffaloe meat, dried and fresh, with which they made a feast that lasted till late at night and caused much music and merriment.

Wednesday, 17th. The weather was pleasant; we passed a low ground covered with small timber on the south, and barren hills on the north which came close to the river; the wind from the northwest then became so strong that we could not move after ten o'clock until late in the afternoon, when we were forced to use the towline, and we therefore made only six miles. We all went out hunting and examining the country. The goats, of which we see large flocks coming to the north bank of the river, spend the summer, says Mr. Gravelines, in the plains east of the Missouri, and at the present season are returning to the Black mountains, where they subsist on leaves and shrubbery during the winter and resume their migrations in the spring. We also saw buffaloe, elk, and deer, and a number of snakes; a beaver house, too, was seen, and we caught a whippoorwill of a small and uncommon kind. The leaves are fast falling; the river wider than usual and full of sandbars, and on the sides of the hills are large stones, and some rock of a brownish colour in the southern bend below us. Our latitude by observation was 46° 23′ 57″.

Thursday, 18th. After three miles we reached the mouth of Le Boulet or Cannonball river; this stream rises in the Black mountains, and falls into the Missouri on the south; its channel is about one hundred and forty yards wide, though the water is now confined within forty, and its name is derived from the numbers of perfectly round large stones on the shore and in the bluffs just above. We here met with two Frenchmen in the employ of Mr. Gravelines, who had been robbed by the Mandans of their traps, furs, and other articles, and were descending the river in a periogue, but they turned

## UP THE MISSOURI

back with us in expectation of obtaining redress through our means. At eight miles is a creek on the north about twenty-eight yards wide, rising in the northeast, and called Chewah or Fish river; one mile above this is another creek on the south; we encamped on a sandbar to the south, at the distance of thirteen miles, all of which we had made with oars and poles. Great numbers of goats are crossing the river and directing their course to the westward; we also saw a herd of buffaloe and of elk; a pelican, too, was killed and six fallow deer, having found, as the Ricaras informed us, that there are none of the black-tail species as high up as this place. The country is in general level and fine, with broken short high grounds, low timbered mounds on the river, and a rugged range of hills at a distance.

Friday, 19th. We set sail with a fine morning and a southeast wind, and at two and a half miles passed a creek on the north side; at eleven and a half miles we came to a lake or large pond on the same side, in which were some swans. On both banks of the Missouri are low grounds which have much more timber than lower down the river; the hills are at one or two miles distance from the banks, and the streams which rise in them are brackish and the mineral salts appear on the sides of the hills and edges of the runs. In walking along the shore we counted fifty-two herds of buffaloe and three of elk at a single view. Besides these we also observed elk, deer, pelicans, and wolves. After seventeen and a half miles we encamped on the north, opposite to the uppermost of a number of round hills forming a cone at the top, one being about ninety, another sixty feet in height, and some of less elevation. Our chief tells us that the calumet bird lives in the holes formed by the filtration of the water from the top of these hills through the sides. Near to one of these moles, on a point of a hill ninety feet above the plain, are the remains of an old village which is high, strong, and has been fortified; this our chief tells us is the remains of one of the Mandan villages, and are the first ruins which we have seen of that nation in ascending the Missouri; opposite to our camp is a deep bend to the south, at the extremity of which is a pond.

Saturday, 20th. We proceeded early with a southeast wind, which continued high all day, and came to a creek on the north at two miles distance twenty yards wide. At eight miles we reached the lower point of an island in the middle of the river, though there is no current on the south. This island is covered with willows and extends about two miles, there being a small creek coming in from the south at its lower extremity. After making twelve miles we encamped on the south, at the upper part of a bluff containing stone-coal of an inferior quality; immediately below this bluff and on the declivity of a hill are the remains of a village covering six or eight acres formerly occupied by the Mandans, who, says our Ricara chief, once lived in a number of villages on each side of the river, till the Sioux forced them forty miles higher; whence, after a few years residence, they moved to their present position. The country through which we passed has wider bottoms and more timber than those we have been accustomed to see, the hills rising at a distance and by gradual ascents. We have seen great numbers of elk, deer, goats, and buffaloe, and the usual attendants of these last, the wolves, who follow their movements and feed upon those who die by accident or who are too poor to keep pace with the herd; we also wounded a white bear, and saw some fresh tracks of those animals, which are twice as large as the track of a man.

Sunday, 21st. Last night the weather was cold, the wind high from the northeast, and the rain which fell froze on the ground. At daylight it began to snow and continued till the afternoon, when it remained cloudy and the ground was covered with snow. We, however, set out early, and just above our camp came to a creek on the south called Chisshetaw, about thirty yards wide and with a considerable quantity of water. Our Ricara chief tells us that at some distance up this river is situated a large rock which is held in great veneration and visited by parties who go to consult it as to their own or their nations' destinies, all of which they discern in some sort of figures or paintings with which it is covered. About two miles off from the mouth of the river the party on shore saw another of the objects of Ricara superstition: it is a large oak tree standing alone in the open prairie, and as it alone has withstood the fire which has consumed everything around, the Indians naturally ascribe to it extraordinary powers. One of their ceremonies is to make a hole in the skin of their necks through which a string is passed and the other end tied to the body of the tree, and after remaining in this way for some time they think they become braver. At two miles from our encampment we came to the ruins of a second Mandan village which was in existence at the same time with that just mentioned. It is situated on the north at the foot of a hill in a beautiful and extensive plain, which is now covered with herds of buffaloe; nearly opposite are remains of a third village on the south of the Missouri, and there is another also about two miles farther on the north, a little off the river. At the distance of seven miles we encamped on the south, and spent a cold night. We procured to-day a buffaloe and an otter only. The river is wide and the sandbars numerous, and a low island near our encampment.

Monday, 22d. In the morning we passed an old Mandan village on the south, near our camp; at four miles another on the same side. About seven o'clock we came to at a camp of eleven Sioux of the Teton tribe, who are almost perfectly naked, having only a piece of skin or cloth round the middle, though we are suffering from the cold. From their appearance, which is warlike, and from their giving two different accounts of themselves, we believe that they are either going to or returning from the Mandans, to which nations the Sioux frequently make excursions to steal horses. As their conduct displeased us, we gave them nothing. At six we reached an island about one mile in length, at the head of which is a Mandan village on the north in ruins, and two miles beyond a bad sandbar. At eight miles are remains of another Mandan village on the south, and at twelve miles encamped on the south. The hunters brought in a buffaloe bull, and mentioned that of about three hundred which they had seen there was not a single female. The beaver is here in plenty, and the two Frenchmen who are returning with us catch several every night.

These villages, which are nine in number, are scattered along each side of the river within a space of twenty miles; almost all that remains of them is the wall which surrounded them, the fallen heaps of earth which covered the houses, and occasionally human skulls and the teeth and bones of men and different animals, which are scattered on the surface of the ground.

Tuesday, 23d. The weather was cloudy and we had some snow; we soon arrived at five lodges where the two Frenchmen had been robbed, but the Indians had left it lately as we found the fires still burning. The country consists as usual of timbered low grounds, with grapes, rushes, and great quantities of a small red acid fruit known among the Indians by a name signifying rabbit-berries, and called by the French graisse de buffle or buffaloe fat. The river, too, is obstructed by many sandbars. At twelve miles we passed an old village on the north which was the former residence of the Ahnahaways, who now live between the Mandans and Minnetarees. After making thirteen miles we encamped on the south.

Wednesday, 24th. The day was again dark and it snowed a little in the morning. At three miles we came to a point on the south where the river, by forcing a channel across a former bend, has formed a large island on the north. On this island we found one of the grand chiefs of the Mandans, who with five lodges was on a hunting excursion. He met his enemy the Ricara chief with great ceremony and apparent cordiality, and smoked with him. After visiting his lodges, the grand chief and his brother came on board our boat for a short time; we then proceeded and encamped on the north, at seven miles from our last night's station and below the old village of the Mandans and Ricaras. Here four Mandans came down from a camp above, and our Ricara chief returned with them to their camp, from which we augur favourably of their pacific views towards each other. The land is low and beautiful and covered with oak and cottonwood, but has been too recently hunted to afford much game.

25th. The morning was cold and the wind gentle from the southeast; at three miles we passed a handsome high prairie on the south, and on an eminence about forty feet above the water, and extending back for several miles in a beautiful plain, was situated an old village of the Mandan nation which has been deserted for many years. A short distance above it, on the continuation of the same rising ground, are two old villages of Ricaras, one on the top of the hill, the other in the level plain, which have been deserted only five years ago. Above these villages is an extensive low ground for several miles, in which are situated, at three or four miles from the Ricara villages, three old villages of Mandans near together. Here the Mandans lived when the Ricaras came to them for protection, and from this they moved to their present situation above. In the low ground the squaws raised their corn, and the timber, of which there was little near the villages, was supplied from the opposite side of the river, where it was and still is abundant.

As we proceeded several parties of Mandans, both on foot and horseback, came along the river to view us, and were very desirous that we should land and talk to them; this we could not do on account of the sandbreaks on the shore, but we sent our Ricara chief to them in a periogue. The wind, too, having shifted to the southwest and being very high it required all our precautions on board, for the river was full of sandbars which made it very difficult to find the channel. We got aground several times and passed a very bad point of rocks, after which we encamped on a sandpoint to the north, above a handsome plain covered with timber, and opposite to a high hill on the south side at the distance of eleven miles. Here we were joined by our Ricara chief, who brought an Indian to the camp, where he remained all night.

26th. We set out early, with a southwest wind, and after putting the Ricara chief on shore to join the Mandans, who were in great numbers along it, we proceeded to the camp of the grand chiefs four miles distant. Here we met a Mr. M'Cracken, one of the northwest or Hudson Bay company, who arrived with another person about nine days ago to trade for horses and buffaloe robes. Two of the chiefs came on board with some of their household furniture, such as earthern pots, and a little corn, and went on with us, the rest of the Indians following on shore. At one mile beyond the camp we passed a small creek, and at three more a bluff of coal of an inferior quality on the south. After making eleven miles we reached an old field where the Mandans had cultivated

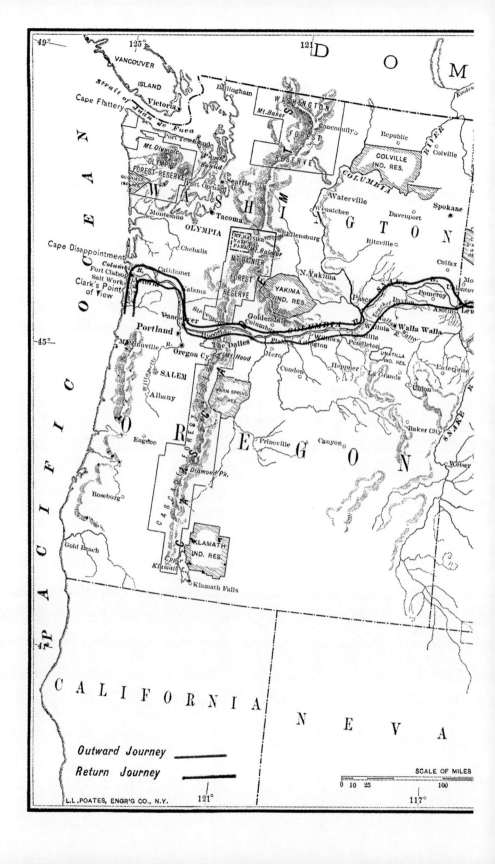

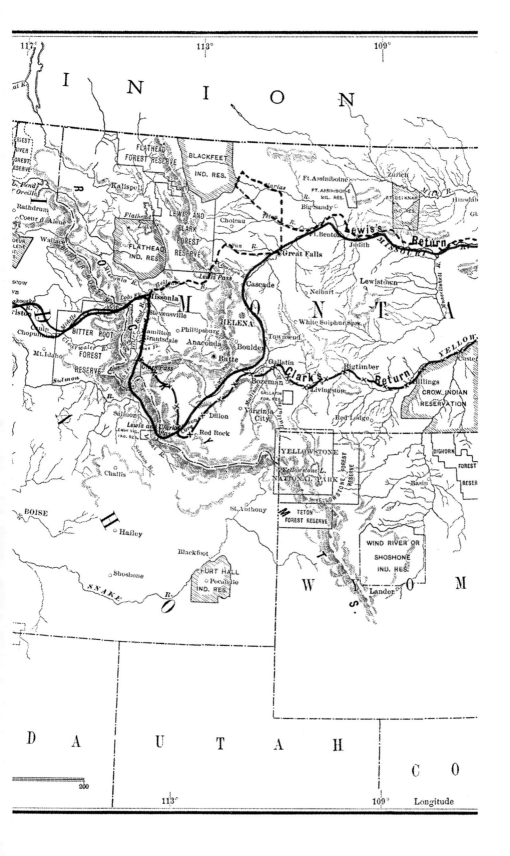

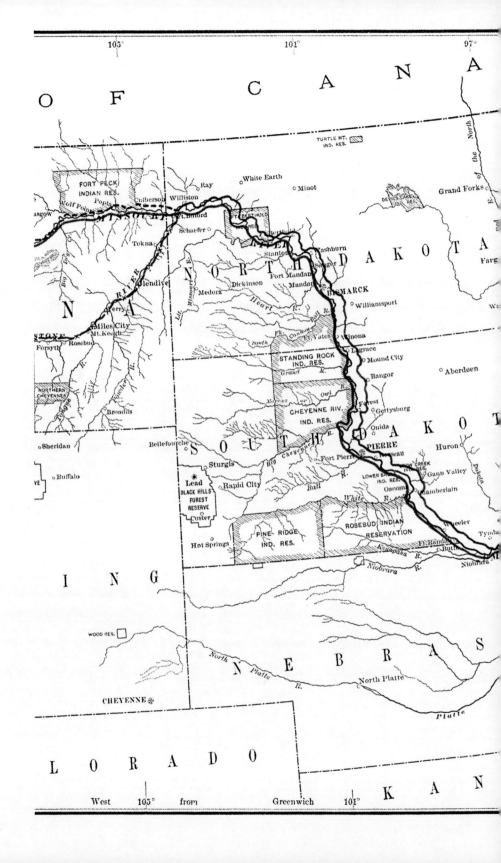

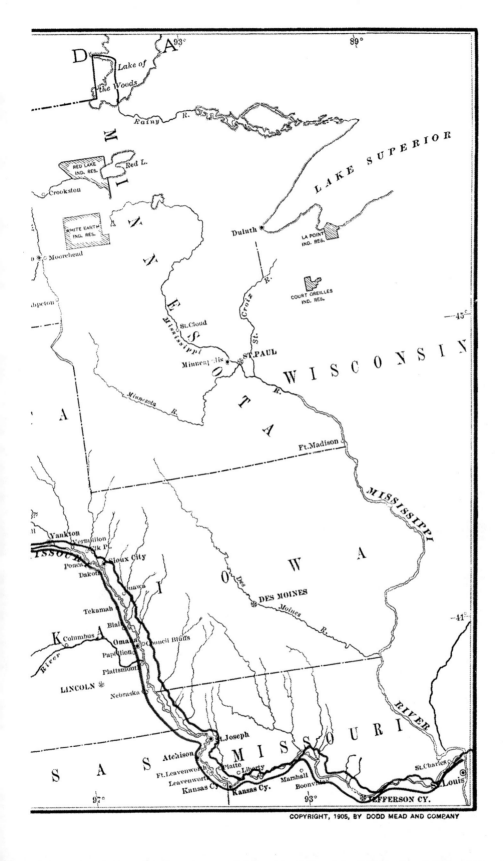

grain last summer, and encamped for the night on the south side, about half a mile below the first village of the Mandans. In the morning we had a willow low ground on the south and highland on the north, which occasionally varied in the course of the day. There is but little wood on this part of the river, which is here subdivided into many channels and obstructed by sandbars. As soon as we arrived a crowd of men, women, and children came down to see us. Captain Lewis returned with the principal chiefs to the village, while the others remained with us during the evening. The object which seemed to surprise them most was a corn mill fixed to the boat which we had occasion to use, and delighted them by the ease with which it reduced the grain to powder. Among others who visited us was the son of the grand chief of the Mandans, who had his two little fingers cut off at the second joints. On inquiring into this accident we found that it was customary to express grief for the death of relations by some corporeal suffering, and that the usual mode was to lose two joints of the little fingers, or sometimes the other fingers. The wind blew very cold in the evening from the southwest. Two of the party are affected with rheumatic complaints.

# CHAPTER V.

Council held with the Mandans—A prairie on fire, and a singular instance of preservation—Peace established between the Mandans and Ricaras—The party encamp for the winter—Indian mode of catching goats—Beautiful appearance of northern lights—Friendly character of the Indians—Some account of the Mandans—The Ahnahaways and the Minnetarees—The party acquire the confidence of the Mandans by taking part in their controversy with the Sioux—Religion of the Mandans, and their singular conception of the term medicine—Their tradition—The sufferings of the party from the severity of the season—Indian game of billiards described—Character of the Missouri, of the surrounding country, and of the rivers, creeks, islands, etc.

SATURDAY, October 27. At an early hour we proceeded, and anchored off the village. Captain Clark went on shore, and after smoking a pipe with the chiefs was desired to remain and eat with them. He declined on account of his being unwell; but his refusal gave great offence to the Indians, who considered it disrespectful not to eat when invited, till the cause was explained to their satisfaction. We sent them some tobacco, and then proceeded to the second village on the north, passing by a bank containing coal, and a second village, and encamped at four miles on the north opposite to a village of Ahnahaways. We here met with a Frenchman named Jesseaume, who lives among the Indians with his wife and children, and who we take as an interpreter. The Indians had flocked to the bank to see us as we passed, and they visited in great numbers the camp, where some of them remained all night. We sent in the evening three young Indians with a present of tobacco for the chiefs of the three upper villages, inviting them to come down in the morning to a council with us. Accordingly, the next day,

Sunday, October 28, we were joined by many of the Minnetarees and Ahnahaways from above, but the wind was so violent from the southwest that the chiefs of the lower villages could not come up, and the council was deferred till tomorrow. In the mean while we entertained our visitors by showing them what was new to them in the boat; all which, as well as our black servant, they called Great Medicine, the meaning of which we afterwards learnt. We also consulted the grand chief of the blandans, Black Cat, and Mr. Jesseaume, as to the names, character, etc., of the chiefs with whom we are to hold the council. In the course of the day we received several presents from the women, consisting of corn, boiled hominy, and garden stuffs; in our turn we gratified the wife of the great chief with a gift of a glazed earthen jar. Our hunter brought us two beaver. In the afternoon we sent the Minnetaree chiefs to smoke for us with the great chief of the Mandans, and told them we would speak in the morning.

Finding that we shall be obliged to pass the winter at this place, we went up the river about one and a half miles to-day, with a view of finding a convenient spot for a fort, but the timber was too scarce and small for our purposes.

Monday, October 29. The morning was fine, and we prepared out presents and speech for the council. After breakfast we were visited by an old chief of the Ahnahaways, who, finding himself growing old and weak, had transferred his power to his son, who is now at war against the Shoshonees. At ten o'clock the chiefs were all assembled under an awning of our sails, stretched so as to exclude the wind, which had become high; that the impression might be the more forcible, the men were all paraded, and the council opened by a discharge from the swivel of the boat. We then delivered a speech, which, like those we had already made, intermingled advice with assurances of friendship and trade. While we were speaking the old Ahnahaway chief grew very restless, and observed that he could not wait long as his camp was exposed to the hostilities of the Shoshonees; he was instantly rebuked with great dignity by one of the chiefs for this violation of decorum at such a moment, and remained quiet during the rest of the council. Towards the end of our speech we introduced the subject of our Ricara chief, with whom we recommended a firm peace; to this they seemed well disposed, and all smoked with him very amicably. We all mentioned the goods which had been taken from the Frenchmen, and expressed a wish that they should be restored. This being over, we proceeded to distribute the presents with great ceremony; one chief of each town was acknowledged by a gift of a flag, a medal with the likeness of the President of the United States, a uniform coat, hat, and feather; to the second chiefs we gave a medal representing some domestic animals, and a loom for weaving; to the third chiefs, medals with the impressions of a farmer sowing grain. A variety of other presents were distributed, but none seemed to give them more satisfaction than an iron corn mill which we gave to the Mandans.

The chiefs who were made to-day are: Shahaka, or Big White, a first chief, and Kagohami, or Little Raven, a second chief of the lower village of the Mandans, called Matootonha; the other chiefs of an inferior quality who were recommended were, I. Ohheenaw, or Big Man, a Chayenne taken prisoner by the Mandans, who adopted him, and he now enjoys great consideration among the tribe. 2. Shotahawrora, or Coal, of the second Mandan village, which is called Rooptahee. We made Poscopsahe, or Black Cat, the first chief of the village, and the grand chief of the whole Mandan nation; his second chief is Kagonomokshe, or Raven man Chief; inferior chiefs of this village were Tawnuheo and Bellahsara, of which we did not learn the translation.

## UP THE MISSOURI 129

In the third village, which is called Mahawha, and where the Arwacahwas reside, we made one first chief, Tetuckopinreha, or White Buffaloe robe unfolded, and recognized two of an inferior order: Minnissurraree, or Neighing Horse, and Locongotiha, or Old woman at a distance.

Of the fourth village, where the Minnetarees live, and which is called Metaharta, we made a first chief, Ompsehara, or Black Moccasin; a second chief, Ohhaw, or Little Fox. Other distinguished chiefs of this village were, Mahnotah, or Big Thief, a man whom we did not see, as he is out fighting, and was killed soon after; and Mahserassa, or Tail of the Calumet Bird. In the fifth village we made a first chief, Eapanopa, or Red Shield; a second chief, Wankerassa, or Two Tailed Calumet Bird, both young chiefs. Other persons of distinction are, Shahakohopinnee, or Little Wolf's Medicine; Ahrattanamockshe, or Wolfman chief, who is now at war, and is the son of the old chief we have mentioned, whose name is Caltahcota, or Cherry on a Bush.

The presents intended for the grand chief of the Minnetarees, who was not at the council, were sent to him by the old chief Caltahcota; and we delivered to a young chief those intended for the chief of the lower village. The council was concluded by a shot from our swivel, and after firing the airgun for their amusement, they retired to deliberate on the answer which they are to give to-morrow.

In the evening the prairie took fire, either by accident or design, and burned with great fury, the whole plain being enveloped in flames; so rapid was its progress that a man and a woman were burnt to death before they could reach a place of safety; another man, with his wife and child, were much burnt, and several other persons narrowly escaped destruction. Among the rest a boy of the half white breed escaped unhurt in the midst of the flames; his safety was ascribed to the great medicine spirit, who had preserved him on account of his being white. But a much more natural cause was the presence of mind of his mother, who, seeing no hopes of carrying off her son, threw him on the ground, and covering him with the fresh hide of a buffaloe, escaped herself from the flames; as soon as the fire had passed, she returned and found him untouched, the skin having prevented the flames from reaching the grass on which he lay.

Tuesday, 30th. We were this morning visited by two persons from the lower village, one the Big White, the chief of the village, the other the Chayenne called the Big Man; they had been hunting, and did not return yesterday early enough to attend the council. At their request we repeated part of our speech of yesterday, and put the medal round the neck of the chief. Captain Clark took a periogue and went up the river in search of a good wintering place, and returned after going seven miles to the lower point of an island on the north side, about one mile in length; he found the banks on the north side high, with coal occasionally, and the country fine on all sides; but the want of wood and the scarcity of game up the river induced us to decide on fixing ourselves lower down during the winter. In the evening our men danced among themselves, to the great amusement of the Indians.

Wednesday, 31 st. A second chief arrived this morning with an invitation from the grand chief of the Mandans to come to his village, where he wished to present some corn to us and to speak with us. Captain Clark walked down to his village. He was first seated with great ceremony on a robe by the side of the chief, who then threw over his shoulders another robe handsomely ornamented. The pipe was then smoked with several of the old men, who were seated around the chief; after some time he began his discourse by observing that he believed what we had told him, and that they should soon enjoy peace, which would gratify him as well as his people, because they could then hunt without fear of being attacked, and the women might work in the fields without looking every moment for the enemy, and at night put off their moccasins, a phrase by which is conveyed the idea of security, when the women could undress at night without fear of attack. As to the Ricaras, he continued, in order to show you that we wish peace with all men, that chief, pointing to his second chief, will go with some warriors back to the Ricaras with their chief now here and smoke with that nation. When we heard of your coming all the nations around returned from their hunting to see you, in hopes of receiving large presents; all are disappointed, and some discontented; for his part he was not much so, though his village was. He added that he would go and see his great father the President. Two of the steel traps stolen from the Frenchmen were then laid before Captain Clark, and the women brought about twelve bushels of corn. After the chief had finished, Captain Clark made an answer to the speech, and then returned to the boat, where he found the chief of the third village and Kagohami (the Little Raven), who smoked and talked about an hour. After they left the boat the grand chief of the Mandans came dressed in the clothes we had given him, with his two children, and begged to see the men dance, in which they willingly gratified him.

Thursday, November 1. Mr.M'Cracken, the trader whom we found here, set out to-day on his return to the British fort and factory on the Assiniboin river, about one hundred and fifty miles from this place. He took a letter from Captain Lewis to the northwest company, inclosing a copy of the passport granted by the British minister in the United States. At ten o'clock the chiefs of the lower village arrived; they requested that we would call at their village for some corn, that they were willing to make peace with the Ricaras, that they had never provoked the war between them, but as the Ricaras had killed some of their chiefs they had retaliated on them; that they had killed them like birds till they were tired of killing them, so that they would send a chief and some warriors to smoke with them. In the evening we dropped down to the lower village, where Captain Lewis went on shore, and Captain Clark proceeded to a point of wood on the north side.

Friday, November 2. He therefore went up to the village, where eleven bushels of corn were presented to him. In the mean time Captain Clark went down with the boats three miles, and having found a good position where there was plenty of timber, encamped and began to fell trees to build our huts. Our Ricara chief set out with one Mandan chief and several Minnetaree and Mandan wrarriors; the wind was from the southeast, and the weather being fine a crowd of Indians came down to visit us.

Saturday, 3d. We now began the building of our cabins, and the Frenchmen, who are to return to St. Louis, are building a pcriogue for the purpose. We sent six men in a periogue to hunt down the river. We were also fortunate enough to engage in our service a Canadian Frenchman, who had been with the Chayenne Indians on the Black mountains and last summer descended thence by the Little Missouri. Mr. Jessaume, our interpreter, also came down with his squaw and children to live at our camp. In the evening we received a visit from Kagohami, or Little Raven, whose wife accompanied him, bringing about sixty weight of dried meat, a robe, and a pot of meal. We gave him in return a piece of tobacco, to his wife an axe and a few small articles, and both of them spent the night at our camp. Two beavers were caught in traps this morning.

## UP THE MISSOURI

Sunday, 4th. We continued our labours; the timber which we employ is large and heavy, and chiefly consists of cottonwood and elm with some ash of an inferior size. Great numbers of the Indians pass our camp on their hunting excursions; the day was clear and pleasant, but last night was very cold and there was a white frost.

Monday, 5th. The Indians are all out on their hunting parties; a camp of Mandans caught within two days one hundred goats a short distance below us. Their mode of hunting them is to form a large strong pen or fold, from which a fence made of bushes gradually widens on each side; the animals are surrounded by the hunters and gently driven towards this pen, in which they imperceptibly find themselves inclosed and are then at the mercy of the hunters. The weather is cloudy and the wind moderate from the northwest. Late at night we were awaked by the sergeant on guard to see the beautiful phenomenon called the northern light; along the northern sky was a large space occupied by a light of a pale but brilliant white colour, which rising from the horizon extended itself to nearly twenty degrees above it. After glittering for some time its colours would be overcast and almost obscured, but again it would burst out with renewed beauty; the uniform colour was pale light, but its shapes were various and fantastic; at times the sky was lined with light-coloured streaks rising perpendicularly from the horizon, and gradually expanding into a body of light in which we could trace the floating columns, sometimes advancing, sometimes retreating, and shaping into infinite forms the space in which they moved. It all faded away before the morning. At daylight,

Tuesday, 6th, the clouds to the north were darkening, and the wind rose high from the northwest at eight o'clock and continued cold during the day. Mr.Gravelines and four others who came with us returned to the Ricaras in a small periogue; we gave him directions to accompany some of the Ricara chiefs to the seat of government in the spring.

Wednesday, 7th. The day was temperate but cloudy and foggy, and we were enabled to go on with our work with much expedition.

Thursday, 8th. The morning again cloudy; our huts advance very well, and we are visited by numbers of Indians who come to let their horses graze near us; in the day the horses are let loose in quest of grass, in the night they are collected and receive an armfull of small boughs of the cottonwood, which, being very juicy, soft, and brittle, form nutritious and agreeable food. The frost this morning was very severe, the weather during the day cloudy and the wind from the northwest. We procured from an Indian a weasel, perfectly white except the extremity of the tail, which was black; great numbers of wild geese are passing to the south, but their flight is too high for us to procure any of them.

November 10. We had again a raw day, a northwest wind, but rose early in hopes of finishing our works before the extreme cold begins. A chief who is a half Pawnee came to us and brought a present of half a buffaloe, in return for which we gave him some small presents and a few articles to his wife and son; he then crossed the river in a buffaloe skin canoe; his wife took the boat on her back and carried it to the village three miles off. Large flocks of geese and brant and also a few ducks are passing towards the south.

Sunday, 11th. The weather is cold. We received the visit of two squaws, prisoners from the Rock mountains, and purchased by Chaboneau. The Mandans at this time are out hunting the buffaloe.

Monday, 12th. The last night had been cold and this morning we had a very hard frost; the wind changeable during the day, and some ice appears on the edges of the rivers; swans, too, are passing to the south. The Big White came down to us, having packed on the back of his squaw about one hundred pounds of very fine meat, for which we gave him as well as the squaw some presents, particularly an axe to the woman, with which she was very much pleased.

Tuesday, 13th. We this morning unloaded the boat and stowed away the contents in a storehouse which we have built. At half past ten ice began to float down the river for the first time. In the course of the morning we were visited by the Black Cat, Poscapsahe, who brought an Assiniboin chief and seven warriors to see us. This man, whose name is Chechawk, is a chief of one out of three bands of Assiniboins who wander over the plains between the Missouri and Assiniboin during the summer, and in the winter carry the spoils of their hunting to the traders on the Assiniboin river, and occasionally come to this place; the whole three bands consist of about eight hundred men. We gave him a twist of tobacco to smoke with his people, and a gold cord for himself; the Sioux also asked for whiskey, which we refused to give them. It snowed all day and the air was very cold.

Wednesday, 14th. The river rose last night half an inch, and is now filled with floating ice. This morning was cloudy, with some snow. About seventy lodges of Assiniboins and some Knistenaux are at the Mandan village, and this being the day of adoption and exchange of property between them all, it is accompanied by a dance, which prevents our seeing more than two Indians to-day. These Knistenaux are a band of Chippeways, whose language they speak; they live on the Assiniboin and Saskashawan rivers, and are about two hundred and forty men. We sent a man down on horseback to see what had become of our hunters, and as we apprehend a failure of provisions we have recourse to our pork this evening.

Two Frenchmen who had been below returned with twenty beaver which they had caught in traps.

Thursday, 15th. The morning again cloudy, and the ice running thicker than yesterday, the wind variable. The man came back with information that our hunters were about thirty miles below, and we immediately sent an order to them to make their way through the floating ice, to assist them in which we sent some tin for the bow of the periogue and a towrope. The ceremony of yesterday seems to continue still, for we were not visited by a single Indian. The swan are still passing to the south.

Friday, 16th. We had a very hard white frost this morning, the trees are all covered with ice, and the weather cloudy. The men this day moved into the huts, although they are not finished. In the evening some horses were sent down to the woods near us in order to prevent their being stolen by the Assiniboins, with whom some difficulty is now apprehended. An Indian came down with four buffaloe robes and some corn, which he offered for a pistol, but was refused.

Saturday, November 17. Last night was very cold, and the ice in the river to-day is thicker than hitherto. We are totally occupied with our huts, but received visits from several Indians.

Sunday, November 18. To-day we had a cold windy morning; the Black Cat came to see us, and occupied us for a long time with questions on the usages of our country. He mentioned that a council had been held yesterday to deliberate on the state of their affairs. It seems that not long ago a party of Sioux fell in with some horses belonging to the Minnetarees and carried them off, but in their flight they were met by some Assiniboins, who killed the Sioux and kept the horses. A Frenchman, too, who had lived many years among the Mandans, was lately killed on his route to the British factory on the Assiniboin; some smaller differences existed between the two nations, all of which being discussed, the council decided that they would not resent the recent insults from the Assiniboins and Knistenaux until they had seen whether we had deceived them or not in our promises of furnishing them with arms and ammunition. They had been disappointed in their hopes of receiving them from Mr. Evans, and were afraid that we too, like him, might tell them what was not true. We advised them to continue at peace, that supplies of every kind would no doubt arrive for them, but that time was necessary to organize the trade. The fact is that the Assiniboins treat the Mandans as the Sioux do the Ricaras; by their vicinity to the British they get all the supplies, which they withhold or give at pleasure to the remoter Indians; the consequence is that, however badly treated, the Mandans and Ricaras are very slow to retaliate lest they should lose their trade altogether.

Monday, 19th. The ice continues to float in the river, the wind high from the northwest, and the weather cold. Our hunters arrived from their excursion below and bring a very fine supply of thirty-two deer, eleven elk, and five buffaloe, all of which was hung in a smokehouse.

Tuesday, 20th. We this day moved into our huts, which are now completed. This place, which we call Fort Mandan, is situated in a point of low ground on the north side of the Missouri, covered with tall and heavy cottonwood. The works consist of two rows of huts or sheds, forming an angle where they joined each other, each row containing four rooms of fourteen feet square and seven feet high, with plank ceiling, and the roof slanting so as to form a loft above the rooms, the highest part of which is eighteen feet from the ground;

the backs of the huts formed a wall of that height, and opposite the angle the place of the wall was supplied by picketing; in the area were two rooms for stores and provisions. The latitude by observation is 47° 21′ 47″, and the computed distance from the mouth of the Missouri sixteen hundred miles.

In the course of the day several Indians came down to partake of our fresh meat, among the rest three chiefs of the second Mandan village. They inform us that the Sioux on the Missouri above the Chayenne river threaten to attack them this winter; that these Sioux are much irritated at the Ricaras for having made peace through our means with the Mandans, and have lately ill treated three Ricaras who carried the pipe of peace to them by beating them and taking away their horses. We gave them assurances that we would protect them from all their enemies.

November 21. The weather was this day fine, the river clear of ice and rising a little; we are now settled in our new winter habitation, and shall wait with much anxiety the first return of spring to continue our journey.

The villages near which we are established are five in number, and are the residence of three distinct nations: the Mandans, the Ahnahaways, and the Minnetarees. The history of the Mandans, as we received it from our interpreters and from the chiefs themselves, and as it is attested by existing monuments, illustrates more than that of any other nation the unsteady movements and the tottering fortunes of the American nations. Within the recollection of living witnesses, the Mandans were settled forty years ago in nine villages, the ruins of which we passed about eighty miles below, and situated seven on the west and two on the east side of the Missouri. The two, finding themselves wasting away before the small-pox and the Sioux, united into one village and moved up the river opposite to the Ricaras. The same causes reduced the remaining seven to five villages, till at length they emigrated in a body to the Ricara nation, where they formed themselves into two villages and joined those of their countrymen who had gone before them. In their new residence they were still insecure, and at length the three villages ascended the Missouri to their present position. The two who had emigrated together still settled in the two villages on the northwest side of the Missouri, while the single village took a position on the southeast side. In this situation they were found by those who visited them in 1796, since which the two villages have united into one. They are now in two villages, one on the southeast of the Missouri, the other on the opposite side and at the distance of three miles across. The first, in an open plain, contains about forty or fifty lodges, built in the same way as those of the Ricaras; the second, the same number; and both may raise about three hundred and fifty men.

On the same side of the river, and at the distance of four miles from the lower Mandan village, is another called Mahaha. It is situated in a high plain at the mouth of Knife river, and is the residence of the Ahnahaways. This nation, whose name indicates that they were "people whose village is on a hill," formerly resided on the Missouri, about thirty miles below where they now live. The Assiniboins and Sioux forced them to a spot five miles higher, where the greatest part of them were put to death, and the rest emigrated to their present situation in order to obtain an asylum near the Minnetarees. They are called by the French, Soulier Noir, or Shoe Indians; by the Mandans, Wattasoons, and their whole force is about fifty men.

On the south side of the same Knife river, half a mile above the Mahaha and in the same open plain with it, is a village of Minnetarees surnamed Metaharta, who are about one hundred and fifty men in number. On the opposite side of Knife river, and one and a half mile above this village, is a second of Minnetarees, who may be considered as the proper Minnetaree nation. It is situated in a beautiful low plain, and contains four hundred and fifty warriors. The accounts which we received of the Minnetarees were contradictory. The Mandans say that this people came out of the water to the east, and settled near them in their former establishment in nine villages; that they were very numerous, and fixed themselves in one village on the southern side of the Missouri. A quarrel about a buffaloe divided the nation, of which two bands went into the plains and were known by the name of Crow and Paunch Indians, and the rest moved to their present establishment. The Minnetarees proper assert, on the contrary, that they grew where they now live, and will never emigrate from the spot, the great spirit having declared that if they moved they would all die. They also say that the Minnetarees Metaharta, that is Minnetarees of the Willows, whose language with very little variation is their own, came many years ago from the plains and settled near them, and perhaps the two traditions may be reconciled by the natural presumption that these Minnetarees were the tribe known to the Mandans below, and that they ascended the river for the purpose of rejoining the Minnetarees proper. These Minnetarees are part of the great nation called Fall Indians, who occupy the intermediate country between the Missouri and the Saskaskawan, and who are known by the name of Minnetarees of the Missouri, and Minnetarees of Fort de Prairie: that is, residing near or rather frequenting the establishment in the prarie on the Saskaskawan. These Minnetarees, indeed, told us that they had relations on the Saskaskawan whom they had never known till they met them in war, and having

## UP THE MISSOURI

engaged in the night were astonished at discovering that they were fighting with men who spoke their own language. The name of Grosventres, or Bigbellies, is given to these Minnetarees, as well as to all the Fall Indians. The inhabitants of these five villages, all of which are within the distance of six miles, live in harmony with each other. The Ahnahaways understand in part the language of the Minnetarees; the dialect of the Mandans differs widely from both, but their long residence together has insensibly blended their manners and occasioned some approximation in language, particularly as to objects of daily occurrence and obvious to the senses.

November 22. The morning was fine and the day warm. We purchased from the Mandans a quantity of corn of a mixed colour, which they dug up in ears from holes made near the front of their lodges, in which it is buried during the winter. This morning the sentinel informed us that an Indian was about to kill his wife near the fort; we went down to the house of our interpreter, where we found the parties, and after forbidding any violence, inquired into the cause of his intending to commit such an atrocity. It appeared that some days ago a quarrel had taken place between him and his wife, in consequence of which she had taken refuge in the house where the two squaws of our interpreter lived; by running away she forfeited her life, which might have been lawfully taken by the husband. About two days ago she had returned to the village, but the same evening came back to the fort much beaten and stabbed in three places, and the husband now came for the purpose of completing his revenge. He observed that he had lent her to one of our serjeants for a night, and that if he wanted her he would give her to him altogether; we gave him a few presents and tried to persuade him to take his wife home; the grand chief, too, happened to arrive at the same moment, and reproached him with his violence, till at length they went off together, but by no means in a state of much apparent love.

November 23. Again we had a fair and warm day, with the wind from the southeast; the river is now at a stand, having risen four inches in the whole.

November 24. The wind continued from the same quarter and the weather was warm; we were occupied in finishing our huts and making a large rope of elk-skin to draw our boat on the bank.

Sunday, November 25. The weather is still fine, warm and pleasant, and the river falls one inch and a half. Captain Lewis went on an excursion to the villages accompanied by eight men. A Minnetaree chief, the first who has visited us, came down to the fort; his name was Waukerassa, but as both the interpreters had gone with Captain Lewis, we were obliged to confine our civilities to some presents, with which he was much pleased; we now completed our huts, and fortunately, too, for the next day,

Monday, November 26, before daylight, the wind shifted to the northwest, and blew very hard, with cloudy weather and a keen cold air, which confined us much and prevented us from working; the night continued very cold, and,

Tuesday, 27th, the weather cloudy, the wind continuing from the northwest and the river crowded with floating ice. Captain Lewis returned with two chiefs, Mahnotah an Ahnahaway, and Minnessurraree a Minnetaree, and a third warrior; they explained to us that the reason of their not having come to see us was that the Mandans had told them that we meant to combine with the Sioux and cut them off in the course of the winter, a suspicion increased by the strength of the fort and the circumstance of our interpreters having both removed there with their families; these reports we did not fail to disprove to their entire satisfaction, and amused them by every attention, particularly by the dancing of the men, which diverted them highly. All the Indians whom Captain Lewis had visited were very well disposed and received him with great kindness, except a principal chief of one of the upper villages, named Mahpahpaparapassatoo, or Horned Weasel, who made use of the civilized indecorum of refusing to be seen, and when Captain Lewis called he was told the chief was not at home. In the course of the day seven of the northwest company's traders arrived from the Assiniboin river, and one of their interpreters having undertaken to circulate among the Indians unfavourable reports, it became necessary to warn them of the consequences if they did not desist from such proceedings. The river fell two inches to-day, and the weather became very cold.

Wednesday, 28th. About eight o'clock last evening it began to snow and continued till daybreak, after which it ceased till seven o'clock, but then resumed and continued during the day, the weather being cold and the river full of floating ice; about eight o'clock Poscopsahe came down to visit us, with some warriors; we gave them presents and entertained them with all that might amuse their curiosity, and at parting we told them that we had heard of the British trader, Mr. Laroche, having attempted to distribute medals and flags among them, but that those emblems could not be received from any other than the American nation without incurring the displeasure of their great father the President. They left us much pleased with their treatment. The river fell one inch to-day.

Thursday, 29th. The wind is again from the northwest, the weather cold, and the snow which fell yesterday and last night is thirteen inches in depth. The river closed during the night at the village above and fell two feet, but this afternoon it began to rise a little. Mr. Laroche, the principal of the seven traders, came with one of his men to see us; we told him that we should not permit him to give medals and flags to the Indians; he declared that he had no such intention, and we then suffered him to make use of one of our interpreters, on his stipulating not to touch any subject but that of his traffic with them. An unfortunate accident occurred to Sergeant Pryor, who, in taking down the boat's mast, dislocated his shoulder, nor was it till after four trials that we replaced it.

Friday, 30th. About eight o'clock an Indian came to the opposite bank of the river calling out that he had something important to communicate, and on sending for him he told us that five Mandans had been met about eight leagues to the southwest by a party of Sioux, who had killed one of them, wounded two, and taken nine horses; that four of the Wattasoons were missing, and that the Mandans expected an attack. We thought this an excellent opportunity to discountenance the injurious reports against us, and to fix the wavering confidence of the nation. Captain Clark therefore instantly crossed the river with twenty-three men, strongly armed, and circling the town approached it from behind. His unexpected appearance surprised and alarmed the chiefs, who came out to meet him and conducted him to the village. He then told them that, having heard of the outrage just committed, he had come to assist his dutiful children; that if they would assemble their warriors and those of the nation, he would lead them against the Sioux and avenge the blood of their countrymen. After some minutes conversation, Oheenaw the Chayenne arose. "We now see," said he, "that what you have told us is true, since as soon as our enemies threaten to attack us you come to protect us, and are ready to chastise those who have spilt our blood. We did indeed listen to your good talk, for when you told us that the other nations were inclined to peace with us, we went out carelessly in small parties, and some have been killed by the Sioux and Ricaras. But I knew that the Ricaras were liars, and I told their chief who accompanied you that his whole nation were liars and bad men; that we had several times made a peace with them which they were the first to break; that whenever we pleased we might shoot them like buffaloe, but that we had no wish to kill them; that we would not suffer them to kill us nor steal our horses; and that although we agreed to make peace with them, because our two fathers desired it, yet we did not believe that they would be faithful long. Such, father, was my language to them in your presence, and you see that instead of listening to your good counsels they have spilt our blood. A few days ago two Ricaras came here and told us that two of their villages were making moccasins, that the Sioux were stirring them up against us, and that we ought to take care of our horses; yet these very Ricaras we sent home as soon as the news reached us to-day, lest our people should kill them in the first moment of grief for their murdered relatives. Four of the Wattasoons, whom we expected back in sixteen days, have been absent twenty-four, and we fear have fallen. But, father, the snow is now deep, the weather cold, and our horses cannot travel through the plains; the murderers have gone off; if you will conduct us in the spring, when the snow has disappeared, we will assemble all the surrounding warriors and follow you."

Captain Clark replied that we were always willing and able to defend them; that he was sorry that the snow prevented their marching to meet the Sioux, since he wished to show them that the warriors of their great father would chastise the enemies of his obedient children who opened their ears to his advice; that if some Ricaras had joined the Sioux, they should remember that there were bad men in every nation, and that they should not be offended at the Ricaras till they saw whether these ill-disposed men were countenanced by the whole tribe; that the Sioux possessed great influence over the Ricaras, whom they supplied with military stores, and sometimes led them astray because they were afraid to oppose them; but that this should be the less offensive since the Mandans themselves were under the same apprehensions from the Assiniboins and Knistenaux, and that while they were thus dependent, both the Ricaras and Mandans ought to keep on terms with their powerful neighbours, whom they may afterwards set at defiance when we shall supply them with arms and take them under our protection.

After two hours conversation Captain Clark left the village. The chief repeatedly thanked him for the fatherly protection he had given them, observing that the whole village had been weeping all night and day for the brave young man who had been slain, but now they would wipe their eyes and weep no more, as they saw that their father would protect them. He then crossed the river on the ice and returned on the north side to the fort. The day as well as the evening was cold, and the river rose to its former height.

Saturday, December 1. The wind was from the northwest, and the whole party engaged in picketing the fort. About ten o'clock the half-brother of the man who had been killed came to inform us that six Sharhas or Chayenne Indians had arrived, bringing a pipe of peace, and that their nation was three days march behind them. Three Pawnees had accompanied the Sharhas, and the Mandans, being afraid of the Sharhas on account of their being at peace with the Sioux, wished to put both them and the three Pawnees to death; but the chiefs had forbidden it, as it would be contrary to our wishes. We gave him a present of tobacco, and although from his connexion with the sufferer he was more embittered against the Pawnees than any other Mandan, yet he seemed perfectly satisfied with our pacific counsels and advice. The Mandans, we observe, call all the Ricaras by the name of Pawnees, the name of Ricaras being that by which the nation distinguishes itself.

In the evening we were visited by a Mr. Henderson, who came from the Hudson bay company to trade with the Minnetarees. He had been about eight days on his route in a direction nearly south, and brought with him tobacco, beads, and other merchandize to trade for furs, and a few guns which are to be exchanged for horses.

Sunday, December 2. The latter part of the evening was warm, and a thaw continued till the morning, when the wind shifted to the north. At eleven o'clock the chiefs of the lower village brought down four of the Sharhas. We explained to them our intentions, and advised them to remain at peace with each other; we also gave them a flag, some tobacco, and a speech for their nation. These were accompanied by a letter to Messrs. Tabeau and Gravelines at the Ricara village, requesting them to preserve peace if possible, and to declare the part which we should be forced to take if the Ricaras and Sioux made war on those whom we had adopted. After distributing a few presents to the Sharhas and Mandans and showing them our curiosities, we dismissed them, apparently well pleased at their reception.

Monday, December 3. The morning was fine, but in the afternoon the weather became cold, with the wind from the northwest. The father of the Mandan who was killed brought us a present of dried pumpkins and some pemitigon, for which we gave him some small articles. Our offer of assistance to avenge the death of his son seemed to have produced a grateful respect from him, as well as from the brother of the deceased, which pleased us much.

Tuesday, 4th. The wind continues from the northwest, the weather cloudy and raw, and the river rose one inch. Oscapsahe and two young chiefs pass the day with us. The whole religion of the Mandans consists in the belief of one great spirit presiding over their destinies. This being must be in the nature of a good genius, since it is associated with the healing art, and the great spirit is synonymous with great medicine, a name also applied to everything which they do not comprehend. Each individual selects for himself the particular object of his devotion, which is termed his medicine, and is either some invisible being or more commonly some animal, which thenceforward becomes his protector or his intercessor with the great spirit, to propitiate whom every attention is lavished, and every personal consideration is sacrificed. "I was lately owner of seventeen horses," said a Mandan to us one day, "but I have offered them all up to my medicine and am now poor." He had in reality taken all his wealth, his horses, into the plain, and turning them loose committed them to the care of his medicine and abandoned them forever. The horses, less religious, took care of themselves, and the pious votary travelled home on foot. Their belief in a future state is connected with this tradition of their origin: The whole nation resided in one large village under ground near a subterraneous lake; a grapevine extended its roots down to their habitation and gave them a view of the light; some of the most adventurous climbed up the vine and were delighted with the sight of the earth, which they found covered with buffaloe and rich with every kind of fruits; returning with the grapes they had gathered, their countrymen were so pleased with the taste of them that the whole nation resolved to leave their dull residence for the charms of the upper region; men, women, and children ascended by means of the vine; but when about half the nation had reached the surface of the earth, a corpulent woman who was clambering up the vine broke it with her weight, and closed upon herself and the rest of the nation the light of the sun. Those who were left on earth made a village below where we saw the nine villages; and when the Mandans die they expect to return to the original seats of their forefathers, the good reaching the ancient village by means of the lake, which the burden of the sins of the wicked will not enable them to cross.

Wednesday, 5th. The morning was cold and disagreeable, the wind from the southeast accompanied with snow; in the evening there was snow again, and the wind shifted to the northeast; we were visited by several Indians with a present of pumpkins, and by two of the traders of the northwest company.

Thursday, 6th. The wind was violent from the north-northwest with some snow, the air keen and cold. At eight o'clock A. M. the thermometer stood at 10° above zero, and the river rose an inch and a half in the course of the day.

Friday, December 7. The wind still continued from the northwest and the day is very cold; Shahaka, the chief of the lower village, came to apprise us that the buffaloe were near, and that his people were waiting for us to join them in the chase. Captain Clark, with fifteen men, went out and found the Indians engaged in killing the buffaloe; the hunters, mounted on horseback and armed with bows and arrows, encircle the herd and gradually drive them into a plain or an open place fit for the movements of horse; they then ride in among them, and singling out a buffaloe, a female being preferred, go as close as possible and wound her with arrows till they think they have given the mortal stroke, when they pursue another till the quiver is exhausted. If, which rarely happens, the wounded buffaloe attacks the hunter, he evades his blow by the agility of his horse, which is trained for the combat with great dexterity. When they have killed the requisite number they collect their game, and the squaws and attendants come up from the rear and skin and dress the animals. Captain Clark killed ten buffaloe, of which five only were brought to the fort, the rest, which could not be conveyed home, being seized by the Indians, among whom the custom is that whenever a buffaloe is found dead without an arrow or any particular mark, he is the property of the finder, so that often a hunter secures scarcely any of the game he kills if the arrow happens to fall off; whatever is left out at night falls to the share of the wolves, who are the constant and numerous attendants of the buffaloe. The river closed opposite the fort last night, an inch and a half in thickness. In the morning the thermometer stood at 1° below zero. Three men were badly frostbitten in consequence of their exposure.

Saturday, 8th. The thermometer stood at 12° below zero, that is at forty-two degrees below the freezing point; the wind was from the northwest. Captain Lewis, with fifteen men, went out to hunt the buffaloe, great numbers of which darkened the prairies for a considerable distance; they did not return till after dark, having killed eight buffaloe and one deer. The hunt was, however, very fatiguing, as they were obliged to make a circuit at a distance of more than seven miles; the cold, too, was so excessive that the air was filled with icy particles resembling a fog, and the snow generally six or eight inches deep and sometimes eighteen, in consequence of which two of the party were hurt by falls, and several had their feet frostbitten.

Sunday, 9th. The wind was this day from the east, the thermometer at 7° above zero, and the sun shone clear; two chiefs visited us, one in a sleigh drawn by a dog and loaded with meat.

Monday, 10th. Captain Clark, who had gone out yesterday with eighteen men to bring in the meat we had killed the day before and to continue the hunt, came in at twelve o'clock. After killing nine buffaloe and preparing that already dead, he had spent a cold disagreeable night on the snow, with no covering but a small blanket, sheltered by the hides of the buffaloe they had killed. We observe large herds of buffaloe crossing the river on the ice. The men who were frostbitten are recovering, but the weather is still exceedingly cold, the wind being from the north and the thermometer at 10° and 11° below zero; the rise of the river is one inch and a half.

Tuesday, 11th. The weather became so intensely cold that we sent for all the hunters who had remained out with Captain Clark's party, and they returned in the evening, several of them frostbitten. The wind was from the north and the thermometer at sunrise stood at 21° below zero, the ice in the atmosphere being so thick as to render the weather hazy and give the appearance of two suns reflecting each other. The river continues at a stand. Pocapsahe made us a visit to-day.

Wednesday, December 12. The wind is still from the north, the thermometer being at sunrise 38° below zero. One of the Ahnahaways brought us down the half of an antelope killed near the fort; we had been informed that all these animals return to the Black mountains, but there are great numbers of them about us at this season which we might easily kill, but are unwilling to venture out before our constitutions are hardened gradually to the climate. We measured the river on the ice, and find it five hundred yards wide immediately opposite the fort.

Thursday, 13th. Last night was clear and a very heavy frost covered the old snow, the thermometer at sunrise being 20° below zero, and followed by a fine day. The river falls.

Friday, 14th. The morning was fine, and the weather having moderated so far that the mercury stood at zero, Captain Lewis went down with a party to hunt; they proceeded about eighteen miles, but the buffaloe having left the banks of the river they saw only two, which were so poor as not to be worth killing, and shot two deer. Notwithstanding the snow we were visited by a large number of the Mandans.

Saturday, 15th. Captain Lewis, finding no game, returned to the fort hunting on both sides of the river, but with no success. The wind being from the north, the mercury at sunrise 8° below zero, and the snow of last night an inch and a half in depth. The Indian chiefs continue to visit us to-day with presents of meat.

Sunday, 16th. The morning is clear and cold, the mercury at sunrise 22° below zero. A Mr. Haney, with two other persons from the British establishment on the Assiniboin, arrived in six days with a letter from Mr. Charles Chabouilles, one of the company, who with much politeness offered to render us any service in his power.

Monday, 17th. The weather to-day was colder than any we had yet experienced, the thermometer at sunrise being 45° below zero, and about eight o'clock it fell to 74° below the freezing point. From Mr. Haney, who is a very sensible intelligent man, we obtained much geographical information with regard to the country between the Missouri and Mississippi, and the various tribes of Sioux who inhabit it.

Tuesday, 18th. The thermometer at sunrise was 32° below zero. The Indians had invited us yesterday to join their chace to-day, but the seven men whom we sent returned in consequence of the cold, which was so severe last night that we were obliged to have the sentinel relieved every half hour. The northwest traders, however, left us on their return home.

Wednesday, 19th. The weather moderated and the river

rose a little, so that we were enabled to continue the picketing of the fort. Notwithstanding the extreme cold, we observe the Indians at the village engaged out in the open air at a game which resembled billiards more than anything we had seen, and which we inclined to suspect may have been acquired by ancient intercourse with the French of Canada. From the first to the second chief's lodge, a distance of about fifty yards, was covered with timber smoothed and joined so as to be as level as the floor of one of our houses, with a battery at the end to stop the rings; these rings were of clay-stone and flat like the chequers for drafts, and the sticks were about four feet long, with two short pieces at one end in the form of a mace, so fixed that the whole will slide along the board. Two men fix themselves at one end, each provided with a stick and one of them with a ring; they then run along the board, and about half way slide the sticks after the ring.

Thursday, 20th. The wind was from the N. W., the weather moderate, the thermometer 24° above zero at sunrise. We availed ourselves of this change to picket the fort near the river.

Friday, 21st. The day was fine and warm, the wind N. W. by W. The Indian who had been prevented a few days ago from killing his wife came with both his wives to the fort, and was very desirous of reconciling our interpreter, a jealousy against whom on account of his wife's taking refuge in his house had been the cause of his animosity. A woman brought her child with an abscess in the lower part of the back, and offered as much corn as she could carry for some medicine; we administered to it of course very cheerfully.

Saturday, 22d. A number of squaws and men dressed like squaws brought corn to trade for small articles with the men. Among other things we procured two horns of the animal called by the French the Rock mountain sheep, and known to the Mandans by the name of ahsahta. The animal itself is about the size of a small elk or large deer, the horns winding like those of a ram, which they resemble also in texture, though larger and thicker.

Sunday, 23d. The weather was fine and warm like that of yesterday; we were again visited by crowds of Indians of all descriptions, who came either to trade or from mere curiosity. Among the rest Kogahami, the Little Raven, brought his wife and son loaded with corn, and she then entertained us with a favorite Mandan dish, a mixture of pumpkins, beans, corn, and chokecherries with the stones, all boiled together in a kettle, and forming a composition by no means unpalatable.

Monday, 24th. The day continued warm and pleasant, and the number of visitors became troublesome. As a present to three of the chiefs, we divided a fillet of sheepskin which we brought for spunging into three pieces, each of two inches in width; they were delighted at the gift, which they deemed of equal value with a fine horse. We this day completed our fort, and the next morning being Christmas,

Tuesday, 25th, we were awaked before day by a discharge of three platoons from the party. We had told the Indians not to visit us as it was one of our great medicine days, so that the men remained at home and amused themselves in various ways, particularly with dancing, in which they take great pleasure. The American flag was hoisted for the first time in the fort; the best provisions we had were brought out, and this, with a little brandy, enabled them to pass the day in great festivity.

Wednesday, 26th. The weather is again temperate, but no Indians have come to see us. One of the northwest traders, who came down to request the aid of our Minnetaree interpreter, informs us that a party of Minnetarees who had gone in pursuit of the Assiniboins who lately stole their horses had just returned. As is their custom, they came back in small detachments, the last of which brought home eight horses which they had captured or stolen from an Assiniboin camp on Mouse river.

Thursday, 27th. A little fine snow fell this morning and the air was colder than yesterday, with a high northwest wind. We were fortunate enough to have among our men a good blacksmith, whom we set to work to make a variety of articles; his operations seemed to surprise the Indians who came to see us, but nothing could equal their astonishment at the bellows, which they considered as a very great medicine. Having heretofore promised a more particular account of the Sioux, the following may serve as a general outline of their history:

Almost the whole of that vast tract of country comprised between the Mississippi, the Red river of lake Winnepeg, the Saskaskawan, and the Missouri, is loosely occupied by a great nation whose primitive name is Darcota, but who are called Sioux by the French, Sues by the English. Their original seats were on the Mississippi, but they have gradually spread themselves abroad and become subdivided into numerous tribes. Of these, what may be considered as the Darcotas are the Mindawarcarton, or Minowakanton, known to the French by the name of the Gens du Lac, or People of the Lake. Their residence is on both sides of the Mississippi near the falls of St. Anthony, and the probable number of their warriors about three hundred. Above them, on the river St. Peter's, is the Wahpatone, a smaller band of nearly two hundred men; and still farther up the same river, below

Yellowwood river, are the Wahpatootas, or Gens de Feuilles, an inferior band of not more than one hundred men; while the sources of the St. Peter's are occupied by the Sisatoones, a band consisting of about two hundred warriors.

These bands rarely if ever approach the Missouri, which is occupied by their kinsmen the Yanktons and the Tetons. The Yanktons are of two tribes: those of the plains, or rather of the north, a wandering race of about five hundred men, who roam over the plains at the heads of the Jacques, the Sioux, and the Red rivers; and those of the south, who possess the country between the Jacques and Sioux rivers and the Desmoine. But the bands of Sioux most known on the Missouri are the Tetons. The first who are met on ascending the Missouri is the tribe called by the French the Tetons of the Bois Brule or Burntwood, who reside on both sides of the Missouri, about White and Teton rivers, and number two hundred warriors. Above them on the Missouri are the Teton Okandandas, a band of one hundred and fifty men living below the Chayenne river, between which and the Wetarhoo river is a third band, called Teton Minnakenozzo, of nearly two hundred and fifty men; and below the Warreconne is the fourth and last tribe of Tetons, of about three hundred men, and called Teton Saone. Northward of these, between the Assiniboin and the Missouri, are two bands of Assiniboins, one on Mouse river of about two hundred men, and called Assiniboin Menatopa; the other, residing on both sides of White river, called by the French Gens de Feuilles, and amounting to two hundred and fifty men. Beyond these a band of Assiniboins of four hundred and fifty men, and called the Big Devils, wander on the heads of Milk, Porcupine, and Martha's rivers; while still farther to the north are seen two bands of the same nation, one of five hundred and the other of two hundred, roving on the Saskaskawan. Those Assiniboins are recognized by a similarity of language and by tradition as descendants or seceders from the Sioux; though often at war, are still acknowledged as relations. The Sioux themselves, though scattered, meet annually on the Jacques, those on the Missouri trading with those on the Mississippi.

## CHAPTER VI.

The party increase in the favour of the Mandans—Description of a buffaloe dance—Medicine dance—The fortitude with which the Indians bear the severity of the season—Distress of the party for want of provisions—The great importance of the blacksmith in procuring it—Depredations of the Sioux—The homage paid to the medicine stone—Summary act of justice among the Minnetarees-The process by which the Mandans and Ricaras make beads—Character of the Missouri, of the surrounding country, and of the rivers, creeks, islands, etc.

FRIDAY, 28th. The wind continued high last night, the frost severe, and the snow drifting in great quantities through the plains.

Saturday, 29th. There was a frost fell last night nearly one-quarter of an inch in depth, which continued to fall till the sun had gained some height; thé mercury at sunrise stood at 9° below zero; there were a number of Indians at the fort in the course of the day.

Sunday, 30th. The weather was cold, and the thermometer 20° below zero. We killed one deer, and yesterday one of the men shot a wolf. The Indians brought corn, beans, and squashes, which they very readily gave for getting their axes and kettles mended. In their general conduct during these visits they are honest, but will occasionally pilfer any small article.

Monday, 31st. During the night there was a high wind which covered the ice with hillocks of mixed sand and snow; the day was, however, fine, and the Indians came in great numbers for the purpose of having their utensils repaired.

Tuesday, January 1, 1805. The new year was welcomed by two shot from the swivel and a round of small arms. The weather was cloudy but moderate; the mercury, which at

## LEWIS AND CLARK EXPEDITION 159

sunrise was at 18°, in the course of the day rose to 34° above zero; towards evening it began to rain, and at night we had snow, the temperature for which is about zero. In the morning we permitted sixteen men with their music to go up to the first village, where they delighted the whole tribe with their dances, particularly with the movements of one of the Frenchmen who danced on his head. In return they presented the dancers with several buffaloe robes and quantities of corn. We were desirous of showing this attention to the village, because they had received an impression that we had been wanting in regard for them, and because they had in consequence circulated invidious comparisons between us and the northern traders; all these, however, they declared to Captain Clark, who visited them in the course of the morning, were made in jest. As Captain Clark was about leaving the village, two of their chiefs returned from a mission to the Grosventres or wandering Minnetarees. These people were encamped about ten miles above, and while there one of the Ahnahaways had stolen a Minnetaree girl; the whole nation immediately espoused the quarrel, and one hundred and fifty of their warriors were marching down to revenge the insult on the Ahnahaways. The chief of that nation took the girl from the ravisher, and, giving her to the Mandans, requested their intercession. The messengers went out to meet the warriors, and delivered the young damsel into the hands of her countrymen, smoked the pipe of peace with them, and were fortunate enough to avert their indignation and induce them to return. In the evening some of the men came to the fort and the rest slept in the village. Pocapsahe also visited us and brought some meat on his wife's back.

Wednesday, January 2. It snowed last night, and during this day the same scene of gayety was renewed at the second village, and all the men returned in the evening.

Thursday, 3d. Last night it became very cold, and this morning we had some snow; our hunters were sent out for buffaloe, but the game had been frightened from the river by the Indians, so that they obtained only one; they, however, killed a hare and a wolf. Among the Indians who visited us was a Minnetaree who came to seek his wife; she had been much abused and came here for protection, but returned with him, as we had no authority to separate those whom even the Mandan rites had united.

Friday, 4th. The morning was cloudy and warm, the mercury being 28° above zero, but towards evening the wind changed to northwest and the weather became cold. We sent some hunters down the river, but they killed only one buffaloe and a wolf. We received the visit of Kagohami, who is very friendly, and to whom we gave a handkerchief and two files.

Saturday, 5th. We had high and boisterous winds last night and this morning; the Indians continue to purchase repairs with grain of different kinds. In the first village there has been a buffaloe dance for the last three nights, which has put them all into commotion, and the description which we received from those of the party who visited the village and from other sources is not a little ludicrous; the buffaloe dance is an institution originally intended for the benefit of the old men, and practised at their suggestion. When buffaloe becomes scarce they send a man to harangue the village, declaring that the game is far off and that a feast is necessary to bring it back, and if the village be disposed a day and place is named for the celebration of it. At the appointed hour the old men arrive, and seat themselves crosslegged on skins round a fire in the middle of the lodge, with a sort of doll or small image, dressed like a female, placed before them. The young men bring with them a platter of provisions, a pipe

## UP THE MISSOURI 161

of tobacco, and their wives, whose dress on the occasion is only a robe or mantle loosely thrown round the body. On their arrival each youth selects the old man whom he means to distinguish by his favour, and spreads before him the provisions, after which he presents the pipe and smokes with him. Mox senex vir simulacrum parvæ puellæ ostensit. Tunc egrediens cætu, jecit effigium solo et superincumbens, senili ardore veneris complexit. Hoc est signum. Denique uxor e turba recessit, et jactu corporis, fovet amplexus viri solo recubante. Maritus appropinquans senex vir dejecto vultu, et honorem et dignitatem ejus conservare amplexu uxoris illum oravit. Forsitan imprimis ille refellit; dehinc, maritus multis precibus, multis lachrymis, et multis donis vehementer intercessit. Tunc senex amator perculsus miserecordia, tot precibus, tot lachrymis, et tot donis, conjugali amplexu submisit. Multum ille jactatus est, sed debilis et effœtus senectute, frustra jactatus est. Maritus interdum stans juxta guadit multum honore, et ejus dignitati sic conservata. Unus nostrum sodalium multum alacrior et potentior juventute, hac nocte honorem quartuor maritorum custodivit.

Sunday, 6th. A clear cold morning, with high wind; we caught in a trap a large gray wolf, and last night obtained in the same way a fox who had for some time infested the neighbourhood of the fort. Only a few Indians visited us to-day.

Monday, 7th. The weather was again clear and cold, with a high northwest wind, and the thermometer at sunrise 22° below zero; the river fell an inch. Shahaka, the Big White chief, dined with us, and gave a connected sketch of the country as far as the mountains.

Tuesday, 8th. The wind was still from the northwest, the day cold, and we received few Indians at the fort. Besides the buffaloe dance we have just described there is another called medicine dance, an entertainment given by any person desirous of doing honour to his medicine or genius. He announces that on such a day he will sacrifice his horses or other property, and invites the young females of the village to assist in rendering homage to his medicine; all the inhabitants may join in the solemnity, which is performed in the open plain and by daylight, but the dance is reserved for the virgins or at least the unmarried females, who disdain the incumbrance or the ornament of dress. The feast is opened by devoting the goods of the master of the feast to his medicine, which is represented by a head of the animal itself, or by a medicine bag if the deity be an invisible being. The young women then begin the dance, in the intervals of which each will prostrate herself before the assembly to challenge or reward the boldness of the youth, who are often tempted by feeling or the hopes of distinction to achieve the adventure.

Wednesday, 9th. The weather is cold, the thermometer at sunrise 21° below zero. Kagohami breakfasted with us, and Captain Clark with three or four men accompanied him and a party of Indians to hunt, in which they were so fortunate as to kill a number of buffaloe; but they were incommoded by snow, by high and squally winds, and by extreme cold; several of the Indians came to the fort nearly frozen, others are missing, and we are uneasy, for one of our men who was separated from the rest during the chase has not returned. In the morning,

Thursday, 10th, however, he came back just as we were sending out five men in search of him. The night had been excessively cold, and this morning at sunrise the mercury stood at 40° below zero, or 72° below the freezing point. He had, however, made a fire and kept himself tolerably warm. A young Indian, about thirteen years of age, also came in soon after. His father, who came last night to inquire after him very anxiously, had sent him in the afternoon to the fort; he was overtaken by the night, and was obliged to sleep on the snow with no covering except a pair of antelope skin moccasins and leggings and a buffaloe robe; his feet being frozen we put them into cold water, and gave him every attention in our power. About the same time an Indian who had also been missing returned to the fort, and although his dress was very thin, and he had slept on the snow without a fire, he had not suffered the slightest inconvenience. We have indeed observed that these Indians support the rigours of the season in a way which we had hitherto thought impossible. A more pleasing reflection occurred at seeing the warm interest which the situation of these two persons had excited in the village: the boy had been a prisoner and adopted from charity, yet the distress of the father proved that he felt for him the tenderest affection; the man was a person of no distinction, yet the whole village was full of anxiety for his safety, and when they came to us, borrowed a sleigh to bring them home with ease, if they survived, or to carry their bodies if they had perished.

Friday, 11th. We despatched three hunters to join the same number whom we had sent below about seven miles to hunt elk. Like that of yesterday the weather to-day was cold and clear, the thermometer standing at 38° below zero. Poscopsahe and Shotahawrora visited us, and passed the night at the fort.

Saturday, 12th. The weather continues very cold, the mercury at sunrise being 20° below zero. Three of the hunters returned, having killed three elk.

Sunday, 13th. We have a continuation of clear weather and the cold has increased, the mercury having sunk to 34° below zero. Nearly one-half of the Mandan nation passed down the river to hunt for several days; in these excursions men, women, and children, with their dogs, all leave the village together, and after discovering a spot convenient for the game, fix their tents; all the family bear their part in the labour, and the game is equally divided among the families of the tribe. When a single hunter returns from the chase with more than is necessary for his own immediate consumption, the neighbours are entitled by custom to a share of it; they do not, however, ask for it, but send a squaw, who, without saying anything, sits down by the door of the lodge till the master understands the hint, and gives her gratuitously a part for her family. Chaboneau, who with one man had gone to some lodges of Minnetarees near the Turtle mountain, returned with their faces much frostbitten. They had been about ninety miles distant, and procured from the inhabitants some meat and grease, with which they loaded the horses. He informs us that the agent of the Hudson bay company at that place had been endeavouring to make unfavourable impressions with regard to us on the mind of the great chief, and that the N. W. company intend building a fort there. The great chief had in consequence spoken slightly of the Americans, but said that if we would give him our great flag he would come and see us.

Monday, 14th. The Mandans continue to pass down the river on their hunting party, and were joined by six of our men. One of those sent on Thursday returned with information that one of his companions had his feet so badly frostbitten that he could not walk home. In their excursion they had killed a buffaloe, a wolf, two porcupines, and a white hare. The weather was more moderate to-day, the mercury being at 16° below zero and the wind from the S. E.; we had, however, some snow, after which it remained cloudy.

Tuesday, 15th. The morning is much warmer than yesterday and the snow begins to melt, though the wind, after being for some time from the S. E., suddenly shifted to N. W. Between twelve and three o'clock A. M. there was a total eclipse of the moon, from which we obtained a part of the observation necessary for ascertaining the longitude.

We were visited by four of the most distinguished men of the Minnetarees, to whom we showed marked attentions, as we knew that they had been taught to entertain strong prejudices against us; these we succeeded so well in removing that when in the morning,

Wednesday, 16th, about thirty Mandans, among whom six were chiefs, came to see us, the Minnetarees reproached them with their falsehoods, declaring that they were bad men and ought to hide themselves. They had told the Minnetarees that we would kill them if they came to the fort, yet on the contrary they had spent a night there and been treated with kindness by the whites, who had smoked with them and danced for their amusement. Kagohami visited us and brought us a little corn, and soon afterwards one of the first war chiefs of the Minnetarees came, accompanied by his squaw, a handsome woman, whom he was desirous we should use during the night. He favoured us with a more acceptable present: a draft of the Missouri, in his manner, and informed us of his intention to go to war in the spring against the Snake Indians. We advised him to reflect seriously before he committed the peace of his nation to the hazards of war; to look back on the numerous nations whom war has destroyed; that if he wished his nation to be happy he should cultivate peace and intercourse with all his neighbours, by which means they would procure more horses, increase in numbers, and that if he went to war he would displease his great father the President and forfeit his protection. We added that we had spoken thus to all the tribes whom we had met, that they had all opened their ears, and that the President would compel those who did not voluntarily listen to his advice. Although a young man of only twenty-six years of age, this discourse seemed to strike him. He observed that if it would be displeasing to us he would not go to war, since he had horses enough, and that he would advise all the nation to remain at home until we had seen the Snake Indians and discovered whether their intentions were pacific. The party who went down with the horses for the man who was frostbitten returned, and we are glad to find his complaint not serious.

Thursday, 17th. The day was very windy from the north; the morning clear and cold, the thermometer at sunrise being at zero; we had several Indians with us.

Friday, 18th. The weather is fine and moderate. Messrs. Laroche and M'Kenzie, two of the N. W. company's traders, visited us with some of the Minnetarees. In the afternoon two of our hunters returned, having killed four wolves and a blaireau.

Saturday, 19th. Another cloudy day. The two traders set out on their return, and we sent two men with the horses thirty miles below to the hunting camp.

Sunday, 20th. The day fair and cold. A number of Indians visit us with corn to exchange for articles, and to pay for repairs to their household utensils.

Monday, 21st. The weather was fine and moderate. The hunters all returned, having killed during their absence three elk, four deer, two porcupines, a fox, and a hare.

Tuesday, 22d. The cold having moderated and the day pleasant, we attempted to cut the boats out of the ice, but at the distance of eight inches came to water, under which the ice became three feet thick, so that we were obliged to desist.

Wednesday, 23d. The cold weather returned, the mercury

having sunk 2° below zero and the snow fell four inches deep.

Thursday, 24th. The day was colder than any we have had lately, the thermometer being at 12° below zero. The hunters whom we sent out returned unsuccessful, and the rest were occupied in cutting wood to make charcoal.

Friday, 25th. The thermometer was at 25° below zero, the wind from N. W. and the day fair, so that the men were employed in preparing coal and cutting the boats out of the ice. A band of Assiniboins, headed by their chief, called by the French Son of the Little Calf, have arrived at the villages.

Saturday, 26th. A fine, warm day; a number of Indians dine with us, and one of our men is attacked with a violent pleurisy.

Sunday, 27th. Another warm and pleasant day; we again attempted to get the boat out of the ice. The man who has the pleurisy was blooded and sweated, and we were forced to take off the toes of the young Indian who was frostbitten some time since. Our interpreter returned from the villages bringing with him three of Mr. Laroche's horses, which he had sent in order to keep them out of the way of the Assiniboins, who are very much disposed to steal and who have just returned to their camp.

Monday, 28th. The weather to-day is clear and cold; we are obliged to abandon the plan of cutting the boat through the ice, and therefore made another attempt the next day,

Tuesday, 29th, by heating a quantity of stones so as to warm the water in the boat and thaw the surrounding ice; but in this, too, we were disappointed, as all the stones on being put into the fire cracked into pieces. The weather warm and pleasant. The man with the pleurisy is recovering.

Wednesday, 30th. The morning was fair, but afterwards

became cloudy. Mr. Laroche, the trader from the northwest company, paid us a visit in hopes of being able to accompany us on our journey westward, but this proposal we thought it best to decline.

Thursday, 31st. It snowed last night and the morning is cold and disagreeable, with a high wind from the northwest; we sent five hunters down the river. Another man is taken with the pleurisy.

Friday, February 1. A cold, windy day; our hunters returned, having killed only one deer. One of the Minnetaree war chiefs, a young man named Maubuksheahokeah, or Seeing Snake, came to see us and procure a war hatchet; he also requested that we would suffer him to go to war against the Sioux and Ricaras, who had killed a Mandan some time ago; this we refused for reasons which we explained to him. He acknowledged that we were right, and promised to open his ears to our counsels.

Saturday, 2d. The day is fine; another deer was killed. Mr. Laroche, who has been very anxious to go with us, left the fort to-day, and one of the squaws of the Minnetaree interpreter is taken ill.

Sunday, 3d. The weather is again pleasant; disappointed in all our efforts to get the boats free, we occupied ourselves in making iron spikes so as to prize them up by means of long poles.

Monday, 4th. The morning fair and cold, the mercury at sunrise being 18° below zero, and the wind from the northwest. The stock of meat which we had procured in November and December being now nearly exhausted, it became necessary to renew our supply; Captain Clark therefore took eighteen men, and with two sleighs and three horses descended the river for the purpose of hunting, as the buffaloe has disappeared from our neighbourhood and the Indians are themselves suffering for want of meat. Two deer were killed to-day, but they were very lean.

Tuesday, 5th. A pleasant, fair morning, with the wind from northwest; a number of the Indians come with corn for the blacksmith, who, being now provided with coal, has become one of our greatest resources for procuring grain. They seem particularly attached to a battle-axe, of a very inconvenient figure: it is made wholly of iron, the blade extremely thin and from seven to nine inches long; it is sharp at the point and five or six inches on each side, whence they converge towards the eye, which is circular and about an inch in diameter, the blade itself being not more than an inch wide; the handle is straight and twelve or fifteen inches long, the whole weighing about a pound. By way of ornament the blade is perforated with several circular holes. The length of the blade compared with the shortness of the handle renders it a weapon of very little strength, particularly as it is always used on horseback; there is still, however, another form which is even worse, the same sort of handle being fixed to a blade resembling an espontoon.

Wednesday, February 6. The morning was fair and pleasant, the wind N. W. A number of Indian chiefs visited us and withdrew after we had smoked with them, contrary to their custom, for after being once introduced into our apartment they are fond of lounging about during the remainder of the day. One of the men killed three antelopes. Our blacksmith has his time completely occupied, so great is the demand for utensils of different kinds. The Indians are particularly fond of sheet iron, out of which they form points for arrows and instruments for scraping hides, and when the blacksmith cut up an old cambouse of that metal, we obtained for every piece of four inches square seven or eight gallons of corn from the Indians, who were delighted at the exchange.

Thursday, 7th. The morning was fair and much warmer than for some days, the thermometer being at 18° above zero and the wind from the S. E. A number of Indians continue to visit us, but learning that the interpreter's squaws had been accustomed to unbar the gate during the night, we ordered a lock put on it and that no Indian should remain in the fort all night, nor any person admitted during the hours when the gate is closed, that is, from sunset to sunrise.

Friday, 8th. A fair, pleasant morning, with S. E. winds. Pocopsahe came down to the fort with a bow, and apologized for his not having finished a shield which he had promised Captain Lewis, and which the weather had prevented him from completing. This chief possesses more firmness, intelligence, and integrity than any Indian of this country, and he might be rendered highly serviceable in our attempts to civilize the nation. He mentioned that the Mandans are very much in want of meat, and that he himself had not tasted any for several days. To this distress they are often reduced by their own improvidence, or by their unhappy situation. Their principal article of food is buffaloe meat, their corn, beans, and other grain being reserved for summer, or as a last resource against what they constantly dread, an attack from the Sioux, who drive off the game and confine them to their villages. The same fear, too, prevents their going out to hunt in small parties to relieve their occasional wants, so that the buffaloe is generally obtained in large quantities and wasted by carelessness.

Saturday, 9th. The morning was fair and pleasant, the wind from the S. E. Mr. M'Kenzie from the N. W. company establishment visited us.

## UP THE MISSOURI

Sunday, 10th. A slight snow fell in the course of the night, the morning was cloudy, and the northwest wind blew so high that although the thermometer was 18° above zero the day was cooler than yesterday, when it was only 10° above the same point. Mr. M'Kenzie left us, and Chaboneau returned with information that our horses, loaded with meat, were below, but could not cross the ice, not being shod.

Monday, 11th. We sent down a party with sleds, to relieve the horses from their loads; the weather fair and cold, with a N. W. wind. About five o'clock one of the wives of Chaboneau was delivered of a boy; this being her first child she was suffering considerable, when Mr. Jessaume told Captain Lewis that he had frequently administered to persons in her situation a small dose of the rattle of the rattlesnake, which had never failed to hasten the delivery. Having some of the rattle Captain Lewis gave it to Mr. Jessaume, who crumbled two of the rings of it between his fingers, and mixing it with a small quantity of water gave it to her. What effect it may really have had it might be difficult to determine, but Captain Lewis was informed that she had not taken it more than ten minutes before the delivery took place.

Tuesday, 12th. The morning is fair though cold, the mercury being 14° below zero, the wind from the S. E. About four o'clock the horses were brought in much fatigued; on giving them meal bran moistened with water they would not eat it, but preferred the bark of the cottonwood, which, as is already observed, forms their principal food during the winter. The horses of the Mandans are so often stolen by the Sioux, Ricaras, and Assiniboins that the invariable rule now is to put the horses every night in the same lodge with the family. In the summer they ramble in the plains in the vicinity of the camp and feed on the grass, but during cold weather the squaws cut down the cottonwood trees as they are wanted, and the horses feed on the boughs and bark of the tender branches, which are also brought into the lodges at night and placed near them. These animals are very severely treated; for whole days they are pursuing the buffaloe or burdened with the fruits of the chase, during which they scarcely ever taste food, and at night return to a scanty allowance of wood; yet the spirit of this valuable animal sustains him through all these difficulties, and he is rarely deficient either in flesh or vigour.

Wednesday, 13th. The morning was cloudy, the thermometer at 2° below zero, the wind from the southeast. Captain Clark returned last evening with all his hunting party. During their excursion they had killed forty deer, three buffaloe, and sixteen elk; but most of the game was too lean for use, and the wolves, who regard whatever lies out at night as their own, had appropriated a large part of it. When he left the fort on the 4th instant he descended on the ice twenty-two miles to New Mandan island, near some of their old villages, and encamped, having killed nothing and therefore without food for the night.

Early on the 5th the hunters went out and killed two buffaloe and a deer, but the last only could be used, the others being too lean. After breakfast they proceeded down to an Indian lodge and hunted during the day; the next morning, 6th, they encamped forty-four miles from the fort on a sand point near the mouth of a creek on the southwest side, which they call Hunting creek, and during this and the following day hunted through all the adjoining plains with much success, having killed a number of deer and elk, On the 8th, the best of the meat was sent with the horses to the fort, and such parts of the remainder as were fit for use were brought to a point of the river three miles below, and after the bones were taken out, secured in pens built of logs so as to keep off the wolves, ravens, and magpies, who are very numerous and constantly disappoint the hunter of his prey; they then went to the low grounds near the Chisshetaw river, where they encamped, but saw nothing except some wolves on the hills, and a number of buffaloe too poor to be worth hunting. The next morning, 9th, as there was no game, and it would have been inconvenient to send it back sixty miles to the fort, they returned up the river and for three days hunted along the banks and plains, and reached the fort in the evening of the 12th much fatigued, having walked thirty miles that day on the ice and through the snow, in many places knee-deep, the moccasins, too, being nearly worn out; the only game which they saw besides what is mentioned was some growse on the sandbars in the river.

Thursday, 14th. Last night the snow fell three inches deep; the day was, however, fine. Four men were despatched with sleds and three horses to bring up the meat which had been collected by the hunters. They returned, however, with intelligence that about twenty-one miles below the fort a party of upwards of one hundred men, whom they supposed to be Sioux, rushed on them, cut the traces of the sleds and carried off two of the horses, the third being given up by intercessions of an Indian who seemed to possess some authority over them; they also took away two of the men's knifes and a tomahawk, which last, however, they returned. We sent up to the Mandans to inform them of it, and to know whether any of them would join a party which intended to pursue the robbers in the morning. About twelve o'clock two of their chiefs came down and said that all their young men were out hunting, and that there were few guns in the village. Several Indians, however, armed some with bows and arrows, some with spears and battle-axes, and two with fusils, accompanied Captain Lewis, who set out,

Friday, 15th, at sunrise, with twenty-four men. The morning was fine and cool, the thermometer being at 16° below zero. In the course of the day one of the Mandan chiefs returned from Captain Lewis's party, his eye-sight having become so bad that he could not proceed. At this season of the year the reflexion from the ice and snow is so intense as to occasion almost total blindness. This complaint is very common, and the general remedy is to sweat the part affected by holding the face over a hot stone, and receiving the fumes from snow thrown on it. A large red fox was killed to-day.

Saturday, 16th. The morning was warm, mercury at 32° above zero, the weather cloudy; several of the Indians who went with Captain Lewis returned, as did also one of our men whose feet had been frostbitten.

Sunday, 17th. The weather continued as yesterday, though in the afternoon it became fair. Shotawhorora and his son came to see us, with about thirty pounds of dried buffaloe meat and some tallow.

Monday, 18th. The morning was cloudy, with some snow, but in the latter part of the day it cleared up. Mr. M'Kenzie, who had spent yesterday at the fort, now left us. Our stock of meat is exhausted, so that we must confine ourselves to vegetable diet, at least till the return of the party; for this, however, we are at no loss, since both on this and the following day,

Tuesday, 19th, our blacksmith got large quantities of corn from the Indians, who came in great numbers to see us. The weather was fair and warm, the wind from the south.

Wednesday, 20th. The day was delightfully fine, the mercury being at sunrise 2° and in the course of the day 22° above zero, the wind southerly. Kagohami came down to see us early; his village is afflicted by the death of one of their eldest men, who from his account to us must have seen one hundred and twenty winters. Just as he was dying, he requested his grandchildren to dress him in his best robe when he was dead, and then carry him on a hill and seat him on a stone, with his face down the river towards their old villages, that he might go straight to his brother who had passed before him to the ancient village under ground. We have seen a number of Mandans who have lived to a great age, chiefly, however, the men, whose robust exercises fortify the body, while the laborious occupations of the women shorten their existence.

Thursday, 21st. We had a continuation of the same pleasant weather. Oheenaw and Shahaka came down to see us, and mentioned that several of their countrymen had gone to consult their medicine stone as to the prospects of the following year. This medicine stone is the great oracle of the Mandans, and whatever it announces is believed with implicit confidence. Every spring, and on some occasions during the summer, a deputation visits the sacred spot, where there is a thick porous stone twenty feet in circumference, with a smooth surface. Having reached the place, the ceremony of smoking to it is performed by the deputies, who alternately take a whiff themselves and then present the pipe to the stone; after this they retire to an adjoining wood for the night, during which it may be safely presumed that all the embassy do not sleep; and in the morning they read the destinies of the nation in the white marks on the stone, which those who made them are at no loss to decipher. The Minnetarees have a stone of a similar kind, which has the same qualities and the same influence over the nation. Captain Lewis returned from his excursion in pursuit of the Indians. On reaching the place where the Sioux had stolen our horses, they found only one sled, and several pair of moccasins which were recognized to be those of the Sioux. The party then followed the Indian tracks till they reached two old lodges, where they slept, and the next morning pursued the course of the river till they reached some Indian camps, where Captain Clark passed the night some time ago, and which the Sioux had now set on fire, leaving a little corn near the place in order to induce a belief that they were Ricaras. From this point the Sioux tracks left the river abruptly and crossed into the plains; but perceiving that there was no chance of overtaking them, Captain Lewis went down to the pen where Captain Clark had left some meat, which he found untouched by the Indians, and then hunted in the low grounds on the river, till he returned with about three thousand pounds of meat, some drawn in a sled by fifteen of the men, and the rest on horseback, having killed thirty-six deer, fourteen elk, and one wolf.

Friday, 22d. The morning was cloudy and a little snow fell, but in the afternoon the weather became fair. We were visited by a number of Indians, among whom was Shotawhorora, a chief of much consideration among the Mandans, although by birth a Ricara.

Saturday, 23d. The day is warm and pleasant. Having worked industriously yesterday and all this morning, we were enabled to disengage one of the periogues and haul it on shore, and also nearly cut out the second. The father of the boy whose foot had been so badly frozen, and whom we had now cured, came to-day and carried him home in a sleigh.

Sunday, 24th. The weather is again fine. We succeeded in loosening the second periogue and barge, though we found a leak in the latter. The whole of the next day,

Monday, 25th, we were occupied in drawing up the boats on the bank; the smallest one we carried there with no difficulty, but the barge was too heavy for our elk-skin ropes, which constantly broke. We were visited by Orupsehara, or

Black Moccasin, and several other chiefs, who brought us presents of meat on the backs of their squaws, and one of the Minnetarees requested and obtained permission for himself and his two wives to remain all night in the fort. The day was exceedingly pleasant.

Tuesday, 26th. The weather is again fine. By great labour during the day we got all the boats on the bank by sunset, an operation which attracted a great number of Indians to the fort.

Wednesday, 27th. The weather continues fine. All of us employed in preparing tools to build boats for our voyage, as we find that small periogues will be much more convenient than the barge in ascending the Missouri.

Thursday, 28th. The day is clear and pleasant. Sixteen men were sent out to examine the country for trees suitable for boats, and were successful in finding them. Two of the N. W. company traders arrived with letters; they had likewise a root which is used for the cure of persons bitten by mad dogs, snakes, and other venomous animals; it is found on high grounds and the sides of hills, and the mode of using it is to scarify the wound and apply to it an inch or more of the chewed or pounded root, which is to be renewed twice a day; the patient must not, however, chew or swallow any of the root, as an inward application might be rather injurious than beneficial.

Mr. Gravelines, with two Frenchmen and two Indians, arrived from the Ricara nation, with letters from Mr. Anthony Tabeau. This last gentleman informs us that the Ricaras express their determination to follow our advice and to remain at peace with the Mandans and Minnetarees, whom they are desirous of visiting; they also wish to know whether these nations would permit the Ricaras to settle near them, and form a league against their common enemies, the Sioux. On mentioning this to the Mandans they agreed to it, observing that they always desired to cultivate friendship with the Ricaras, and that the Ahnahaways and Minnetarees have the same friendly views.

Mr. Gravelines states that the band of Tetons whom we had seen was well disposed to us, owing to the influence of their chief, the Black Buffaloe; but that the three upper bands of Tetons, with the Sisatoons, and the Yanktons of the north, mean soon to attack the Indians in this quarter, with a resolution to put to death every white man they encounter. Moreover, that Mr. Cameron of St. Peter's has armed the Sioux against the Chippeways, who have lately put to death three of his men. The men who had stolen our horses we found to be all Sioux, who, after committing the outrage, went to the Ricara villages, where they said that they had hesitated about killing our men who were with the horses, but that in future they would put to death any of us they could, as we were bad medicines and deserved to be killed. The Ricaras were displeased at their conduct and refused to give them anything to eat, which is deemed the greatest act of hostility short of actual violence.

Friday, March 1. The day is fine, and the whole party is engaged, some in making ropes and periogues, others in burning coal and making battle-axes to sell for corn.

Saturday, 2d. Mr. Laroche, one of the N. W. company's traders, has just arrived with merchandize from the British establishments on the Assiniboin. The day is fine and the river begins to break up in some places, the mercury being between 28° and 36° above zero, and the wind from the N. E. We were visited by several Indians.

Sunday, 3d. The weather pleasant, the wind from the E. with clouds; in the afternoon the clouds disappeared and the wind came from the N. W. The men are all employed in preparing the boats; we are visited by Poscapsahe and several other Indians with corn. A flock of ducks passed up the river to-day.

Monday, 4th. A cloudy morning with N. W. wind, the latter part of the day clear. We had again some Indian visitors with a small present of meat. The Assiniboins, who a few days since visited the Mandans, returned and attempted to take horses from the Minnetarees, who fired on them,— a circumstance which may occasion some disturbance between the two nations.

Tuesday, 5th. About four o'clock in the morning there was a slight fall of snow, but the day became clear and pleasant, with the mercury 40° above zero. We sent down an Indian and a Frenchman to the Ricara villages with a letter to Mr. Tabeau.

Wednesday, 6th. The day was cloudy and smoky in consequence of the burning of the plains by the Minnetarees; they have set all the neighbouring country on fire in order to obtain an early crop of grass which may answer for the consumption of their horses, and also as an inducement for the buffaloe and other game to visit it. The horses stolen two days ago by the Assiniboins have been returned to the Minnetarees. Ohhaw, second chief of the lower Minnetaree village, came to see us. The river rose a little and overran the ice, so as to render the crossing difficult.

Thursday, 7th. The day was somewhat cloudy, and colder than usual, the wind from the northeast. Shotawhorora visited us with a sick child, to whom some medicine was administered. There were also other Indians, who brought corn and dried buffaloe meat, in exchange for blacksmith's work.

Friday, 8th. The day cold and fair, with a high easterly wind. We were visited by two Indians, who gave us an account of the country and people near the Rocky mountains where they had been.

Saturday, 9th. The morning cloudy and cool, the wind from the north. The grand chief of the Minnetarees, who is called by the French Le Borgne, from his having but one eye, came down for the first time to the fort. He was received with much attention, two guns were fired in honour of his arrival, the curiosities were exhibited to him, and as he said that he had not received the presents which we had sent to him on his arrival, we again gave him a flag, a medal, shirt, armbraces, and the usual presents on such occasions, with all which he was much pleased. In the course of the conversation the chief observed that some foolish young men of his nation had told him there was a person among us who was quite black, and he wished to know if it could be true. We assured him that it was true, and sent for York. The Borgne was very much surprised at his appearance, examined him closely, and spit on his finger and rubbed the skin in order to wash off the paint; nor was it until the negro uncovered his head, and showed his short hair, that the Borgne could be persuaded that he was not a painted white man.

Sunday, 10th. A cold, windy day. Tetuckopinreha, chief of the Ahnahaways, and the Minnetaree chief, Ompschara, passed the day with us, and the former remained during the night. We had occasion to see an instance of the summary justice of the Indians. A young Minnetaree had carried off the daughter of Cagonomokshe, the Raven Man, second chief of the upper village of the Mandans; the father went to the village and found his daughter, whom he brought home, and took with him a horse belonging to the offender; this reprisal satisfied the vengeance of the father and of the nation, as the young man would not dare to reclaim his horse, which from that time became the property of the injured party. The stealing of young women is one of the most common offences against the police of the village, and the punishment of it always measured by the power or the passions of the kindred of the female. A voluntary elopement is of course more rigorously chastised. One of the wives of the Borgne deserted him in favour of a man who had been her lover before the marriage, and who after some time left her, and she was obliged to return to her father's house. As soon as he heard it the Borgne walked there, and found her sitting near the fire; without noticing his wife, he began to smoke with the father, when they were joined by the old men of the village, who, knowing his temper, had followed in hopes of appeasing him. He continued to smoke quietly with them, till, rising to return, he took his wife by the hair, led her as far as the door, and with a single stroke of his tomahawk put her to death before her father's eyes; then turning fiercely upon the spectators, he said that if any of her relations wished to avenge her, they might always find him at his lodge; but the fate of the woman had not sufficient interest to excite the vengeance of the family. The caprice or the generosity of the same chief gave a very different result to a similar incident which occurred some time afterwards. Another of his wives eloped with a young man, who, not being able to support her as she wished, they both returned to the village, and she presented herself before the husband, supplicating his pardon for her conduct. The Borgne sent for the lover; at the moment when the youth expected that he would be put to death, the chief mildly asked them if they still preserved their affection for each other; and on their declaring that want, and not a change of affection, had induced them to return, he gave up his wife to her lover, with the liberal present of three horses, and restored them both to his favour.

Monday, 11th. The weather was cloudy in the morning

and a little snow fell, the wind then shifted from southeast to northwest, and the day became fair. It snowed again in the evening, but the next day,

Tuesday, 12th, was fair, with the wind from the northwest.

Wednesday, 13th. We had a fine day and a southwest wind. Mr. M'Kenzie came to see us, as did also many Indians, who are so anxious for battle-axes that our smiths have not a moment's leisure, and procure us an abundance of corn. The river rose a little to-day, and so continued.

Thursday, 14th. The wind being from the west, and the day fine, the whole party were employed in building boats and in shelling corn.

Friday, 15th. The day is clear, pleasant, and warm. We take advantage of the fine weather to hang all our Indian presents and other articles out to dry before our departure.

Saturday, 16th. The weather is cloudy, the wind from the southeast. A Mr. Garrow, a Frenchmen who has resided a long time among the Ricaras and Mandans, explained to us the mode in which they make their large beads, an art which they are said to have derived from some prisoners of the Snake Indian nation, and the knowledge of which is a secret even now confined to a few among the Mandans and Ricaras. The process is as follows: glass of different colours is first pounded fine and washed, till each kind, which is kept separate, ceases to stain the water thrown over it; some well-seasoned clay, mixed with a sufficient quantity of sand to prevent its becoming very hard when exposed to heat, and reduced by water to the consistency of dough, is then rolled on the palm of the hand till it becomes of the thickness wanted for the hole in the bead; these sticks of clay are placed upright, each on a little pedestal or ball of the same material about an ounce in weight, and distributed over a small earthen platter, which is laid on the fire for a few minutes, when they are taken off to cool; with a little paddle or shovel, three or four inches long and sharpened at the end of the handle, the wet pounded glass is placed in the palm of the hand; the beads are made of an oblong form wrapped in a cylindrical form round the stick of clay, which is laid crosswise over it, and gently rolled backwards and forwards till it becomes perfectly smooth. If it be desired to introduce any other colour, the surface of the bead is perforated with the pointed end of the paddle and the cavity filled with pounded glass of that colour; the sticks with the string of beads are then replaced on their pedestals, and the platter deposited on burning coals or hot embers; over the platter an earthern pot containing about three gallons, with a mouth large enough to cover the platter, is reversed, being completely closed except a small aperture at the top, through which are watched the beads; a quantity of old dried wood formed into a sort of dough or paste is placed round the pot so as almost to cover it, and afterwards set on fire; the manufacturer then looks through the small hole in the pot, till he sees the beads assume a deep red colour, to which succeeds a paler or whitish red, or they become pointed at the upper extremity, on which the fire is removed and the pot suffered to cool gradually; at length it is removed, the beads taken out, the clay in the hollow of them picked out with an awl or needle, and it is then fit for use. The beads thus formed are in great demand among the Indians, and used as pendants to their ears and hair, and are sometimes worn round the neck.

Sunday, 17th. A windy but clear and pleasant day, the river rising a little and open in several places. Our Minnetaree interpreter, Chaboneau, whom we intended taking with us to the Pacific, had some days ago been worked upon by the British traders and appeared unwilling to accompany us except on certain terms, such as his not being subject to our orders, and do duty, or to return, whenever he chose. As we saw clearly the source of his hesitation, and knew that it was intended as an obstacle to our views, we told him that the terms were inadmissable, and that we could dispense with his services; he had accordingly left us with some displeasure. Since then he had made an advance towards joining us, which we showed no anxiety to meet; but this morning he sent an apology for his improper conduct, and agreed to go with us and perform the same duties as the rest of the corps; we therefore took him again into our service.

Monday, 18th. The weather was cold and cloudy, the wind from the north. We were engaged in packing up the goods into eight divisions, so as to preserve a portion of each in case of accident. We hear that the Sioux have lately attacked a party of Assiniboins and Knistenaux, near the Assiniboin river, and killed fifty of them.

Tuesday, 19th. Some snow fell last night, and this morning was cold, windy, and cloudy. Shahaka and Kagohami came down to see us, as did another Indian with a sick child, to whom we gave some medicine. There appears to be an approaching war, as two parties have already gone from the Minnetarees, and a third is preparing.

Wednesday, 20th. The morning was cold and cloudy, the wind high from the north, but the afternoon was pleasant. The canoes being finished, four of them were carried down to the river, at the distance of a mile and a half from where they were constructed.

Thursday, 21st. The remaining periogues were hauled to the same place, and all the men except three, who were left to watch them, returned to the fort. On his way down, which was about six miles, Captain Clark passed along the points of the high hills, where he saw large quantities of pumicestone on the foot, sides, and tops of the hills, which had every appearance of having been at some period on fire. He collected specimens of the stone itself, the pumicestone, and the hard earth, and on being put into the furnace the hard earth melted and glazed, the pumicestone melted, and the hardstone became a pumicestone glazed.

# CHAPTER VII.

Indian method of attacking the buffaloe on the ice—An enumeration of the presents sent to the President of the United States—The party are visited by a Ricara chief—They leave their encampment, and proceed on their journey—Description of the Little Missouri Some account of the Assiniboins—Their mode of burying the dead—Whiteearth river described—Great quantity of salt discovered on its banks—Yellowstone river described—A particular account of the country at the confluence of the Yellowstone and Missouri—Description of the Missouri, the surrounding country, and of the rivers, creeks, islands, etc.

FRIDAY, 22d. This was a clear, pleasant day, with the wind from the S. S. W. We were visited by the second chief of the Minnetarees, to whom we gave a medal and some presents, accompanied by a speech. Mr. M'Kenzie and Mr. Laroche also came to see us. They all took their leave next day.

Saturday, 23d. Soon after their departure a brother of the Borgne, with other Indians, came to the fort. The weather was fine, but in the evening we had the first rain that has fallen during the winter.

Sunday, 24th. The morning cloudy, but the afternoon fair, the wind from the N. E. We are employed in preparing for our journey. This evening swans and wild geese flew towards the N. E.

Monday, 25th. A fine day, the wind S. W. The river rose nine inches, and the ice began breaking away in several places, so as to endanger our canoes which we are hauling down to the fort.

Tuesday, 26th. The river rose only half an inch, and being choked up with ice near the fort, did not begin to run till towards evening. This day is clear and pleasant.

Wednesday, 27th. The wind is still high from the S. W.; the ice, which is occasionally stopped for a few hours, is then thrown over shallow sandbars when the river runs. We had all our canoes brought down, and were obliged to cauk and pitch very attentively the cracks so common in cottonwood.

Thursday, 28th. The day is fair. Some obstacle above has prevented the ice from running. Our canoes are now nearly ready, and we expect to set out as soon as the river is sufficiently clear to permit us to pass.

Friday, 29th. The weather clear, and the wind from N. W. The obstruction above gave way this morning, and the ice came down in great quantities, the river having fallen eleven inches in the course of the last twenty-four hours. We have had few Indians at the fort for the last three or four days, as they are now busy in catching the floating buffaloe. Every spring as the river is breaking up the surrounding plains are set on fire, and the buffaloe tempted to cross the river in search of the fresh grass which immediately succeeds to the burning; on their way they are often insulated on a large cake or mass of ice, which floats down the river. The Indians now select the most favourable points for attack, and, as the buffaloe approaches, dart with astonishing agility across the trembling ice, sometimes pressing lightly a cake of not more than two feet square; the animal is of course unsteady and his footsteps insecure on this new element, so that he can make but little resistance, and the hunter who has given him his death wound paddles his icy boat to the shore and secures his prey.

Saturday, 30th. The day was clear and pleasant, the wind N. W., and the ice running in great quantities. All our Indian presents were again exposed to the air, and the barge made ready to descend the Missouri.

Sunday, 31st. Early this morning it rained, and the

weather continued cloudy during the day; the river rose nine inches, the ice not running so much as yesterday. Several flocks of geese and ducks fly up the river.

Monday, April 1, 1805. This morning there was a thunder storm, accompanied with large hail, to which succeeded rain for about half an hour. We availed ourselves of this interval to get all the boats in the water. At four o'clock P.M. it began to rain a second time, and continued till twelve at night. With the exception of a few drops at two or three different times, this is the first rain we have had since the 15th of October last.

Tuesday, 2d. The wind was high last night and this morning, from N. W., and the weather continued cloudy. The Mandans killed yesterday twenty-one elk, about fifteen miles below, but they were so poor as to be scarcely fit for use.

Wednesday, 3d. The weather is pleasant, though there was a white frost, and some ice on the edge of the water. We were all engaged in packing up our baggage and merchandize.

Thursday, 4th. The day is clear and pleasant, though the wind is high, from N. W. We now packed up in different boxes a variety of articles for the President, which we shall send in the barge. They consisted of a stuffed male and female antelope with their skeletons, a weasel, three squirrels from the Rocky mountains, the skeleton of the prairie wolf, those of the white and gray hare, a male and female blaireau or burrowing dog of the prairie, with a skeleton of the female, two burrowing squirrels, a white weasel, and the skin of the louservia, the horns of the mountain ram or big-horn, a pair of large elk horns, the horns and tail of the black-tailed deer, and a variety of skins, such as those of the red fox, white hare, martin, yellow bear, obtained from the Sioux; also a number of articles of Indian dress, among which was a buffaloe robe representing a battle fought about eight years since between the Sioux and Ricaras against the Mandans and Minnetarees, in which the combatants are represented on horseback. It has of late years excited much discussion to ascertain the period when the art of painting was first discovered; how hopeless all researches of this kind are, is evident from the foregoing fact. It is indebted for its origin to one of the strongest passions of the human heart: a wish to preserve the features of a departed friend, or the memory of some glorious exploit; this inherits equally the bosoms of all men, either civilized or savage. Such sketches, rude and imperfect as they are, delineate the predominant character of the savage nations. If they are peaceable and inoffensive, the drawings usually consist of local scenery and their favourite diversions. If the band are rude and ferocious, we observe tomahawks, scalping-knives, bows, arrows, and all the engines of destruction. A Mandan bow and quiver of arrows, also some Ricara tobacco-seed and an ear of Mandan corn; to these were added a box of plants, another of insects, and three cases containing a burrowing squirrel, a prairie hen and four magpies, all alive.

Friday, 5th. Fair and pleasant, but the wind high from the northwest. We were visited by a number of Mandans, and are occupied in loading our boats in order to proceed on our journey.

Saturday, 6th. Another fine day, with a gentle breeze from the south. The Mandans continue to come to the fort, and in the course of the day informed us of the arrival of a party of Ricaras on the other side of the river. We sent our interpreter to inquire into their reason for coming; and in the morning,

Sunday, 7th, he returned with a Ricara chief and three of his nation. The chief, whose name is Kagohweto, or Brave

Raven, brought a letter from Mr. Tabeau, mentioning the wish of the grand chiefs of the Ricaras to visit the President, and requesting permission for himself and four men to join our boat when it descends; to which we consented, as it will then be manned with fifteen hands and be able to defend itself against the Sioux. After presenting the letter, he told us that he was sent with ten warriors by his nation to arrange their settling near the Mandans and Minnetarees, whom they wished to join; that he considered all the neighbouring nations friendly except the Sioux, whose persecution they would no longer withstand, and whom they hoped to repel by uniting with the tribes in this quarter; he added that the Ricaras intended to follow our advice and live in peace with all nations, and requested that we would speak in their favour to the Assiniboin Indians. This we willingly promised to do, and assured them that their great father would protect them and no longer suffer the Sioux to have good guns, or to injure his dutiful children. We then gave him a small medal, a certificate of his good conduct, a carrot of tobacco, and some wampum, with which he departed for the Mandan village well satisfied with his reception. Having made all our arrangements, we left the fort about five o'clock in the afternoon. The party now consisted of thirty-two persons. Besides ourselves were Serjeants John Ordway, Nathaniel Pryor, and Patrick Gass; the privates were William Bratton, John Colter, John Collins, Peter Cruzatte, Robert Frazier, Reuben Fields, Joseph Fields, George Gibson, Silas Goodrich, Hugh Hall, Thomas P. Howard, Baptiste Lapage, Francis Labiche, Hugh M'Neal, John Potts, John Shields, George Shannon, John B. Thompson, William Werner, Alexander Willard, Richard Windsor, Joseph Whitehouse, Peter Wiser, and Captain Clark's black servant, York. The two interpreters were George Drewyer and Toussaint Chaboneau. The wife of

## UP THE MISSOURI

Chaboneau also accompanied us with her young child, and we hope may be useful as an interpreter among the Snake Indians. She was herself one of that tribe, but having been taken in war by the Minnetarees, by whom she was sold as a slave to Chaboneau, who brought her up and afterwards married her. One of the Mandans likewise embarked with us, in order to go to the Snake Indians and obtain a peace with them for his countrymen. All this party, with the baggage, was stowed in six small canoes and two large periogues. We left the fort with fair, pleasant weather, though the northwest wind was high, and after making about four miles encamped on the north side of the Missouri, nearly opposite the first Mandan village. At the same time that we took our departure, our barge, manned with seven soldiers, two Frenchmen, and Mr. Gravelines as pilot, sailed for the United States loaded with our presents and despatches.

Monday, 8th. The day was clear and cool, the wind from the northwest, so that we travelled slowly. After breakfasting at the second Mandan village we passed the Mahaha at the mouth of Knife river, a handsome stream about eighty yards wide. Beyond this we reached the island which Captain Clark had visited on the 30th October. This island has timber as well as the lowlands on the north, but its distance from the water had prevented our encamping there during the winter. From the head of this island we made three and a half miles to a point of wood on the north, passing a high bluff on the south, and having come about fourteen miles. In the course of the day one of our boats filled and was near sinking; we, however, saved her with the loss of a little biscuit and powder.

Tuesday, April 9. We set off as soon as it was light, and proceeded five miles to breakfast, passing a low ground on the south covered with groves of cottonwood timber. At the distance of six miles we reached on the north a hunting camp of Minnetarees, consisting of thirty lodges, and built in the usual form of earth and timber. Two miles and a quarter farther comes in on the same side Miry creek, a small stream about ten yards wide, which, rising in some lakes near the Mouse river, passes through beautiful level fertile plains, without timber, in a direction nearly southwest, the banks near its entrance being steep and rugged on both sides of the Missouri. Three miles above this creek we came to a hunting party of Minnetarees, who had prepared a park or inclosure and were waiting the return of the antelope. This animal, which in the autumn retires for food and shelter to the Black mountains during the winter, recross the river at this season of the year, and spread themselves through the plains on the north of the Missouri. We halted and smoked a short time with them, and then proceeded on through handsome plains on each side of the river, and encamped at the distance of twenty-three and a half miles on the north side. The day was clear and pleasant, the wind high from the south, but afterwards changed to a western steady breeze. The bluffs which we passed to-day are upwards of one hundred feet high, composed of a mixture of yellow clay and sand, with many horizontal strata of carbonated wood resembling pit-coal, from one to five feet in depth, and scattered through the bluff at different elevations, some as high as eighty feet above the water; the hills along the river are broken, and present every appearance of having been burned at some former period; great quantities of pumicestone and lava, or rather earth, which seems to have been boiled and then hardened by exposure, being seen in many parts of these hills, where they are broken and washed down into gullies by the rain and melting snow. A great number of brants pass up the river; there are some of them perfectly white, except the large feathers of the first and second joint of the wing, which are black, though in every other characteristic they resemble common gray brant; we also saw, but could not procure; an animal that burrows in the ground, and similar in every respect to the burrowing squirrel, except that it is only one-third of its size. This may be the animal whose works we have often seen in the plains and prairies; they resemble the labours of the salamander in the sand hills of South Carolina and Georgia, and like him the animals rarely come above ground; they consist of a little hillock of ten or twelve pounds of loose ground, which would seem to have been reversed from a pot, though no aperture is seen through which it could have been thrown; on removing gently the earth, you discover that the soil has been broken in a circle of about an inch and a half diameter, where the ground is looser, though still no opening is perceptible. When we stopped for dinner the squaw went out, and after penetrating with a sharp stick the holes of the mice, near some driftwood, brought to us a quantity of wild artichokes, which the mice collect and hoard in large numbers; the root is white, of an ovate form, from one to three inches long, and generally of the size of a man's finger, and two, four, and sometimes six roots are attached to a single stalk. Its flavour, as well as the stalk which issues from it, resemble those of the Jerusalem artichoke, except that the latter is much larger. A large beaver was caught in a trap last night, and the musquitoes begin to trouble us.

Wednesday, 10th. We again set off early, with clear, pleasant weather, and halted about ten for breakfast, above a sandbank which was falling in, and near a small willow island. On both sides of the Missouri, after ascending the hills near the water, one fertile unbroken plain extends itself as far as the eye can reach, without a solitary tree or shrub, except in moist situations or in the steep declivities of hills where they are sheltered from the ravages of fire. At the distance of twelve miles we reached the lower point of a bluff on the south, which is in some parts on fire and throws out quantities of smoke which has a strong sulphurous smell, the coal and other appearances in the bluffs being like those described yesterday. At one o'clock we overtook three Frenchmen who left the fort a few days before us in order to make the first attempt on this river of hunting beaver, which they do by means of traps; their efforts promise to be successful, for they have already caught twelve which are finer than any we have ever seen; they mean to accompany us as far as the Yellowstone river in order to obtain our protection against the Assiniboins, who might attack them. In the evening we encamped on a willow point to the south opposite to a bluff, above which a small creek falls in, and just above a remarkable bend in the river to the southwest, which we called the Little Basin. The low grounds which we passed to-day possess more timber than is usual, and are wider; the current is moderate, at least not greater than that of the Ohio in high tides; the banks, too, fall in but little, so that the navigation comparatively with that lower down the Missouri is safe and easy. We were enabled to make eighteen and a half miles; we saw the track of a large white bear; there were also a herd of antelopes in the plains; the geese and swan are now feeding in considerable quantities on the young grass in the low prairies; we shot a prairie hen, and a bald eagle, of which there were many nests in the tall cottonwood trees, but could procure neither of two elk which were in the plain. Our old companions, the musquitoes, have renewed their visit, and gave us much uneasiness.

Thursday, 11th. We set out at daylight, and after passing bare and barren hills on the south, and a plain covered with timber on the north, breakfasted at five miles distance; here we were regaled with a deer brought in by the hunters, which was very acceptable, as we had been for several days without fresh meat, the country between this and fort Mandan being so frequently disturbed by hunters that the game has become scarce. We then proceeded, with a gentle breeze from the south which carried the periogues on very well; the day was, however, so warm that several of the men worked with no clothes except round the waist, which is the less inconvenient as we are obliged to wade in some places owing to the shallowness of the river. At seven miles we reached a large sandbar making out from the north. We again stopped for dinner, after which we went on to a small plain on the north covered with cottonwood, where we encamped, having made nineteen miles. The country around is much the same as that we passed yesterday; on the sides of the hills, and even on the banks of the rivers, as well as on the sandbars, is a white substance which appears in considerable quantities on the surface of the earth, and tastes like a mixture of common salt with glauber salts; many of the streams which come from the foot of the hills are so strongly impregnated with this substance that the water has an unpleasant taste and a purgative effect. A beaver was caught last night by one of the Frenchmen; we killed two geese, and saw some cranes, the largest bird of that kind common to the Missouri and Mississippi, and perfectly white except the large feathers on the two first joints of the wing, which are black. Under a bluff opposite to our encampment we discovered some Indians with horses, whom we supposed were Minnetarees, but the width of the river prevented our speaking to them.

Friday, 12th. We set off early and passed a high range of hills on the south side, our periogues being obliged to go over to the south in order to avoid a sandbank which was rapidly falling in. At six miles we came to at the lower side of the entrance of the Little Missouri, where we remained during the day for the purpose of making celestial observations. This river empties itself on the south side of the Missouri, one thousand six hundred and ninety-three miles from its confluence with the Mississippi. It rises to the west of the Black mountains, across the northern extremity of which it finds a narrow rapid passage along high perpendicular banks, then seeks the Missouri in a northeastern direction, through a broken country with highlands bare of timber, and the low grounds particularly supplied with cottonwood, elm, small ash, box-alder, and an undergrowth of willow, redwood, sometimes called red or swamp willow, the redberry and chokecherry. In its course it passes near the northwest side of the Turtle mountain, which is said to be only twelve or fifteen miles from its mouth in a straight line a little to the south of west, so that both the Little Missouri and Knife river have been laid down too far southwest. It enters the Missouri with a bold current, and is one hundred and thirty-four yards wide, but its greatest depth is two feet and a half, and this, joined to its rapidity and its sandbars, make the navigation difficult except for canoes, which may ascend it for a considerable distance. At the mouth, and as far as we could discern from the hills between the two rivers, about three miles from their junction, the country is much broken, the soil consisting of a deep rich dark-coloured loam, intermixed with a small proportion of fine sand, and covered generally with a short grass resembling blue grass. In its colour, the nature of its bed, and its general appearance it resembles so much the Missouri as to induce a belief that the countries they water are similar in point of soil. From the Mandan villages to this place the country is hilly and irregular, with the same appearance of glauber salts and carbonated wood, the low grounds smooth, sandy, and partially covered with cottonwood and small ash; at some distance back there are extensive plains of a good soil, but without timber or water.

We found great quantities of small onions, which grow single, the bulb of an oval form, white, about the size of a bullet, with a leaf resembling that of the shive. On the side of a neighbouring hill there is a species of dwarf cedar; it spreads its limbs along the surface of the earth, which it almost conceals by its closeness and thickness, and is sometimes covered by it, having always a number of roots on the under side, while on the upper are a quantity of shoots which, with their leaves, seldom rise higher than six or eight inches; it is an evergreen, its leaf more delicate than that of the common cedar, though the taste and smell is the same.

The country around has been so recently hunted that the game are extremely shy, so that a white rabbit, two beaver, a deer, and a bald eagle were all that we could procure. The weather had been clear, warm, and pleasant in the morning, but about three we had a squall of high wind and rain, with some thunder, which lasted till after sunset, when it again cleared off.

Saturday, 13th. We set out at sunrise, and at nine o'clock, having the wind in our favour, went on rapidly past a timbered low ground on the south, and a creek on the north at the distance of nine miles, which we called Onion creek, from the quantity of that plant which grows in the plains near it; this creek is about sixteen yards wide at a mile and a half above its mouth; it discharges more water than is usual for creeks of that size in this country, but the whole plain which it waters is totally destitute of timber. The Missouri itself widens very remarkably just above the junction with the Little Missouri; immediately at the entrance of the latter it is not more than two hundred yards wide, and so shallow that it may be passed in canoes with setting poles, while a few miles above it is upwards of a mile in width. Ten miles beyond Onion creek we came to another, discharging itself on the north in the centre of a deep bend; on ascending it for about a mile and a half, we found it to be the discharge of a pond or small lake, which seemed to have been once the bed of the Missouri; near this lake were the remains of forty-three temporary lodges which seem to belong to the Assiniboins, who are now on the river of the same name. A great number of swan and geese were also in it, and from this circumstance we named the creek Goose creek, and the lake by the same name; these geese we observe do not build their nests on the ground or in sandbars, but in the tops of lofty cottonwood trees. We saw some elk and buffaloe to-day but at too great a distance to obtain any of them, though a number of the carcases of the latter animal are strewed along the shore, having fallen through the ice and been swept along when the river broke up. More bald eagles are seen on this part of the Missouri than we have previously met with; the small or common hawk, common in most parts of the United States, are also found here; great quantities of geese are feeding in the prairies, and one flock of white brant or geese, with black wings, and some gray brant with them, pass up the river, and from their flight they seem to proceed much farther to the northwest. We killed two antelopes, which were very lean, and caught last night two beaver. The French hunters, who had procured seven, thinking the neighbourhood of the Little Missouri a convenient hunting ground for that animal, remained behind there. In the evening we encamped in a beautiful plain on the north, thirty feet above the river, having made twenty-two and a half miles.

Sunday, 14th. We set off early, with pleasant and fair

## UP THE MISSOURI 199

weather; a dog joined us, which we suppose had strayed from the Assiniboin camp on the lake. At two and a half miles we passed timbered low grounds and a small creek; in these low grounds are several uninhabited lodges, built with the boughs of the elm, and the remains of two recent encampments, which, from the hoops of small kegs found in them, we judged could belong to Assiniboins only, as they are the only Missouri Indians who use spirituous liquors; of these they are so passionately fond that it forms their chief inducement to visit the British on the Assiniboin, to whom they barter for kegs of rum their dried and pounded meat, their grease, and the skins of large and small wolves and small foxes. The dangerous exchange is transported to their camps with their friends and relations, and soon exhausted in brutal intoxication. So far from considering drunkenness as disgraceful, the women and children are permitted and invited to share in these excesses with their husbands and fathers, who boast how often their skill and industry as hunters has supplied them with the means of intoxication. In this, as in their other habits and customs, they resemble the Sioux, from whom they are descended. The trade with the Assiniboins and Knistenaux is encouraged by the British, because it procures provision for their *engages* on their return from Rainy lake to the English river and the Athabasky country where they winter, these men being obliged during that voyage to pass rapidly through a country but scantily supplied with game. We halted for dinner near a large village of burrowing squirrels, who we observe generally select a southeasterly exposure, though they are sometimes found in the plains. At ten and a quarter miles we came to the lower point of an island, which, from the day of our arrival there, we called Sunday island; here the river washes the bases of the hills on both sides and above the island, which with its sandbar extends a mile and a half; two small creeks fall in from the south; the uppermost of these, which is the largest, we called Chaboneau's creek, after our interpreter, who once encamped on it several weeks with a party of Indians. Beyond this no white man had ever been except two Frenchmen, one of whom, Lapage, is with us, and who, having lost their way, straggled a few miles farther, though to what point we could not ascertain; about a mile and a half beyond this island we encamped on a point of woodland on the north, having made in all fourteen miles.

The Assiniboins have so recently left the river that game is scarce and shy. One of the hunters shot at an otter last evening; a buffaloe, too, was killed, and an elk, both so poor as to be almost unfit for use; two white bear were also seen, and a muskrat swimming across the river. The river continues wide, and of about the same rapidity as the ordinary current of the Ohio. The low grounds are wide, the moister parts containing timber, the upland extremely broken, without wood, and in some places seem as if they had slipped down in masses of several acres in surface. The mineral appearances of salts, coal, and sulphur, with the burnt hill and pumicestone, continue, and a bituminous water, about the colour of strong lye, with the taste of glauber salts and a slight tincture of allum. Many geese were feeding in the prairies, and a number of magpies, who build their nests much like those of the blackbird, in trees, and composed of small sticks, leaves, and grass, open at top; the egg is of a bluish brown colour, freckled with reddish brown spots. We also killed a large hooting owl resembling that of the United States, except that it was more booted and clad with feathers. On the hills are many aromatic herbs, resembling in taste, smell, and appearance the sage, hysop, wormwood, southern wood, juniper, and dwarf cedar; a plant, also, about two or three feet high, similar to the camphor in smell and taste, and another plant of the same size, with a long, narrow, smooth, soft leaf, of an agreeable smell and flavour, which is a favorite food of the antelope, whose necks are often perfumed by rubbing against it.

Monday, 15th. We proceeded under a fine breeze from the south, and clear, pleasant weather. At seven miles we reached the lower point of an island, in a bend to the south, which is two miles in length. Captain Clark, who went about nine miles northward from the river, reached the high grounds, which, like those we have seen, are level plains without timber; here he observed a number of drains, which, descending from the hills, pursue a northeast course, and probably empty into the Mouse river, a branch of the Assiniboin, which from Indian accounts approaches very near to the Missouri at this place. Like all the rivulets of this neighbourhood, these drains were so strongly impregnated with mineral salts that they are not fit to drink. He saw also the remains of several camps of Assiniboins. The low grounds on both sides of the river are extensive, rich, and level. In a little pond on the north we heard for the first time this season the croaking of frogs, which exactly resembles that of the small frogs in the United States; there are also in these plains great quantities of geese, and many of the grouse, or prairie hen, as they are called by the N. W. company traders; the note of the male, as far as words can represent it, is cook, cook, cook, coo, coo, coo, the first part of which both male and female use when flying; the male, too, drums with his wings when he flies in the same way, though not so loud, as the pheasant; they appear to be mating. Some deer, elk, and goats were in the low grounds, and buffaloe on the sand beaches, but they were uncommonly shy; we also saw a black bear and two white ones. At fifteen miles we passed on the north side a small creek twenty yards wide, which we called Goatpen creek, from a park or enclosure for the purpose of catching that animal which those who went up the creek found, and which we presume to have been left by the Assiniboins. Its water is impregnated with mineral salts, and the country through which it flows consists of wide and very fertile plains, but without any trees. We encamped at the distance of twenty-three miles, on a sand point to the south; we passed in the evening a rock in the middle of the river, the channel of which, a little above our camp, is confined within eighty yards.

Tuesday, 16th. The morning was clear, the wind light from the S. E. The country presents the same appearance of low plains and meadows on the river, bounded a few miles back by broken hills, which end in high, level, fertile lands; the quantity of timber is, however, increasing. The appearances of minerals continue as usual, and to-day we found several stones which seemed to have been wood, first carbonated and then petrified by the water of the Missouri, which has the same effect on many vegetable substances. There is indeed reason to believe that the strata of coal in the hills cause the fire and appearances which they exhibit of being burned. Whenever these marks present themselves in the bluffs on the river the coal is seldom seen, and when found in the neighbourhood of the strata of burnt earth, the coal, with the sand and sulphurous matter usually accompanying it, is precisely at the same height and nearly of the same thickness with those strata. We passed three small creeks, or rather runs, which rise in the hills to the north. Numbers of geese, and a few ducks, chiefly of the mallard and blue-winged teal, many buffaloe, elk, and deer were also observed, and in the timbered low grounds this morning we were surprised to observe a great quantity of old hornets'

## UP THE MISSOURI

nests. We encamped in a point of woods on the south, having come eighteen miles, though the circuits which we were obliged to make round sandbars very much increased the real distance.

Wednesday, April 17. We set off early, the weather being fine, and the wind so favourable as to enable us to sail the greater part of the course. At ten and three-quarter miles we passed a creek ten yards wide, on the south; at eighteen miles a little run on the north, and at night encamped in a woody point on the south. We had travelled twenty-six miles, through a country similar to that of yesterday, except that there were greater appearances of burnt hills, furnishing large quantities of lava and pumicestone, of the last of which we observe some pieces floating down the river, as we had previously done, as low as the Little Missouri. In all the copses of wood are the remains of the Assiniboin encampments. Around us are great quantities of game, such as herds of buffaloe, elk, antelope, some deer and wolves, the tracks of bears; a curlue was also seen, and we obtained three beaver, the flesh of which is more relished by the men than any other food which we have. Just before we encamped we saw some tracks of Indians, who had passed twenty-four hours before and left four rafts, and whom we supposed to be a band of Assiniboins on their return from war against the Indians on the Rocky mountains.

Thursday, 18th. We had again a pleasant day, and proceeded on with a westerly wind, which, however, changed to N. W., and blew so hard that we were obliged to stop at one o'clock and remain four hours, when it abated, and we then continued our course.

We encamped about dark on a woody bank, having made thirteen miles. The country presented. the usual variety of highlands interspersed with rich plains. In one of these we observed a species of pea bearing a yellow flower which is now in blossom, the leaf and stalk resembling the common pea. It seldom rises higher than six inches, and the root is perennial. On the rosebushes we also saw a quantity of the hair of the buffaloe, which had become perfectly white by exposure and resembled the wool of the sheep, except that it was much finer and more soft and silky. A buffaloe which we killed yesterday had shed his long hair, and that which remained was about two inches long, thick, fine, and would have furnished five pounds of wool, of which we have no doubt an excellent cloth may be made. Our game to-day was a beaver, a deer, an elk, and some geese. The river has been crooked all day and bearing towards the south.

On the hills we observed considerable quantities of dwarf juniper, which seldom grows higher than three feet. We killed in the course of the day an elk, three geese, and a beaver. The beaver on this part of the Missouri are in greater quantities, larger and fatter, and their fur is more abundant and of a darker colour, than any we had hitherto seen; their favorite food seems to be the bark of the cottonwood and willow, as we have seen no other species of tree that has been touched by them, and these they gnaw to the ground through a diameter of twenty inches.

The next day, Friday, 19th, the wind was so high from northwest that we could not proceed, but being less violent on

Saturday, 20th, we set off about seven o'clock, and had nearly lost one of the canoes as we left the shore by the falling in of a large part of the bank. The wind, too, became again so strong that we could scarcely make one mile an hour, and the sudden squalls so dangerous to the small boats that we stopped for the night among some willows on the north, not being able to advance more than six and a half miles. In walking through the neighbouring plains we found a fine fertile soil, covered with cottonwood, some box alder, ash, red elm, and an undergrowth of willow, rosebushes, honeysuckle, red willow, gooseberry, currant, and serviceberries, and along the foot of the hills great quantities of hysop. Our hunters procured elk and deer, which are now lean, and six beaver, which are fatter and more palatable. Along the plain there were also some Indian camps; near one of these was a scaffold about seven feet high, on which were two sleds with their harness, and under it the body of a female, carefully wrapped in several dressed buffaloe skins; near it lay a bag made of buffaloe skin, containing a pair of moccasins, some red and blue paint, beaver's nails, scrapers for dressing hides, some dried roots, several plaits of sweet grass, and a small quantity of Mandan tobacco. These things, as well as the body itself, had probably fallen down by accident, as the custom is to place them on the scaffold. At a little distance was the body of a dog, not yet decayed, who had met this reward for having dragged thus far in the sled the corpse of his mistress, to whom, according to the Indian usage, he had been sacrificed.

Sunday, 21st. Last night there was a hard, white frost, and this morning the weather cold, but clear and pleasant; in the course of the day, however, it became cloudy and the wind rose. The country is of the same description as within the few last days. We saw immense quantities of buffaloe, elk deer, antelopes, geese, and some swan and ducks, out of which we procured three deer, four buffaloe calves, which last are equal in flavour to the most delicious veal; also two beaver and an otter. We passed one large and two small creeks on the south side, and reached at sixteen miles the mouth of Whiteearth river, coming in from the north. This river, before it reaches the low grounds near the Missouri, is a fine bold stream sixty yards wide, and is deep and navigable, but it is so much choked up at the entrance by the mud of the Missouri that its mouth is not more than ten yards wide. Its course, as far as we could discern from the neighbouring hills, is nearly due north, passing through a beautiful and fertile valley, though without a tree or bush of any description. Half a mile beyond this river we encamped on the same side, below a point of highland, which from its appearance we call Cut bluff.

Monday, 22d. The day clear and cold; we passed a high bluff on the north and plains on the south, in which were large herds of buffaloe, till breakfast, when the wind became so strong ahead that we proceeded with difficulty even with the aid of the towline. Some of the party now walked across to the Whiteearth river, which here, at the distance of four miles from its mouth, approaches very near to the Missouri. It contains more water than is usual in streams of the same size at this season, with steep banks about ten or twelve feet high, and the water is much clearer than that of the Missouri; the salts which have been mentioned as common on the Missouri are here so abundant that in many places the ground appears perfectly white, and from this circumstance it may have derived its name; it waters an open country and is navigable almost to its source, which is not far from the Saskaskawan, and, judging from its size and course, it is probable that it extends as far north as the fiftieth degree of latitude. After much delay in consequence of the high wind, we succeeded in making eleven miles, and encamped in a low ground on the south covered with cottonwood and rabbitberries. The hills of the Missouri near this place exhibit large irregular broken masses of rocks and stones, some of which, although two hundred feet above the water, seem at some remote period to have been subject to its influence, being apparently worn smooth by the agitation of the water.

These rocks and stones consist of white and gray granite, a brittle black rock, flint, limestone, freestone, some small specimens of an excellent pebble, and occasionally broken stratas of a black-coloured stone like petrified wood, which make good whetstones. The usual appearances of coal, or carbonated wood, and pumicestone still continue, the coal being of a better quality, and when burnt affords a hot and lasting fire, emitting very little smoke or flame. There are large herds of deer, elk, buffaloe, and antelopes in view of us; the buffaloe are not so shy as the rest, for they suffer us to approach within one hundred yards before they run, and then stop and resume their pasture at a very short distance. The wolves to-day pursued a herd of them, and at length caught a calf that was unable to keep up with the rest, the mothers on these occasions defending their young as long as they can retreat as fast as the herd, but seldom returning any distance to seek for them.

Tuesday, 23d. A clear and pleasant morning, but at nine o'clock the wind became so high that the boats were in danger of upsetting; we therefore were forced to stop at a place of safety till about five in the afternoon, when, the wind being lower, we proceeded, and encamped on the north at the distance of thirteen and a half miles; the party on shore brought us a buffaloe calf and three black-tailed deer; the sand on the river has the same appearances as usual, except that the quantity of wood increases.

Wednesday, 24th. The wind blew so high during the whole day that we were unable to move; such, indeed, was its violence that, although we were sheltered by high timber, the waves wet many articles in the boat. The hunters went out, and returned with four deer, two elk, and some young wolves of the small kind. The party are very much afflicted with sore eyes, which we presume are occasioned by the vast quantities of sand which are driven from the sandbars in such clouds as often to hide from us the view of the opposite bank. The particles of this sand are so fine and light that it floats for miles in the air like a column of thick smoke, and is so penetrating that nothing can be kept free from it, and we are compelled to eat, drink, and breathe it very copiously. To the same cause we attribute the disorder of one of our watches, although her cases are double and tight, since, without any defect in its works that we can discover, it will not run for more than a few minutes without stopping.

Thursday, 25th. The wind moderated this morning, but was still high; we therefore set out early, the weather being so cold that the water froze on the oars as we rowed, and about ten o'clock the wind increased so much that we were obliged to stop. This detention from the wind, and the reports from our hunters of the crookedness of the river, induced us to believe that we were at no great distance from the Yellowstone river. In order, therefore, to prevent delay as much as possible, Captain Lewis determined to go on by land in search of that river and make the necessary observations, so as to be enabled to proceed on immediately after the boats should join him; he therefore landed about eleven o'clock on the south side, accompanied by four men; the boats were prevented from going until five in the afternoon, when they went on a few miles farther, and encamped for the night at the distance of fourteen and a half miles.

Friday, 26th. We continued our voyage in the morning, and by twelve o'clock encamped, at eight miles distance, at the junction of the Missouri and Yellowstone rivers, where we were soon joined by Captain Lewis.

On leaving us yesterday he pursued his route along the foot of the hills, which he ascended at the distance of eight miles; from these the wide plains watered by the Missouri

## UP THE MISSOURI

and the Yellowstone spread themselves before the eye, occasionally varied with the wood of the banks, enlivened by the irregular windings of the two rivers, and animated by vast herds of buffaloe, deer, elk, and antelope. The confluence of the two rivers was concealed by the wood, but the Yellowstone itself was only two miles distant to the south. He therefore descended the hills and encamped on the bank of the river, having killed as he crossed the plain four buffaloes; the deer alone are shy and retire to the woods, but the elk, antelope, and buffaloe suffered him to approach them without alarm, and often followed him quietly for some distance. This morning he sent a man up the river to examine it, while he proceeded down to the junction. The ground on the lower side of the Yellowstone near its mouth is flat, and for about a mile seems to be subject to inundation, while that at the point of junction, as well as on the opposite side of the Missouri, is at the usual height of ten or eighteen feet above the water, and therefore not overflown. There is more timber in the neighbourhood of this place, and on the Missouri as far below as the Whiteearth river, than on any other part of the Missouri on this side of the Chayenne; the timber consists principally of cottonwood, with some small elm, ash, and box alder. On the sandbars and along the margin of the river grows the small-leafed willow; in the low grounds adjoining are scattered rosebushes three or four feet high, the redberry, serviceberry, and redwood. The higher plains are either immediately on the river, in which case they are generally timbered, and have an undergrowth like that of the low grounds, with the addition of the broad-leafed willow, gooseberry, chokecherry, purple currant, and honeysuckle; or they are between the low grounds and the hills, and for the most part without wood or anything except large quantities of wild hysop; this plant rises about two feet high, and, like

Vol. I.    14

the willow of the sandbars, is a favourite food of the buffaloe, elk, deer, grouse, porcupine, hare, and rabbit. This river, which had been known to the French as the Roche jaune, or as we have called it the Yellowstone, rises, according to Indian information, in the Rocky mountains; its sources are near those of the Missouri and the Platte, and it may be navigated in canoes almost to its head. It runs first through a mountainous country, but in many parts fertile and well timbered; it then waters a rich delightful land, broken into valleys and meadows, and well supplied with wood and water, till it reaches near the Missouri open meadows and low grounds, sufficiently timbered on its borders. In the upper country its course is represented as very rapid, but during the two last and largest portions its current is much more gentle than that of the Missouri, which it resembles also in being turbid, though with less sediment. The man who was sent up the river reported in the evening that he had gone about eight miles, that during that distance the river winds on both sides of a plain four or five miles wide, that the current was gentle and much obstructed by sandbars, that at five miles he had met with a large timbered island, three miles beyond which a creek falls in on the S. E. above a high bluff, in which are several strata of coal. The country, as far as he could discern, resembled that of the Missouri, and in the plain he met several of the bighorn animals, but they were too shy to be obtained. The bed of the Yellowstone, as we observed it near the mouth, is composed of sand and mud, without a stone of any kind. Just above the confluence we measured the two rivers, and found the bed of the Missouri five hundred and twenty yards wide, the water occupying only three hundred and thirty, and the channel deep; while the Yellowstone, including its sandbar, occupied eight hundred and fifty-eight yards, with two hundred and ninety-seven

## UP THE MISSOURI 211

yards of water; the deepest part of the channel is twelve feet, but the river is now falling and seems to be nearly at its summer height.

April 27. We left the mouth of the Yellowstone. From the point of junction a wood occupies the space between the two rivers, which at the distance of a mile come within two hundred and fifty yards of each other. There a beautiful low plain commences, and, widening as the rivers recede, extends along each of them for several miles, rising about half a mile from the Missouri into a plain twelve feet higher than itself. The low plain is a few inches above high water mark, and where it joins the higher plain there is a channel of sixty or seventy yards in width, through which a part of the Missouri when at its greatest height passes into the Yellowstone. At two and a half miles above the junction, and between the high and low plain, is a small lake two hundred yards wide, extending for a mile parallel with the Missouri along the edge of the upper plain. At the lower extremity of this lake, about four hundred yards from the Missouri, and twice that distance from the Yellowstone, is a situation highly eligible for a trading establishment; it is in the high plain which extends back three miles in width, and seven or eight miles in length, along the Yellowstone, where it is bordered by an extensive body of woodland, and along the Missouri with less breadth, till three miles above it is circumscribed by the hills within a space four yards in width. A sufficient quantity of limestone for building may easily be procured near the junction of the rivers; it does not lie in regular stratas, but is in large irregular masses, of a light colour, and apparently of an excellent quality. Game, too, is very abundant, and as yet quite gentle. Above all, its elevation recommends it as preferable to the land at the confluence of the rivers, which their variable channels may render very insecure. The N. W. wind rose so high at eleven o'clock that we were obliged to stop till about four in the afternoon, when we proceeded till dusk. On the south a beautiful plain separates the two rivers, till at about six miles there is a timbered piece of low ground, and a little above it bluffs, where the country rises gradually from the river, the situations on the north more high and open. We encamped on that side, the wind, the sand which it raised, and the rapidity of the current having prevented our advancing more than eight miles; during the latter part of the day the river becomes wider and crowded with sandbars; although the game is in such plenty, we kill only what is necessary for our subsistence. For several days past we have seen great numbers of buffaloe lying dead along the shore, and some of them partly devoured by the wolves; they have either sunk through the ice during the winter, or been drowned in attempting to cross, or else, after crossing to some high bluff, found themselves too much exhausted either to ascend or swim back again, and perished for want of food; in this situation we found several small parties of them. There are geese, too, in abundance, and more bald-eagles than we have hitherto observed, the nests of these last being always accompanied by those of two or three magpies, who are their inseparable attendants.

## CHAPTER VIII.

Unusual appearance of salt—The formidable character of the white bear—Porcupine river described—Beautiful appearance of the surrounding country—Immense quantities of game—Milk river described—Extraordinary character of Bigdry river—An instance of uncommon tenacity of life in a white bear—Narrow escape of one of the party from that animal—A still more remarkable instance—Muscleshell river described.

SUNDAY, 28th. The day was clear and pleasant, and the wind having shifted to southeast we could employ our sails, and went twenty-four miles to a low ground on the north opposite to steep bluffs. The country on both sides is much broken, the hills approaching nearer to the river, and forming bluffs, some of a white and others of a red colour, and exhibiting the usual appearances of minerals, and some burnt hills, though without any pumicestone; the salts are in greater quantities than usual, and the banks and sandbars are covered with a white incrustation like frost. The low grounds are level, fertile, and partially timbered, but are not so wide as for a few days past. The woods are now green, but the plains and meadows seem to have less verdure than those below; the only streams which we met to-day are two small runs on the north and one on the south, which rise in the neighbouring hills, and have very little water. At the distance of eighteen miles the Missouri makes a considerable bend to the southeast. The game is very abundant, the common and mule, or black-tailed, deer, elk, buffaloe, antelope, brown bear, beaver, and geese. The beaver have committed great devastation among the trees, one of which, nearly three feet in diameter, had been gnawed through by them.

Monday, 29th. We proceeded early, with a moderate

wind. Captain Lewis, who was on shore with one hunter, met about eight o'clock two white bears. Of the strength and ferocity of this animal the Indians had given us dreadful accounts; they never attack him but in parties of six or eight persons, and even then are often defeated with the loss of one or more of the party. Having no weapons but bows and arrows, and the bad guns with which the traders supply them, they are obliged to approach very near to the bear, and as no wound except through the head or heart is mortal, they frequently fall a sacrifice if they miss their aim. He rather attacks than avoids a man, and such is the terror which he has inspired that the Indians who go in quest of him paint themselves and perform all the superstitious rites customary when they make war on a neighbouring nation. Hitherto those we had seen did not appear desirous of encountering us, but although to a skilful rifleman the danger is very much diminished, yet the white bear is still a terrible animal. On approaching these two, both Captain Lewis and the hunter fired and each wounded a bear; one of them made his escape; the other turned upon Captain Lewis and pursued him seventy or eighty yards, but being badly wounded he could not run so fast as to prevent him from reloading his piece, which he again aimed at him, and a third shot from the hunter brought him to the ground; he was a male, not quite full grown, and weighed about three hundred pounds; the legs are somewhat longer than those of the black bear, and the talons and tusks much larger and longer. The testicles are also placed much farther forward, and suspended in separate pouches from two to four inches asunder, while those of the black bear are situated back between the thighs and in a single pouch, like those of the dog; its colour is a yellowish brown, the eyes small, black, and piercing, the front of the fore legs near the feet is usually black, and the fur is finer, thicker, and deeper than that of the black bear; add to which, it is a more furious animal, and very remarkable for the wounds which it will bear without dying.

We are surrounded with deer, elk, buffaloe, antelopes, and their companions the wolves, who have become more numerous and make great ravages among them. The hills are here much more rough and high, and almost overhang the banks of the river. There are greater appearances of coal than we have hitherto seen, the stratas of it being in some places six feet thick, and there are stratas of burnt earth, which are always on the same level with those of coal. In the evening, after coming twenty-five miles, we encamped at the entrance of a river which empties itself into a bend on the north side of the Missouri. This stream, which we called Martha's river, is about fifty yards wide, with water for fifteen yards; the banks are of earth, and steep though not high, and the bed principally of mud. Captain Clark, who ascended it for three miles, found that it continued of the same width, with a gentle current, and pursuing its course about north 30° west, through an extensive, fertile, and beautiful valley, but without a single tree. The water is clear, and has a brownish yellow tint; at this place the highlands, which yesterday and to-day had approached so near the river, became lower, and, receding from the water, left a valley seven or eight miles wide.

Tuesday, 30th. The wind was high from the north during last evening, and continued so this morning; we, however, continued, and found the river more winding than usual and with a number of sand islands and bars, on one of which last we encamped, at the distance of twenty-four miles. The low grounds are fertile and extensive, but with very little timber, and that cottonwood, very bad of its kind, being too small for planks, and broken and dead at the top and unsound in the centre of the trunk. We passed some ancient lodges of driftwood which do not appear to have been lately inhabited. The game continues abundant; we killed the largest male elk we have yet seen; on placing it in its natural erect position, we found that it measured five feet three inches from the point of the hoof to the top of the shoulder. The antelopes are yet lean, and the females are with young. This fleet and quick-sighted animal is generally the victim of its curiosity: when they first see the hunters, they run with great velocity; if he lies down on the ground and lifts up his arm, his hat, or his foot, the antelope returns on a light trot to look at the object, and sometimes goes and returns two or three times, till they approach within reach of the rifle; so, too, they sometimes leave their flock to go and look at the wolves, who crouch down, and if the antelope be frightened at first, repeat the same manœuvre, and sometimes relieve each other, till they decoy it from the party, when they seize it. But generally the wolves take them as they are crossing the rivers, for, although swift of foot, they are not good swimmers.

Wednesday, May 1. The wind was in our favour and we were enabled to use the sails till twelve o'clock, when the wind became so high and squally that we were forced to come to, at the distance of ten miles, on the south, in a low ground stocked with cottonwood, and remain there during the day, one of the canoes being separated from us, and not able to cross over in consequence of the high waves. The country around is more pleasant than that through which we had passed for several days, the hills being lower, the low grounds wider and better supplied with timber, which consists principally of cottonwood; the undergrowth, willow on the banks and sandbars, rosebushes, redwillow, and the broad-leafed willow, in the low plains, while the high country on both sides is one extensive plain without wood, though the soil is a dark, rich, mellow loam. Our hunters killed a buffaloe, an elk, a goat, and two beaver, and also a bird of the plover kind.

Thursday, 2d. The wind continued high during the night, and at daylight it began to snow and did not stop till ten o'clock, when the ground was covered an inch deep, forming a striking contrast with the vegetation, which is now considerably advanced, some flowers having put forth, and the cottonwood leaves as large as a dollar. The wind lulled about five o'clock in the afternoon, and we then proceeded along wide fertile low grounds and high level plains, and encamped at the distance of four miles. Our game to-day was deer, elk, and buffaloe; we also procured three beaver, who are quite gentle as they have not been hunted, but when the hunters are in pursuit they never leave their huts during the day; this animal we esteem a great delicacy, particularly the tail, which when boiled resembles in flavour the flesh tongues and sounds of the codfish, and is generally so large as to afford a plentiful meal for two men. One of the hunters, in passing near an old Indian camp, found several yards of scarlet cloth suspended on the bough of a tree, as a sacrifice to the deity by the Assiniboins, the custom of making these offerings being common among that people, as indeed among all the Indians on the Missouri. The air was sharp this evening; the water froze on the oars as we rowed, and in the morning,

Friday, 3d, the weather became quite cold; the ice was a quarter of an inch thick in the kettle, and the snow still continued on the hills, though it has melted from the plains. The wind, too, continued high from the west, but not so violently as to prevent our going on. At two miles from our encampment we passed a curious collection of bushes, about thirty feet high and ten or twelve in diameter, tied in the form of a fascine and standing on end in the middle of the low ground; this, too, we supposed to have been left by the Indians as a religious sacrifice. At twelve o'clock, the usual hour, we halted for dinner. The low grounds on the river are much wider than common, sometimes extending from five to nine miles to the highlands, which are much lower than heretofore, not being more than fifty or sixty feet above the lower plain; through all this valley traces of the ancient bed of the river are everywhere visible, and since the hills have become lower, the stratas of coal, burnt earth, and pumicestone have in a great measure ceased, there being in fact none to-day. At the distance of fourteen miles we reached the mouth of a river on the north, which, from the unusual number of porcupines near it, we called Porcupine river. This is a bold and beautiful stream one hundred and twelve yards wide, though the water is only forty yards at its entrance; Captain Clark, who ascended it several miles and passed it above where it enters the highlands, found it continued nearly of the same width and about knee deep, and as far as he could distinguish for twenty miles from the hills its course was from a little to the east of north. There was much timber on the low grounds; he found some limestone, also, on the surface of the earth in the course of his walk, and saw a range of low mountains, at a distance to the west of north, whose direction was northwest, the adjoining country being everywhere level, fertile, open, and exceedingly beautiful. The water of this river is transparent, and is the only one that is so of all those that fall into the Missouri; before entering a large sandbar through which it discharges itself its low grounds are formed of a stiff blue and black clay, and its banks, which are from eight to ten feet high and seldom if ever overflow, are composed of the same materials. From the quantity of water which this river contains, its direction, and the nature of the country through which it passes, it is

## UP THE MISSOURI

not improbable that its sources may be near the main body of the Saskaskawan, and as in high water it can be no doubt navigated to a considerable distance, it may be rendered the means of intercourse with the Athabasky country, from which the northwest company derive so many of their valuable furs.

A quarter of a mile beyond this river a creek falls in on the south, to which, on account of its distance from the mouth of the Missouri, we gave it the name of Two-thousand mile creek; it is a bold stream, with a bed thirty yards wide. Three miles and a half above Porcupine river we reached some high timber on the north, and encamped just above an old channel of the river, which is now dry. We saw vast quantities of buffaloe, elk, deer, principally of the long-tailed kind, antelopes, beaver, geese, ducks, brant, and some swan. The porcupines, too, are numerous, and so careless and clumsy that we can approach very near without disturbing them as they are feeding on the young willows; towards evening we also found for the first time the nest of a goose among some driftwood, all that we have hitherto seen being on the top of a broken tree on the forks, and invariably from fifteen to twenty feet or more in height.

Saturday, 4th. We were detained till nine in order to repair the rudder of one of the boats, and when we set out the wind was ahead; at six and a half miles we passed a small creek in a deep bend on the south, with a sand island opposite to it, and then passing along an extensive plain which gradually rises from the north side of the river, encamped at the distance of eighteen miles in a point of woodland on the north. The river is this day wider than usual, and crowded with sandbars on all sides; the country is level, fertile, and beautiful, the low grounds extensive, and contain a much greater portion of timber than is common; indeed, all the fore part of the day the river was bordered with timber on both sides, a circumstance very rare on the Missouri, and the first that has occurred since we left the Mandans. There are as usual vast quantities of game, and extremely gentle; the male buffaloe particularly will scarcely give way to us, and as we approach will merely look at us for a moment, as something new, and then quietly resume their feeding. In the course of the day we passed some old Indian hunting camps, one of which consisted of two large lodges, fortified with a circular fence twenty or thirty feet in diameter, and made of timber laid horizontally, the beams overlaying each other to the height of five feet, and covered with the trunks and limbs of trees that have drifted down the river; the lodges themselves are formed by three or more strong sticks, about the size of a man's leg or arm and twelve feet long, which are attached at the top by a whith of small willows, and spreading out so as to form at the base a circle of ten or fourteen feet in diameter; against these are placed pieces of driftwood and fallen timber, usually in three ranges one on the other, and the interstices are covered with leaves, bark, and straw, so as to form a conical figure about ten feet high, with a small aperture in one side for the door. It is, however, at best a very imperfect shelter against the inclemencies of the seasons.

Sunday, 5th. We had a fine morning, and the wind being from the east we used our sails. At the distance of five miles we came to a small island, and twelve miles farther encamped on the north, at the distance of seventeen miles. The country, like that of yesterday, is beautiful in the extreme. Among the vast quantities of game around us we distinguish a small species of goose, differing considerably from the common Canadian goose, its neck, head, and beak being much thicker, larger, and shorter in proportion to its size, which is nearly a third smaller, the noise, too, resembling more that of the brant or of a young goose that has not yet fully acquired its note; in other respects, its colour, habits, and the number of feathers in the tail, the two species correspond; this species also associates in flocks with the large geese, but we have not seen it pair off with them. The white brant is about the size of the common brown brant, or two-thirds of the common goose, than which it is also six inches shorter from the extremity of the wings, though the beak, head, and neck are larger and stronger; the body and wings are of a beautiful pure white, except the black feathers of the first and second joints of the wings; the beak and legs are of a reddish or flesh-coloured white, the eye of a moderate size, the pupil of a deep sea-green incircled with a ring of yellowish brown; the tail consists of sixteen feathers, equally long; the flesh is dark, and, as well as its note, differs but little from those of the common brant, whom in form and habits it resembles, and with whom it sometimes unites in a common flock; the white brant also associate by themselves in large flocks, but as they do not seem to be mated or paired off, it is doubtful whether they reside here during the summer for the purpose of rearing their young.

The wolves are also very abundant, and are of two species. First, the small wolf, or burrowing dog of the prairies, which are found in almost all the open plains. It is of an intermediate size between the fox and dog, very delicately formed, fleet and active. The ears are large, erect, and pointed; the head long and pointed, like that of the fox; the tail long and bushy; the hair and fur of a pale reddish brown colour, though much coarser than that of the fox; the eye of a deep sea-green colour, small and piercing; the talons rather longer than those of the wolf of the Atlantic states, which animal, as far as we can perceive, is not to be found on this side of the river Platte. These wolves usually associate in bands of ten or twelve, and are rarely if ever seen alone, not being able singly to attack a deer or antelope. They live and rear their young in burrows, which they fix near some pass or spot much frequented by game, and sally out in a body against any animal which they think they can overpower, but on the slightest alarm retreat to their burrows, making a noise exactly like that of a small dog.

The second species is lower, shorter in the legs, and thicker than the Atlantic wolf; their colour, which is not affected by the seasons, is of every variety of shade, from a gray or blackish brown to a cream-coloured white. They do not burrow, nor do they bark, but howl, and they frequent the woods and plains, and skulk along the skirts of the buffaloe herds, in order to attack the weary or wounded.

Captain Clark and one of the hunters met this evening the largest brown bear we have seen. As they fired he did not attempt to attack, but fled with a most tremendous roar, and such was its extraordinary tenacity of life that, although he had five balls passed through his lungs and five other wounds, he swam more than half across the river to a sandbar, and survived twenty minutes. He weighed between five and six hundred pounds at least, and measured eight feet seven inches and a half from the nose to the extremity of the hind feet, five feet ten inches and a half round the breast, three feet eleven inches round the neck, one foot eleven inches round the middle of the foreleg, and his talons, five on each foot, were four inches and three-eighths in length. It differs from the common black bear in having its talons much longer and more blunt; its tail shorter; its hair of a reddish or bay brown, longer, finer, and more abundant; his liver, lungs, and heart much larger, even in proportion to his size, the heart particularly being equal to that of a large ox; his maw ten times larger; his testicles pendant from the belly, and in separate pouches, four inches apart; besides fish and flesh, he feeds on roots and every kind of wild fruit.

The antelope are now lean and with young, so that they may readily be caught at this season, as they cross the river from S. W. to N. E.

Monday, 6th. The morning being fair and the wind favourable, we set sail, and proceeded on very well the greater part of the day. The country continues level, rich, and beautiful; the low grounds wide and, comparatively with the other parts of the Missouri, well supplied with wood. The appearances of coal, pumicestone, and burnt earth have ceased, though the salts of tartar or vegetable salts continue on the banks and sandbars, and sometimes in the little ravines at the base of the low hills. We passed three streams on the south: the first, at the distance of one mile and a half from our camp, was about twenty-five yards wide, but although it contained some water in standing pools, it discharges none; this we called Littledry creek, about eight miles beyond which is Bigdry creek, fifty yards wide, without any water; the third is six miles farther, and has the bed of a large river two hundred yards wide, yet without a drop of water; like the other two, this stream, which we called Bigdry river, continues its width undiminished as far as we can discern. The banks are low, the channel formed of a fine brown sand, intermixed with a small proportion of little pebbles of various colours, and the country around flat and without trees. They had recently discharged their waters, and from their appearance and the nature of the country through which they pass, we concluded that they rose in the Black mountains, or in the level low plains which are probably between this place and the mountains; that the country, being nearly of the same kind and of the same latitude, the rains of spring melting the snows about the same time, conspire with them to throw at once vast quantities of water down these channels, which are then left dry during the summer, autumn, and winter, when there is very little rain. We had to-day a slight sprinkling, but it lasted a very short time. The game is in such plenty that it has become a mere amusement to supply the party with provisions. We made twenty-five miles to a clump of trees on the north, where we passed the night.

Tuesday, 7th. The morning was pleasant, and we proceeded at an early hour. There is much driftwood floating, and, what is contrary to our expectation, although the river is rising the water is somewhat clearer than usual. At eleven o'clock the wind became so high that one of the boats was nearly sunk, and we were obliged to stop till one, when we proceeded on, and encamped on the south, above a large sandbar projecting from the north, having made fifteen miles. On the north side of the river are the most beautiful plains we have yet seen; they rise gradually from the low grounds on the water to the height of fifty or sixty feet, and then extend in an unbroken level as far as the eye can reach; the hills on the south are more broken and higher, though at some distance back the country becomes level and fertile. There are no more appearances of burnt earth, coal, or pumicestone, though that of salt still continues, and the vegetation seems to have advanced but little since the twenty-eighth of last month; the game is as abundant as usual. The bald-eagles, of whom we see great numbers, probably feed on the carcases of dead animals, for on the whole Missouri we have seen neither the blue-crested fisher nor the fishing-hawks to supply them with their favourite food, and the water of the river is so turbid that no bird which feeds exclusively on fish can procure a subsistence.

## UP THE MISSOURI

Wednesday, 8th. A light breeze from the east carried us sixteen miles, till we halted for dinner at the entrance of a river on the north. Captain Clark, who had walked on the south, on ascending a high point opposite to its entrance discovered a level and beautiful country which it watered; that its course for twelve or fifteen miles was N. W., when it divided into two nearly equal branches, one pursuing a direction nearly north, the other to the W. of N. W.; its width at the entrance is one hundred and fifty yards, and on going three miles up, Captain Lewis found it to be of the same breadth, and sometimes more; it is deep, gentle, and has a large quantity of water; its bed is principally of mud, the banks abrupt, about twelve feet in height, and formed of a dark, rich loam and blue clay; the low grounds near it are wide and fertile, and possess a considerable proportion of cottonwood and willow. It seems to be navigable for boats and canoes, and this circumstance, joined to its course and the quantity of water, which indicates that it passes through a large extent of country, we are led to presume that it may approach the Saskashawan and afford a communication with that river. The water has a peculiar whiteness, such as might be produced by a tablespoon full of milk in a dish of tea, and this circumstance induced us to call it Milk river. In the evening we had made twenty-seven miles, and encamped on the south. The country on that side consists in general of high broken hills, with much gray, black, and brown granite scattered over the surface of the ground. At a little distance from the river there is no timber on either side, the wood being confined as below to the margin of the river, so that, unless the contrary is particularly mentioned, it is always understood that the upland is perfectly naked, and that we consider the low grounds well timbered if even a fifth be covered with wood. The wild liquorice is found in great abundance on these hills, as is also the white apple. As usual we are surrounded by buffaloe, elk, common and black-tailed deer, beaver, antelopes, and wolves. We observed a place where an Indian had recently taken the hair off an antelope's skin, and some of the party thought they distinguished imperfectly some smoke and Indian lodges up Milk river, marks which we are by no means desirous of realizing, as the Indians are probably Assiniboins, and might be very troublesome.

Thursday, 9th. We again had a favourable wind, and sailed along very well. Between four and five miles we passed a large island in a deep bend to the north, and a large sandbar at the upper point. At fifteen and a quarter miles we reached the bed of a most extraordinary river which presents itself on the south; though as wide as the Missouri itself, that is about half a mile, it does not discharge a drop of water and contains nothing but a few standing pools. On ascending it three miles we found an eminence from which we saw the direction of the channel, first south for ten or twelve miles, then, turning to the east of southeast as far as we could see, it passes through a wide valley without timber, and the surrounding country consists of waving low hills interspersed with some handsome level plains; the banks are abrupt, and consist of a black or yellow clay or of a rich sandy loam, but though they do not rise more than six or eight feet above the bed, they exhibit no appearance of being overflowed; the bed is entirely composed of a light brown sand, the particles of which, like those of the Missouri, are extremely fine. Like the dry rivers we passed before, this seemed to have discharged its waters recently, but the watermark indicated that its greatest depth had not been more than two feet; this stream, if it deserve the name, we called Bigdry river. About a mile below is a large creek on the same side, which is also perfectly dry; the mineral salts and quartz are in large quantities near this neighbourhood. The sand of the Missouri from its mouth to this place has been mixed with a substance which we had presumed to be a granulated talk, but which is most probably this quartz. The game is now in great quantities, particularly the elk and buffaloe, which last is so gentle that the men are obliged to drive them out of the way with sticks and stones. The ravages of the beaver are very apparent; in one place the timber was entirely prostrated for a space of three acres in front on the river and one in depth, and a great part of it removed, although the trees were in large quantities, and some of them as thick as the body of a man. At the distance of twenty-four miles we encamped, after making twenty-five and a half miles, at the entrance of a small creek in a bend on the north, to which we gave the name of Werner's creek, after one of our men.

For several days past the river has been as wide as it generally is near its mouth, but as it is much shallower, crowded with sandbars, and the colour of the water has become much clearer, we do not yet despair of reaching the Rock mountains, for which we are very anxious.

Friday, 10th. We had not proceeded more than four and a quarter miles when the violence of the wind forced us to halt for the day under some timber in a bend on the south side. The wind continued high, the clouds thick and black, and we had a slight sprinkling of rain several times in the course of the day. Shortly after our landing a dog came to us, and as this induced us to believe that we are near the hunting grounds of the Assiniboins, who are a vicious, ill-disposed people, it was necessary to be on our guard; we therefore inspected our arms, which we found in good order, and sent several hunters to scour the country, but they returned in the evening having seen no tents nor any recent tracks of Indians. Biles and imposthumes are very common among the party, and sore eyes continue in a greater or less degree with all of us; for the imposthumes we use emollient poultices, and apply to the eyes a solution of two grains of white vitriol and one of sugar of lead with one ounce of water.

Saturday, 11th. The wind blew very hard in the night, but having abated this morning we went on very well, till in the afternoon the wind arose and retarded our progress; the current was too strong, the river very crooked, and the banks as usual constantly precipitating themselves in large masses into the water. The highlands are broken, and approach nearer the river than they do below. The soil, however, of both hills and low grounds appears as fertile as that farther down the river; it consists of a black-looking loam with a small portion of sand, which cover the hills and bluffs to the depth of twenty or thirty feet, and when thrown in the water dissolves as readily as loaf-sugar, and effervesces like marle. There are also great appearances of quartz and mineral salts; the first is most commonly seen in the faces of the bluffs, the second is found on the hills as well as the low grounds, and in the gullies which come down from the hills; it lies in a crust of two or three inches in depth, and may be swept up with a feather in large quantities. There is no longer any appearance of coal, burnt earth, or pumicestone. We saw and visited some high hills on the north side about three miles from the river, whose tops were covered with the pitch-pine; this is the first pine we have seen on the Missouri, and it is like that of Virginia, except that the leaves are somewhat longer; among this pine is also a dwarf cedar, sometimes between three or four feet high, but generally spreading itself like a vine along the surface of the earth, which it covers very closely, putting out roots from the under side. The fruit and smell resemble those of the common red cedar, but the leaf is finer and more delicate. The tops of the hills where these plants grow have a soil quite different from that just described; the basis of it is usually yellow or white clay, and the general appearance light-coloured, sandy, and barren, some scattering tufts of sedge being almost its only herbage. About five in the afternoon one of our men, who had been afflicted with biles and suffered to walk on shore, came running to the boats with loud cries and every symptom of terror and distress; for some time after we had taken him on board he was so much out of breath as to be unable to describe the cause of his anxiety, but he at length told us that about a mile and a half below he had shot a brown bear, which immediately turned and was in close pursuit of him; but the bear being badly wounded could not overtake him. Captain Lewis with seven men immediately went in search of him and having found his track followed him by the blood for a mile, and found him concealed in some thick brushwood, and shot him with two balls through the skull. Though somewhat smaller than that killed a few days ago, he was a monstrous animal and a most terrible enemy; our man had shot him through the centre of the lungs, yet he had pursued him furiously for half a mile, then returned more than twice that distance, and with his talons had prepared himself a bed in the earth two feet deep and five feet long, and was perfectly alive when they found him, which was at least two hours after he received the wound. The wonderful power of life which these animals possess render them dreadful; their very track in the mud or sand, which we have sometimes found eleven inches long and seven and a quarter wide, exclusive of the talons, is alarming; and we had rather encounter two Indians than meet a single brown bear. There is no chance of killing them by a single shot unless the ball goes through the brains, and this is very difficult on account of two large muscles which cover the side of the forehead, and the sharp projection of the centre of the frontal bone, which is also thick. Our encampment was on the south at the distance of sixteen miles from that of last night. The fleece and skin of the bear were a heavy burden for two men, and the oil amounted to eight gallons.

Sunday, 12th. The weather being clear and calm, we set out early. Within a mile we came to a small creek, about twenty yards wide, emptying itself on the south. At eleven and three-quarter miles we reached a point of woodland on the south, opposite to which is a creek of the same width as the last, but with little water, which we called Pine creek. At eighteen and three-quarter miles we came to on the south opposite to the lower point of a willow island, situated in a deep bend of the river to the southeast; here we remained during the day, the wind having risen at twelve so high that we could not proceed; it continued to blow violently all night, with occasional sprinklings of rain from sunset till midnight. On both sides of the river the country is rough and broken, the low grounds becoming narrower; the tops of the hills on the north exhibit some scattered pine and cedar; on the south the pine has not yet commenced, though there is some cedar on the sides of the hills and in the little ravines. The chokecherry, the wild hysop, sage, fleshy-leafed thorn, and particularly the aromatic herb on which the antelope and hare feed, are to be found on the plains and hills. The soil of the hills has now altered its texture considerably; their bases, like that of the river plains, is as usual a rich, black loam, while from the middle to the summits they are composed of a light brown-coloured earth, poor and sterile, and intermixed with a coarse white sand.

Monday, 13th. The wind was so strong that we could

not proceed until about one o'clock, when we had to encounter a current rather stronger than usual. In the course of a mile and a half we passed two small creeks on the south, one of eighteen the other of thirty yards width, but neither of them containing any water, and encamped on the south at a point of woodland, having made only seven miles. The country is much the same as yesterday, with little timber in the low grounds, and a small quantity of pine and cedar on the northern hills. The river, however, continues to grow clearer, and this as well as the increased rapidity induces us to hope for some change of country. The game is as usual so abundant that we can get without difficulty all that is necessary.

Tuesday, 14th. There was some fog on the river this morning, which is a very rare occurrence. At the distance of a mile and a half we reached an island in a bend on the north, which continued for about half a mile, when at the head of it a large creek comes in on the north, to which we gave the name of Gibson's creek. At seven and a half miles is a point of rocks on the south, above a creek on the same side, which we called Sticklodge creek; five miles farther is a large creek on the south, which like the two others has no running water; and at sixteen and a half miles a timbered point on the north, where we encamped for the night. The country is like that of yesterday, except that the low grounds are wider; there are also many high black bluffs along the banks; the game, too, is in great abundance. Towards evening the men in the hindmost canoes discovered a large brown bear lying in the open grounds, about three hundred paces from the river; six of them, all good hunters, immediately went to attack him, and concealing themselves by a small eminence came unperceived within forty paces of him; four of the hunters now fired, and each lodged a ball in his body, two of them directly through the lungs; the furious animal sprung up and ran open-mouthed upon them; as he came near, the two hunters who had reserved their fire gave him two wounds, one of which, breaking his shoulder, retarded his motion for a moment; but before they could reload he was so near that they were obliged to run to the river, and before they reached it he had almost overtaken them; two jumped into the canoe; the other four separated, and concealing themselves in the willows fired as fast as each could reload; they struck him several times, but instead of weakening the monster each shot seemed only to direct him towards the hunter, till at last he pursued two of them so closely that they threw aside their guns and pouches and jumped down a perpendicular bank of twenty feet into the river; the bear sprang after them, and was within a few feet of the hindmost when one of the hunters on shore shot him in the head and finally killed him; they dragged him to the shore, and found that eight balls had passed through him in different directions; the bear was old and the meat tough, so that they took the skin only, and rejoined us at camp, where we had been as much terrified by an accident of a different kind. This was the narrow escape of one of our canoes containing all our papers, instruments, medicine, and almost every article indispensable for the success of our enterprise. The canoe being under sail, a sudden squall of wind struck her obliquely and turned her considerably. The man at the helm, who was unluckily the worst steersman of the party, became alarmed, and instead of putting her before the wind luffed her up into it. The wind was so high that it forced the brace of the square-sail out of the hand of the man who was attending it, and instantly upset the canoe, which would have been turned bottom upwards but for the resistance made by the awning. Such was the confusion on board, and the waves ran so high, that it was half a minute before she righted, and

## UP THE MISSOURI

then nearly full of water, but by baling out she was kept from sinking until they rowed ashore; besides the loss of the lives of three men who, not being able to swim, would probably have perished, we should have been deprived of nearly everything necessary for our purposes, at a distance of between two and three thousand miles from any place where we could supply the deficiency.

Wednesday, 15th. As soon as a slight shower of rain had passed we spread out the articles to dry, but the weather was so damp and cloudy that they derived little benefit from exposure. Our hunters procured us deer, buffaloe, and beaver.

Thursday, 16th. The morning was fair and we were enabled to dry and repack our stores; the loss we sustained is chiefly in the medicines, many articles of which are completely spoiled, and others considerably injured. At four o'clock we embarked, and after making seven miles encamped on the north near some wood; the country on both sides is broken, the low grounds narrower and with less timber, though there are some scattered pine and cedar on the steep declivities of the hills, which are now higher than usual. A white bear tore the coat of one of the men which he had left on shore, and two of the party wounded a large panther who was feasting on a deer. We caught some lean antelopes as they were swimming the river, and killed two buffaloe.

Friday, 17th. We set out early and proceeded on very well; the banks being firm and the shore bold we were enabled to use the towline, which, whenever the banks will permit it, is the safest and most expeditious mode of ascending the river, except under a sail with a steady breeze. At the distance of ten and a half miles we came to the mouth of a small creek on the south, below which the hills approach the river, and continue near it during the day; three miles farther is a large creek on the north, and again six and three-quarter miles beyond it another large creek to the south, which contain a small quantity of running water of a brackish taste. The last we called Rattlesnake creek, from our seeing that animal near it. Although no timber can be observed on it from the Missouri, it throws out large quantities of driftwood, among which were some pieces of coal brought down by the stream. We continued on one mile and a quarter, and encamped on the south, after making twenty and a half miles. The country in general is rugged, the hills high, with their summits and sides partially covered with pine and cedar, and their bases on both sides washed by the river; like those already mentioned, the lower part of these hills is a dark rich loam, while the upper region for one hundred and fifty feet consists of a whitish brown sand, so hard as in many places to resemble stone, though in fact very little stone or rock of any kind is to be seen on the hills. The bed of the Missouri is much narrower than usual, being not more than between two and three hundred yards in width, with an uncommonly large proportion of gravel; but the sandbars and low points covered with willows have almost entirely disappeared; the timber on the river consists of scarcely anything more than a few scattered cottonwood trees. The saline incrustations along the banks and the foot of the hills are more abundant than usual. The game is in great quantities, but the buffaloe are not so numerous as they were some days ago; two rattlesnakes were seen to-day, and one of them killed; it resembles those of the middle Atlantic states, being about two feet six inches long, of a yellowish brown on the back and sides, variegated with a row of oval dark brown spots lying transversely on the back from the neck to the tail, and two other rows of circular spots of the same colour on the sides along the edge of the scuta; there are one hundred and seventy-six scuta on the belly, and seventeen on the tail. Captain Clark saw in his excursions a fortified Indian camp which appeared to have been recently occupied, and was, we presumed, made by a party of Minnetarees who went to war last March.

Late at night we were roused by the sergeant of the guard in consequence of a fire which had communicated to a tree overhanging our camp. The wind was so high that we had not removed the camp more than a few minutes when a large part of the tree fell precisely on the spot it had occupied, and would have crushed us if we had not been alarmed in time.

Saturday, 18th. The wind continued high from the west, but by means of the towline we were able to make nineteen miles, the sandbars being now few in number, the river narrow and the current gentle; the willow has in a great measure disappeared, and even the cottonwood, almost the only timber remaining, is growing scarce. At twelve and three-quarter miles we came to a creek on the north, which was perfectly dry. We encamped on the south opposite the lower point of an island.

Sunday, 19th. The last night was disagreeably cold, and in the morning there was a very heavy fog, which obscured the river so much as to prevent our seeing the way. This is the first fog of any degree of thickness which we have experienced; there was also last evening a fall of dew, the second which we have seen since entering this extensive open country. About eight o'clock the fog dispersed, and we proceeded with the aid of the towline; the island near which we were encamped was three-quarters of a mile in length. The country resembles that of yesterday, high hills closely bordering the river. In the afternoon the river became crooked, and contained more sawyers or floating timber than we have seen in the same space since leaving the Platte. Our game consisted of deer, beaver, and elk; we also killed a brown bear, who, although shot through the heart, ran at his usual pace nearly a quarter of a mile before he fell. At twenty-one miles is a willow island half a mile in length, on the north side, a quarter of a mile beyond which is a shoal of rapid water under a bluff; the water continued very strong for some distance beyond it; at half a mile we came to a sandbar on the north, from which to our place of encampment was another half mile, making in all twenty-two and a quarter miles. The saline substances which we have mentioned continue to appear, and the men are much afflicted with sore eyes and imposthumes.

Monday, 20th. As usual we set out early, and the banks being convenient for that purpose, we used the towline; the river is narrow and crooked, the water rapid, and the country much like that of yesterday. At the distance of two and a quarter miles we passed a large creek with but little water, to which we gave the name of Blowingfly creek, from the quantity of those insects found in its neighbourhood. They are extremely troublesome, infesting our meat whilst cooking and at our meals. After making seven miles we reached by eleven o'clock the mouth of a large river on the south, and encamped for the day at the upper point of its junction with the Missouri. This stream, which we suppose to be that called by the Minnetarees the Muscleshell river, empties into the Missouri two thousand two hundred and seventy miles above the mouth of the latter river, and in latitude 47° 0′ 24″ 6 north. It is one hundred and ten yards wide, and contains more water than streams of that size usually do in this country; its current is by no means rapid, and there is every appearance of its being susceptible of navigation by canoes for a considerable distance; its bed is chiefly formed of coarse sand and gravel, with an occasional mixture of black mud; the banks abrupt and nearly twelve feet high, so that they are secure from being overflowed; the water is of a greenish yellow cast and much more transparent than that of the Missouri, which itself, though clearer than below, still retains its whitish hue and a portion of its sediment. Opposite to the point of junction the current of the Missouri is gentle and two hundred and twenty-two yards in width, the bed principally of mud (the little sand remaining being wholly confined to the points), and still too deep to use the settingpole. If this be, as we suppose, the Muscleshell, our Indian information is that it rises in the first chain of the Rocky mountains not far from the sources of the Yellowstone, whence in its course to this place it waters a high broken country, well timbered, particularly on its borders, and interspersed with handsome fertile plains and meadows. We have reason, however, to believe, from their giving a similar account of the timber where we now are, that the timber of which they speak is similar to that which we have seen for a few days past, which consists of nothing more than a few straggling small pine and dwarf cedar on the summits of the hills, nine-tenths of the ground being totally destitute of wood, and covered with a short grass, aromatic herbs, and an immense quantity of prickly pears, though the party who explored it for eight miles represented low grounds on the river as well supplied with cottonwood of a tolerable size, and of an excellent soil. They also reported that the country is broken and irregular like that near our camp; that about five miles up, a handsome river about fifty yards wide, which we named after Chaboneau's wife, Sahcajahweah, or Birdwoman's river, discharges itself into the Muscleshell on the north or upper side. Another party found at the foot of the southern hills, about four miles from the Missouri, a fine bold spring, which in this country is so rare that since we left the Mandans we have found only one of a similar kind, and that was under the bluffs on the south side of the Missouri, at some distance from it, and about five miles below the Yellowstone; with this exception all the small fountains, of which we have met a number, are impregnated with the salts which are so abundant here, and with which the Missouri is itself most probably tainted, though to us who have been so much accustomed to it the taste is not perceptible. Among the game to-day we observed two large owls, with remarkably long feathers resembling ears on the sides of the head, which we presume are the hooting owls, though they are larger and their colours are brighter than those common in the United States.

Tuesday, 21st. The morning being very fine we were able to employ the rope, and made twenty miles to our camp on the north. The shores of the river are abrupt, bold, and composed of a black and yellow clay, the bars being formed of black mud and a small proportion of fine sand, the current strong. In its course the Missouri makes a sudden and extensive bend towards the south, to receive the waters of the Muscleshell. The neck of land thus formed, though itself high, is lower than the surrounding country, and makes a waving valley extending for a great distance to the northward, with a fertile soil which, though without wood, produces a fine turf of low grass, some herbs, and vast quantities of prickly pear. The country on the south is high, broken, and crowned with some pine and dwarf cedar; the leaf of this pine is longer than that of the common pitch or red pine of Virginia, the cone is longer and narrower, the imbrications wider and thicker, and the whole frequently covered with rosin. During the whole day the bends of the river are short and sudden, and the points covered with some cottonwood, large or broad-leaved willow, and a small quantity of redwood, the undergrowth consisting of wild roses and the bushes of the small honeysuckle.

The mineral appearances on the river are as usual. We do not find the grouse or prairie hen so abundant as below, and think it probable that they retire from the river to the plains during this season.

The wind had been moderate during the fore part of the day, but continued to rise towards evening, and about dark veered to northwest, and blew a storm all night. We had encamped on a bar on the north, opposite the lower point of an island, which from this circumstance we called Windy island; but we were so annoyed by clouds of dust and sand that we could neither eat nor sleep, and were forced to remove our camp at eight o'clock to the foot of an adjoining hill, which shielded us in some degree from the wind; we procured elk, deer, and buffaloe.

Wednesday, 22d. The wind blew so violently that it was deemed prudent to wait till it had abated, so that we did not leave the camp till ten o'clock, when we proceeded, principally by the towline. We passed Windy island, which is about three-quarters of a mile in length, and five and a half miles above it a large island in a bend to the north; three miles beyond this we came to the entrance of a creek twenty yards wide, though with little water, which we called Grouse creek, from observing near its mouth a quantity of the prairie hen with pointed tails, the first we have seen in such numbers for several days. The low grounds are somewhat wider than usual and apparently fertile, though the short and scanty grass on the hills does not indicate much richness of soil. The country around is not so broken as that of yesterday, but is still waving, the southern hills possessing more pine than usual, and some appearing on the northern hills, which are accompanied by the usual salt and mineral appearances.

The river continues about two hundred and fifty yards wide, with fewer sandbars, and the current more gentle and regular. Game is no longer in such abundance since leaving the Muscleshell. We have caught very few fish on this side of the Mandans, and these were the white catfish, of two to five pounds. We killed a deer and a bear; we have not seen in this quarter the black bear, common in the United States and on the lower parts of the Missouri, nor have we discerned any of their tracks, which may easily be distinguished by the shortness of its talons from the brown, grizzly, or white bear, all of which seem to be of the same family, which assumes those colours at different seasons of the year. We halted earlier than usual, and encamped on the north, in a point of woods, at the distance of sixteen and a half miles.

## CHAPTER IX.

The party continue their route—Description of Judith river—Indian mode of taking the buffaloe—Slaughter river described—Phenomena of nature—Of walls on the banks of the Missouri—The party encamp on the banks of the river to ascertain which of the streams constitute the Missouri—Captain Lewis leaves the party to explore the northern fork, and Captain Clark explores the southern—The surrounding country described in the route of Captain Lewis—Narrow escape of one of his party.

THURSDAY, 23d. Last night the frost was severe, and this morning the ice appeared along the edges of the river, and the water froze on our oars. At the distance of a mile we passed the entrance of a creek on the north, which we named Teapot creek; it is fifteen yards wide, and although it has running water at a small distance from its mouth, yet it discharges none into the Missouri, resembling, we believe, most of the creeks in this hilly country, the waters of which are absorbed by the thirsty soil near the river. They indeed afford but little water in any part, and even that is so strongly tainted with salts that it is unfit for use, though all the wild animals are very fond of it. On experiment it was found to be moderately purgative, but painful to the intestines in its operation. This creek seems to come from a range of low hills, which run from east to west for seventy miles, and have their eastern extremity thirty miles to the north of Teapot creek. Just above its entrance is a large assemblage of the burrowing squirrels, on the north side of the river. At nine miles we reached the upper point of an island in a bend on the south, and opposite the centre of the island, a small dry creek on the north. Half a mile farther a small creek falls in on the same side, and six and a half miles beyond this another on the south. At four and a half we passed a small island in a deep bend to the north, and on the same side, in a deep northeastern bend of the river, another small island. None of these creeks, however, possessed any water, and at the entrances of the islands the two first are covered with tall cottonwood timber, and the last with willows only. The river has become more rapid, the country much the same as yesterday, except that there is rather more rocks on the face of the hills, and some small spruce pine appears among the pitch. The wild roses are very abundant and now in bloom; they differ from those of the United States only in having the leaves and the bush itself of a somewhat smaller size. We find the musquitoes troublesome, notwithstanding the coolness of the morning. The buffaloe is scarce to-day, but the elk, deer, and antelope are very numerous. The geese begin to lose the feathers of the wings, and are unable to fly. We saw five bears, one of which we wounded, but in swimming from us across the river, he became entangled in some driftwood and sank. We formed our camp on the north, opposite to a hill and a point of wood in a bend to the south, having made twenty-seven miles.

Friday, 24th. The water in the kettles froze one-eighth of an inch during the night; the ice appears along the margin of the river, and the cottonwood trees, which have lost nearly all their leaves by the frost, are putting forth other buds. We proceeded, with the line principally, till about nine o'clock, when a fine breeze sprung up from the S. E. and enabled us to sail very well, notwithstanding the rapidity of the current. At one mile and a half is a large creek thirty yards wide, and containing some water, which it empties on the north side, over a gravelly bed, intermixed with some stone. A man who was sent up to explore the country returned in the evening, after having gone ten miles directly towards the ridge of mountains to the north, which is the source of this as well as of Teapot creek. The air of these highlands is so pure that objects appear much nearer than they really are, so that although our man went ten miles without thinking himself by any means half way to the mountains, they do not from the river appear more than fifteen miles distant; this stream we called Northmountain creek. Two and a half miles higher is a creek on the south which is fifteen yards wide, but without any water, and to which we gave the name of Littledog creek, from a village of burrowing squirrels opposite to its entrance, that being the name given by the French watermen to those animals. Three miles from this a small creek enters on the north, five beyond which is an island a quarter of a mile in length, and two miles farther a small river; this falls in on the south, is forty yards wide, and discharges a handsome stream of water; its bed rocky with gravel and sand, and the banks high; we called it Southmountain creek, as from its direction it seemed to rise in a range of mountains about fifty or sixty miles to the S. W. of its entrance. The low grounds are narrow, and without timber, the country high and broken; a large portion of black rock and brown sandy rock appears in the face of the hills, the tops of which are covered with scattered pine, spruce, and dwarf cedar; the soil is generally poor, sandy near the tops of the hills, and nowhere producing much grass, the low grounds being covered with little else than the hysop, or southern wood, and the pulpy-leafed thorn. Game is more scarce, particularly beaver, of which we have seen but few for several days, and the abundance or scarcity of which seems to depend on the greater or less quantity of timber. At twenty-four and a half miles we reached a point of woodland on the south, where we observed that the trees had no leaves, and encamped for the night. The high country through which we have passed for some days, and where we now are, we suppose to be a continuation of what the French traders called the Cote Noire or Black hills. The country thus denominated consists of high, broken, irregular hills and short chains of mountains, sometimes one hundred and twenty miles in width, sometimes narrower, but always much higher than the country on either side. They commence about the head of the Kanzas, where they diverge, the first ridge going westward, along the northern shore of the Arkansaw; the second approaches the Rock mountains obliquely, in a course a little to the W. of N. W., and after passing the Platte above its forks, and intersecting the Yellowstone near the Bigbend, crosses the Missouri at this place, and probably swell the country as far as the Saskashawan, though as they are represented much smaller here than to the south, they may not reach that river.

Saturday, 25th. Two canoes which were left behind yesterday to bring on the game did not join us till eight o'clock this morning, when we set out with the towline, the use of which the banks permitted. The wind was, however, ahead, the current strong, particularly round the points against which it happened to set, and the gullies from the hills having brought down quantities of stone, these projected into the river, forming barriers for forty or fifty feet round, which it was very difficult to pass. At the distance of two and three-quarter miles we passed a small island in a deep bend on the south, and on the same side a creek twenty yards wide, but with no running water. About a mile farther is an island between two and three miles in length, separated from the northern shore by a narrow channel, in which is a sand island at the distance of half a mile from its lower extremity. To this large island we gave the name of Teapot island, two miles above which is an island a mile long, and situated on the south. At three and a half miles is another small island, and one mile beyond it a second, three-quarters of a mile in length, on the north side. In the middle of the river two miles above this is an island with no timber, and of the same extent as this last. The country on each side is high, broken, and rocky, the rock being either a soft brown sandstone, covered with a thin stratum of limestone, or else a hard, black, rugged granite, both usually in horizontal stratas, and the sandrock overlaying the other. Salts and quartz, as well as some coal and pumicestone, still appear; the bars of the river are composed principally of gravel; the river low grounds are narrow, and afford scarcely any timber, nor is there much pine on the hills. The buffaloe have now become scarce; we saw a polecat this evening, which was the first for several days; in the course of the day we also saw several herds of the big-horned animals among the steep cliffs on the north, and killed several of them. At the distance of eighteen miles we encamped on the south, and the next morning,

Sunday, 26th, proceeded on at an early hour by means of the towline, using our oars merely in passing the river, to take advantage of the best banks. There are now scarcely any low grounds on the river, the hills being high and in many places pressing on both sides to the verge of the water. The black rock has given place to a very soft sandstone, which seems to be washed away fast by the river, and being thrown into the river renders its navigation more difficult than it was yesterday; above this sandstone, and towards the summits of the hills, a hard freestone of a yellowish brown colour shows itself in several stratas of unequal thickness, frequently overlaid or incrusted by a thin stratum of limestone, which seems to be formed of concreted shells. At eight and a quarter miles we came to the mouth of a creek on the north, thirty yards wide, with some running water and a rocky bed; we called it Windsor creek, after one of the party. Four and three-quarter miles beyond this we came to another creek in a bend to the north, which is twenty yards wide, with a handsome little stream of water; there is, however, no timber on either side of the river, except a few pines on the hills. Here we saw for the first time since we left the Mandans several soft-shelled turtles, though this may be owing rather to the season of the year than to any scarcity of the animal. It was here that, after ascending the highest summits of the hills on the north side of the river, Captain Lewis first caught a distant view of the Rock mountains, the object of all our hopes, and the reward of all our ambition. On both sides of the river, and at no great distance from it, the mountains followed its course; above these, at the distance of fifty miles from us, an irregular range of mountains spread themselves from west to northwest from his position. To the north of these a few elevated points, the most remarkable of which bore north 65° west, appeared above the horizon, and as the sun shone on the snows of their summits he obtained a clear and satisfactory view of those mountains which close on the Missouri the passage to the Pacific. Four and a half miles beyond this creek we came to the upper point of a small sand island. At the distance of five miles, between high bluffs, we passed a very difficult rapid, reaching quite across the river, where the water is deep, the channel narrow, and gravel obstructing it on each side; we had great difficulty in ascending it, although we used both the rope and the pole and doubled the crews. This is the most considerable rapid on the Missouri, and in fact the only place where there is a sudden descent; as we were labouring over them, a female elk with its fawn swam down through the waves, which ran very high, and obtained for the place the name of the Elk rapids. Just above them is a small low ground of cottonwood trees, where, at twenty-two and a quarter miles, we fixed our encampment, and were joined by Captain Lewis, who had been on the hills during the afternoon,

The country has now become desert and barren; the appearances of coal, burnt earth, pumicestone, salts, and quartz continue as yesterday; but there is no timber, except the thinly scattered pine and spruce on the summits of the hills, or along the sides. The only animals we have observed are the elk, the bighorn, and the hare, common in this country. In the plain where we lie are two Indian cabins made of sticks, and during the last few days we have passed several others in the points of timber on the river.

Monday, 27th. The wind was so high that we did not start till ten o'clock, and even then were obliged to use the line during the greater part of the day. The river has become very rapid, with a very perceptible descent; its general width is about two hundred yards; the shoals, too, are more frequent, and the rocky points at the mouth of the gullies more troublesome to pass; great quantities of this stone lie in the river and on its banks, and seem to have fallen down as the rain washed away the clay and sand in which they were imbedded. The water is bordered by high rugged bluffs, composed of irregular but horizontal stratas of yellow and brown or black clay, brown and yellowish white sand, soft yellowish white sandstone, hard dark brown freestone, and also large round kidney-formed irregular separate masses of a hard black ironstone, imbedded in the clay and sand; some coal or carbonated wood also makes its appearance in the cliffs, as do also its usual attendants the pumicestone and burnt earth. The salts and quartz are less abundant, and generally speaking the country is if possible more rugged and barren than that we passed yesterday, the only growth of the hills being a few pine, spruce, and dwarf cedar, interspersed with an occasional contrast, once in the course of some miles, of several acres of level ground, which supply a scanty subsistence for a few little cottonwood trees.

Soon after setting out we passed a small untimbered island on the south; at about seven miles we reached a considerable bend which the river makes towards the southeast, and in the evening, after making twelve and a half miles, encamped on the south near two dead cottonwood trees, the only timber for fuel which we could discover in the neighbourhood.

Tuesday, 28th. The weather was dark and cloudy; the air smoky, and there fell a few drops of rain. At ten o'clock we had again a slight sprinkling of rain, attended with distant thunder, which is the first we have heard since leaving the Mandans. We employed the line generally, with the addition of the pole at the ripples and rocky points, which we find more numerous and troublesome than those we passed yesterday. The water is very rapid round these points, and we are sometimes obliged to steer the canoes through the points of sharp rocks rising a few inches above the surface of the water, and so near to each other that if our ropes give way the force of the current drives the sides of the canoe against them, and must inevitably upset them or dash them to pieces. These cords are very slender, being almost all made of elkskin, and much worn and rotted by exposure to the weather; several times they gave way, but fortunately always in places where there was room for the canoe to turn without striking the rock; yet with all our precautions, it was with infinite risk and labour that we passed these points. An Indian pole for building floated down the river, and was worn at one end as if dragged along the ground in travelling; several other articles were also brought down by the current, which indicate that the Indians

## UP THE MISSOURI 249

are probably at no great distance above us, and, judging from a football which resembles those used by the Minnetarees near the Mandans, we conjecture that they must be a band of the Minnetarees of fort de Prairie. The appearance of the river and the surrounding country continued as usual, till towards evening, at about fifteen miles, we reached a large creek on the north thirty-five yards wide, discharging some water, and named after one of our men Thompson's creek. Here the country assumed a totally different aspect; the hills retired on both sides from the river, which now spreads to more than three times its former size, and is filled with a number of small handsome islands covered with cottonwood. The low grounds on the river are again wide, fertile, and enriched with trees; those on the north are particularly wide, the hills being comparatively low and opening into three large vallies, which extend themselves for a considerable distance towards the north; these appearances of vegetation are delightful after the dreary hills over which we have passed, and we have now to congratulate ourselves at having escaped from the last ridges of the Black mountains. On leaving Thompson's creek we passed two small islands, and at twenty-three miles distance encamped among some timber on the north, opposite to a small creek, which we named Bull creek. The bighorn is in great quantities, and must bring forth their young at a very early season, as they are now half grown. One of the party saw a large bear also, but being at a distance from the river, and having no timber to conceal him, he would not venture to fire.

Wednesday, 29th. Last night we were alarmed by a new sort of enemy. A buffaloe swam over from the opposite side and to the spot where lay one of our canoes, over which he clambered to the shore; then taking fright, he ran full speed up the bank towards our fires, and passed within eighteen inches of the heads of some of the men, before the sentinel could make him change his course; still more alarmed, he ran down between four fires and within a few inches of the heads of a second row of men, and would have broken into our lodge if the barking of the dog had not stopped him. He suddenly turned to the right and was out of sight in a moment, leaving us all in confusion, everyone seizing his rifle and inquiring the cause of the alarm. On learning what had happened, we had to rejoice at suffering no more injury than the damage to some guns which were in the canoe which the buffaloe crossed.

In the morning early we left our camp, and proceeded as usual by the cord. We passed an island and two sandbars, and at the distance of two and a half miles we came to a handsome river which discharges itself on the south, and which we ascended to the distance of a mile and a half. We called it Judith's river; it rises in the Rock mountains, in about the same place with the Muscleshell and near the Yellowstone river. Its entrance is one hundred yards wide from one bank to the other, the water occupying about seventy-five yards, and in greater quantity than that of the Muscleshell river, and, though more rapid, equally navigable, there being no stones or rocks in the bed, which is composed entirely of gravel and mud, with some sand; the water, too, is clearer than any which we have yet seen, and the low grounds, as far as we could discern, wider and more woody than those of the Missouri; along its banks we observed some box-alder, intermixed with the cottonwood and the willow, the undergrowth consisting of rosebushes, honeysuckles, and a little red willow. There was a great abundance of the argalea or bighorned animals in the high country through which it passes, and a great number of the beaver in its waters; just above the entrance of it we saw the fires of one hundred and twenty-six lodges, which appeared to have been deserted about twelve or fifteen days, and on the other side of the Missouri a large encampment, apparently made by the same nation. On examining some moccasins which we found there, our Indian woman said that they did not belong to her own nation, the Snake Indians, but she thought that they indicated a tribe on this side of the Rocky mountain, and to the north of the Missouri; indeed, it is probable that these are the Minnetarees of fort de Prairie. At the distance of six and a half miles the hills again approach the brink of the river, and the stones and rocks washed down from them form a very bad rapid, with rocks and ripples more numerous and difficult than those we passed on the 27th and 28th; here the same scene was renewed, and we had again to struggle and labour to preserve our small craft from being lost. Near this spot are a few trees of the ash, the first we have seen for a great distance, and from which we named the place Ash rapids. On these hills there is but little timber, but the salts, coal, and other mineral appearances continue. On the north we passed a precipice about one hundred and twenty feet high, under which lay scattered the fragments of at least one hundred carcases of buffaloes, although the water, which had washed away the lower part of the hill, must have carried off many of the dead. These buffaloes had been chased down the precipice in a way very common on the Missouri, and by which vast herds are destroyed in a moment. The mode of hunting is to select one of the most active and fleet young men, who is disguised by a buffaloe skin round his body, the skin of the head, with the ears and horns, fastened on his own head in such a way as to deceive the buffaloe; thus dressed, he fixes himself at a convenient distance between a herd of buffaloe and any of the river precipices, which sometimes extend for some miles. His companions in the meantime get in the rear and side of the herd, and at a given signal show themselves, and advance towards the buffaloe; they instantly take the alarm, and finding the hunters beside them, they run towards the disguised Indian or decoy, who leads them on at full speed toward the river, when suddenly securing himself in some crevice of the cliff which he had previously fixed on, the herd is left on the brink of the precipice; it is then in vain for the foremost to retreat, or even to stop; they are pressed on by the hindmost rank, who, seeing no danger but from the hunters, goad on those before them till the whole are precipitated and the shore is strewed with their dead bodies. Sometimes in this perilous seduction the Indian is himself either trodden under foot by the rapid movements of the buffaloe, or, missing his footing in the cliff, is urged down the precipice by the falling herd. The Indians then select as much meat as they wish, and the rest is abandoned to the wolves, and creates a most dreadful stench. The wolves who had been feasting on these carcases were very fat, and so gentle that one of them was killed with an esponton. Above this place we came to for dinner at the distance of seventeen miles, opposite to a bold running river of twenty yards wide, and falling in on the south. From the objects we had just passed we called this stream Slaughter river. Its low grounds are narrow, and contain scarcely any timber. Soon after landing it began to blow and rain, and as there was no prospect of getting wood for fuel farther on, we fixed our camp on the north, three-quarters of a mile above Slaughter river. After the labours of the day we gave to each man a dram, and such was the effect of long abstinence from spirituous liquors that from the small quantity of half a gill of rum several of the men were considerably affected by it, and all very much exhilarated. Our game to-day consisted of an elk and two beaver.

Thursday, 30th. The rain, which commenced last evening, continued with little intermission till eleven this morning, when, the high wind which accompanied it having abated, we set out. More rain has now fallen than we have had since the 1st of September last, and many circumstances indicate our approach to a climate differing considerably from that of the country through which we have been passing. The air of the open country is astonishingly dry and pure. Observing that the case of our sextant, though perfectly seasoned, shrank and the joints opened, we tried several experiments, by which it appeared that a tablespoon full of water exposed in a saucer to the air would evaporate in thirty-six hours, when the mercury did not stand higher than the temperate point at the greatest heat of the day. The river, notwithstanding the rain, is much clearer than it was a few days past; but we advance with great labour and difficulty, the rapid current, the ripples, and rocky points rendering the navigation more embarrassing than even that of yesterday, in addition to which the banks are now so slippery, after the rain, that the men who draw the canoes can scarcely walk, and the earth and stone constantly falling down the high bluffs make it dangerous to pass under them; still, however, we are obliged to make use of the cord, as the wind is strong ahead, the current too rapid for oars and too deep for the pole. In this way we passed, at the distance of five and a half miles, a small rivulet in a bend on the north; two miles farther an island on the same side, half a mile beyond which came to a grove of trees at the entrance of a run in a bend to the south, and encamped for the night on the northern shore. The eight miles which we made to-day cost us much trouble. The air was cold, and rendered more disagreeable by the rain, which fell in several slight showers in the course of the day; our cords, too, broke several times, but fortunately without injury to the boats. On ascending the hills near the river, one of the party found that there was snow mixed with the rain on the heights; a little back of these the country becomes perfectly level on both sides of the river. There is now no timber on the hills, and only a few scattering cottonwood, ash, box-alder, and willows, along the water. In the course of the day we passed several encampments of Indians, the most recent of which seemed to have been evacuated about five weeks since, and from the several apparent dates we supposed that they were made by a band of about one hundred lodges who were travelling slowly up the river. Although no part of the Missouri from the Minnetarees to this place exhibit signs of permanent settlements, yet none seem exempt from the transient visits of hunting parties. We know that the Minnetarees of the Missouri extend their excursions on the south side of the river as high as the Yellowstone; and the Assiniboins visit the northern side, most probably as high as Porcupine river. All the lodges between that place and the Rocky mountains we supposed to belong to the Minnetarees of fort de Prairie, who live on the south fork of the Saskashawan.

Friday, 31st. We proceeded in two periogues, leaving the canoes to bring on the meat of two buffaloes killed last evening. Soon after we set off it began to rain, and though it ceased at noon, the weather continued cloudy during the rest of the day. The obstructions of yesterday still remain, and fatigue the men excessively; the banks are so slippery in some places, and the mud so adhesive, that they are unable to wear their moccasins; one-fourth of the time they are obliged to be up to their armpits in the cold water, and sometimes walk for several yards over the sharp fragments of rocks which have fallen from the hills; all this, added to the burden of dragging the heavy canoes, is very painful, yet the men bear it with great patience and good humour. Once the rope of one of the periogues, the only one we had made of hemp, broke short, and the periogue swung and just touched a point of rock which almost overset her. At nine miles we came to a high wall of black rock rising from the water's edge on the south, above the cliffs of the river; this continued about a quarter of a mile, and was succeeded by a high open plain, till three miles farther a second wall, two hundred feet high, rose on the same side. Three miles farther a wall of the same kind, about two hundred feet high and twelve in thickness, appeared to the north. These hills and river cliffs exhibit a most extraordinary and romantic appearance; they rise in most places nearly perpendicular from the water, to the height of between two and three hundred feet, and are formed of very white sandstone, so soft as to yield readily to the impression of water, in the upper part of which lie imbedded two or three thin horizontal stratas of white freestone, insensible to the rain, and on the top is a dark rich loam, which forms a gradually ascending plain, from a mile to a mile and a half in extent, when the hills again rise abruptly to the height of about three hundred feet more. In trickling down the cliffs, the water has worn the soft sandstone into a thousand grotesque figures, among which with a little fancy may be discerned elegant ranges of freestone buildings, with columns variously sculptured, and supporting long and elegant galleries, while the parapets are adorned with statuary; on a nearer approach they represent every form of elegant ruins: columns, some with pedestals and capitals entire, others mutilated and prostrate, and some rising pyramidally over each other till they terminate in a sharp point. These are varied by niches, alcoves, and the customary appearances of desolated magnificence; the illusion is increased by the number of martins who have built their globular nests in the niches and hover over these columns; as in our country, they are accustomed to frequent large stone structures. As we advance, there seems no end to the visionary enchantment which surrounds us. In the midst of this fantastic scenery are vast ranges of walls, which seem the productions of art, so regular is the workmanship; they rise perpendicularly from the river, sometimes to the height of one hundred feet, varying in thickness from one to twelve feet, being equally broad at the top as below. The stones of which they are formed are black, thick, and durable, and composed of a large portion of earth, intermixed and cemented with a small quantity of sand, and a considerable proportion of talk or quartz. These stones are almost invariably regular parallelipeds, of unequal sizes in the wall, but equally deep, and laid regularly in ranges over each other like bricks, each breaking and covering the interstice of the two on which it rests; but though the perpendicular interstice be destroyed, the horizontal one extends entirely through the whole work; the stones, too, are proportioned to the thickness of the wall in which they are employed, being largest in the thickest walls. The thinner walls are composed of a single depth of the paralleliped, while the thicker ones consist of two or more depths; these walls pass the river at several places, rising from the water's edge much above the sandstone bluffs, which they seem to penetrate; thence they cross, in a straight line on either side of the river, the plains, over which they tower to the height of from ten to seventy feet, until they lose themselves in the second range of hills; sometimes they run parallel in several ranges near to each other, sometimes intersect each other at right angles, and have the appearance of walls of ancient houses or gardens.

The face of some of these river hills is composed of very excellent freestone of a light yellowish brown colour, and among the cliffs we found a species of pine which we had not yet seen, and differing from the Virginia pitchpine in having a shorter leaf and a longer and more pointed cone. The coal appears only in small quantities, as do the burnt earth and pumicestone; the mineral salts have abated. Among the animals are a great number of the bighorn, a few buffaloe and elk, and some mule-deer, but none of the common deer, nor any antelopes. We saw, but could not procure, a beautiful fox, of a colour varied with orange, yellow, white, and black, rather smaller than the common fox of this country, and about the same size as the red fox of the United States.

The river to-day has been from about one hundred and fifty to two hundred and fifty yards wide, with but little timber. At the distance of two miles and a half from the last stone wall is a stream on the north side, twenty-eight yards in width, and with some running water. We encamped just above its mouth, having made eighteen miles.

Saturday, June 1. The weather was cloudy, with a few drops of rain. As we proceeded, by the aid of our cord, we found the river cliffs and bluffs not so high as yesterday, and the country more level. The timber, too, is in greater abundance on the river, though there is no wood on the high ground; coal, however, appears in the bluffs. The river is from two hundred to two hundred and fifty feet wide, the current more gentle, the water becoming still clearer, and fewer rocky points and shoals than we met yesterday, though those which we did encounter were equally difficult to pass. Game is by no means in such plenty as below; all that we obtained were one bighorn and a mule-deer, though we saw in the plains a quantity of buffaloe, particularly near a small lake about eight miles from the river to the south. Notwithstanding the wind was ahead all day, we dragged the canoes along the distance of twenty-three miles. At fourteen and a quarter miles we came to a small island opposite a bend of the river to the north; two and a half miles, to the upper point of a small island on the north; five miles, to another island on the south side and opposite to a bluff. In the next two miles we passed an island on the south, a second beyond it on the north, and reached, near a high bluff on the north, a third, on which we encamped. In the plains near the river are the chokecherry, yellow and red currant bushes, as well as the wild rose and prickly pear, both of which are now in bloom. From the tops of the river hills, which are lower than usual, we enjoyed a delightful view of the rich fertile plains on both sides, in many places extending from the river cliffs to a great distance back. In these plains we meet occasionally large banks of pure sand, which were driven apparently by the southwest winds, and there deposited. The plains are more fertile some distance from the river than near its banks, where the surface of the earth is very generally strewed with small pebbles, which appear to be smoothed and worn by the agitation of the waters with which they were no doubt once covered. A mountain, or part of the North mountain, approaches the river within eight or ten miles, bearing north from our encampment of last evening; and this morning a range of high mountains, bearing S. W. from us and apparently running to the westward, are seen at a great distance, covered with snow. In the evening we had a little more rain.

Sunday, 2d. The wind blew violently last night, and a slight shower of rain fell, but this morning was fair. We set out at an early hour, and, although the wind was ahead, by means of the cord went on much better than for the last two days, as the banks were well calculated for towing. The current of the river is strong but regular, its timber increases in quantity, the low grounds become more level and extensive, and the bluffs on the river are lower than usual. In the course of the day we had a small shower of rain, which lasted a few minutes only. As the game is very abundant we think it necessary to begin a collection of hides for the purpose of making a leathern boat, which we intend constructing shortly, The hunters, who were out the greater part of the day, brought in six elk, two buffaloe, two mule-deer, and a bear. This last animal had nearly cost us the lives of two of our hunters, who were together when he attacked them; one of them narrowly escaped being caught, and the other, after running a considerable distance, concealed himself in some thick bushes, and while the bear was in quick pursuit of his hiding place, his companion came up and fortunately shot the animal through the head.

At six and a half miles we reached an island on the northern side; one mile and a quarter thence is a timbered low ground on the south; and in the next two and three-quarter miles we passed three small islands, and came to a dark bluff on the south; within the following mile are two small islands on the same side. At three and a quarter miles we reached the lower part of a much larger island near a northern point, and as we coasted along its side, within two miles passed a smaller island, and half a mile above reached the head of another. All these islands are small, and most of them contain some timber. Three-quarters of a mile beyond the last, and at the distance of eighteen miles from our encampment, we came to for the night in a handsome low cottonwood plain on the south, where we remained for the purpose of making some celestial observations during the night, and of examining in the morning a large river which comes in opposite to us. Accordingly, at an early hour,

Monday, 3d, we crossed, and fixed our camp in the point formed by the junction of the river with the Missouri. It now became an interesting question which of these two streams is what the Minnetarees call Ahmateahza or the Missouri, which they described as approaching very near to the Columbia. On our right decision much of the fate of the expedition depends; since if, after ascending to the Rocky mountains or beyond them, we should find that the river we were following did not come near the Columbia, and be obliged to return, we should not only lose the travelling season, two months of which had already elapsed, but probably dishearten the men so much as to induce them either to abandon the enterprise, or yield us a cold obedience instead of the warm and zealous support which they had hitherto afforded us. We determined, therefore, to examine well before we decided on our future course; and for this purpose despatched two canoes with three men up each of the streams, with orders to ascertain the width, depth, and rapidity of the current, so as to judge of their comparative bodies of water. At the same time parties were sent out by land to penetrate the country, and discover from the rising grounds, if possible, the distant bearings of the two rivers; and all were directed to return towards evening. While they were gone we ascended together the high grounds in the fork of these two rivers, whence we had a very extensive prospect of the surrounding country. On every side it was spread into one vast plain covered with verdure, in which innumerable herds of buffaloe were roaming, attended by their enemies the wolves; some flocks of elk also were seen, and the solitary antelopes were scattered with their young over the face of the plain. To the south was a range of lofty mountains, which we supposed to be a continuation of the South mountain, stretching themselves from southeast to northwest, and terminating abruptly about southwest from us. These were partially covered with snow; but at a great distance behind them was a more lofty ridge completely covered with snow, which seemed to follow the same direction as the first, reaching from west to the north of northwest, where their snowy tops were blended with the horizon. The direction of the rivers could not, however, be long distinguished, as they were soon lost in the extent of the plain. On our return we continued our examination; the width of the north branch is two hundred yards, that of the south is three hundred and seventy-two. The north, although narrower and with a gentler current, is deeper than the south; its waters, too, are of the same whitish brown colour, thickness, and turbidness; they run in the same boiling and rolling manner which has uniformly characterized the Missouri; and its bed is composed of some gravel, but principally mud. The south fork is deeper, but its waters are perfectly transparent; its current is rapid, but the surface smooth and unruffled; and its bed, too, is composed of round and flat smooth stones like those of rivers issuing from a mountainous country. The air and character of the north fork so much resemble those of the Missouri that almost all the party believe that to be the true course to be pursued. We, however, although we have given no decided opinion, are inclined to think otherwise, because, although this branch does give the colour and character to the Missouri, yet these very circumstances induce an opinion that it rises in and runs through an open plain country, since if it came from the mountains it would be clearer, unless, which from the position of the country is improbable, it passed through a vast extent of low ground after leaving them. We thought it probable that it did not even penetrate the Rocky mountains, but drew its sources from the open country towards the lower and middle parts of the Saskashawan, in a direction north of this place. What embarrasses us most is, that the Indians, who appeared to be well acquainted with the geography of the country, have not mentioned this northern river; for "the river which scolds at all others," as it is termed, must be, according to their account, one of the rivers which we have passed; and if this north fork be the Missouri, why have they not designated the south branch, which they must also have passed, in order to reach the great falls which they mention on the Missouri. In the evening our parties returned, after ascending the rivers in canoes for some distance, then continuing on foot, just leaving themselves time to return by night. The north fork was less rapid, and therefore afforded the easiest navigation; the shallowest water of the north was five feet deep, that of the south six feet. At two and a half miles up the north fork is a small river coming in on the left or western side, sixty feet wide, with a bold current three feet in depth. The party by land had gone up the south fork in a straight line, somewhat north of west, for seven miles, where they discovered that this little river came within one hundred yards of the south fork, and on returning down it found it a handsome stream, with as much timber as either of the larger rivers, consisting of the narrow and wide-leafed cottonwood, some birch and box-alder, and undergrowth of willows, rosebushes, and currants; they also saw on this river a great number of elk and some beaver.

All these accounts were, however, very far from deciding the important question of our future route, and we therefore determined each of us to ascend one of the rivers during a day and a half's march, or farther, if necessary for our satisfaction. Our hunters killed two buffaloe, six elk, and four deer to-day. Along the plains near the junction, are to be found the prickly pear in great quantities; the chokecherry is also very abundant in the river low grounds, as well as the ravines along the river bluffs; the yellow and red currants are not yet ripe; the gooseberry is beginning to ripen, and the

## UP THE MISSOURI 263

wild rose, which now covers all the low grounds near the rivers, is in full bloom. The fatigues of the last few days have occasioned some falling off in the appearance of the men, who, not having been able to wear moccasins, had their feet much bruised and mangled in passing over the stones and rough ground. They are, however, perfectly cheerful, and have an undiminished ardour for the expedition.

Tuesday, June 4. At the same hour this morning Captain Lewis and Captain Clark set out to explore the two rivers; Captain Lewis with six men crossed the north fork near the camp, below a small island, from which he took a course N. 30° W. for four and a half miles to a commanding eminence. Here we observed that the North mountain, changing its direction parallel to the Missouri, turned towards the north and terminated abruptly at the distance of about thirty miles, the point of termination bearing N. 48° E. The South mountain, too, diverges to the south, and terminates abruptly, its extremity bearing S. 8° W. distant about twenty miles; to the right of, and retreating from this extremity, is a separate mountain at the distance of thirty-five miles in a direction S. 38° W., which, from its resemblance to the roof of a barn, we called the Barn mountain. The north fork, which is now on the left, makes a considerable bend to the northwest, and on its western border a range of hills about ten miles long, and bearing from this spot N. 60° W. runs parallel with it; north of this range of hills is an elevated point of the river bluff on its south side, bearing N. 72° W. about twelve miles from us; towards this he directed his course across a high, level, dry, open plain, which in fact embraces the whole country to the foot of the mountains. The soil is dark, rich, and fertile, yet the grass by no means so luxuriant as might have been expected, for it is short, and scarcely more than sufficient to cover the ground. There are vast quantities of prickly pears, and myraids of grasshoppers, which afford food for a species of curlew which is in great numbers in the plain. He then proceeded up the river to the point of observation they had fixed on; from which he went two miles N. 15° W. to a bluff point on the north side of the river; thence his course was N. 30° W. for two miles to the entrance of a large creek on the south. The part of the river along which he passed is from forty to sixty yards wide, the current strong, the water deep and turbid, the banks falling in; the salts, coal, and mineral appearances are as usual, and in every respect, except as to size, this river resembles the Missouri. The low grounds are narrow, but well supplied with wood; the bluffs are principally of dark brown yellow, and some white clay, with freestone in some places. From this point the river bore N. 20° E. to a bluff on the south, at the distance of twelve miles; towards this he directed his course, ascending the hills, which are about two hundred feet high, and passing through plains for three miles, till he found the dry ravines so steep and numerous that he resolved to return to the river and follow its banks. He reached it about four miles from the beginning of his course, and encamped on the north in a bend, among some bushes which sheltered the party from the wind; the air was very cold, the northwest wind high, and the rain wet them to the skin. Besides the game just mentioned, he observed buffaloe, elk, wolves, foxes, and we got a blaireau and a weasel, and wounded a large brown bear, whom it was too late to pursue. Along the river are immense quantities of roses, which are now in full bloom, and which make the low grounds a perfect garden.

Wednesday, 5th. The rain fell during the greater part of the last night, and in the morning the weather was cloudy and cold, with a high northwest wind. At sunrise he proceeded up the river eight miles to the bluff on the left side, towards which he had been directing his course yesterday. Here he found the bed of a creek twenty-five yards wide at the entrance, with some timber, but no water, notwithstanding the rain; it is, indeed, astonishing to observe the vast quantities of water absorbed by the soil of the plains, which, being opened in large crevices, presents a fine rich loam; at the mouth of this stream (which he called Lark creek) the bluffs are very steep and approach the river, so that he ascended them, and crossing the plains reached the river, which from the last point bore N. 50° W.; four miles from this place it extended north two miles. Here he discovered a lofty mountain standing alone, at the distance of more than eighty miles in the direction of N. 30° W., and which, from its conical figure, he called Tower mountain. He then proceeded on these two hills, and afterwards in different courses six miles, when he again changed for a western course across a deep bend along the south side; in making this passage over the plains he found them like those of yesterday, level and beautiful, with great quantities of buffaloe, and some wolves, foxes, and antelopes, and intersected near the river by deep ravines. Here, at the distance of from one to nine miles from the river, he met the largest village of barking squirrels which we had yet seen, for he passed a skirt of their territory for seven miles. He also saw near the hills a flock of the mountain cock, or a large species of heath hen, with a long pointed tail, which the Indians below had informed us were common among the Rock mountains. Having finished his course of ten miles west across a bend, he continued two miles N. 80° W., and from that point discovered some lofty mountains to the northwest of Tower mountain and bearing N. 65° W. at eighty or one hundred miles distance; here he encamped on the north side in a handsome low ground, on which were several old stick lodges. There had been but little timber on the river in the fore part of the day, but now there is a greater quantity than usual. The river itself is about eighty yards wide, from six to ten feet deep, and has a strong steady current. The party had killed five elk and a muledeer; and, by way of experiment, roasted the burrowing squirrels, which they found to be well-flavoured and tender.

Thursday, 6th. Captain Lewis was now convinced that this river pursued a direction too far north for our route to the Pacific, and therefore resolved to return, but waited till noon to take a meridian altitude. The clouds, however, which had gathered during the latter part of the night, continued and prevented the observation; part of the men were sent forward to a commanding eminence, six miles S. 70° W., from which they saw, at the distance of about fifteen miles S. 80° W., a point of the south bluff of the river, which thence bore northwardly. In their absence two rafts had been prepared, and when they returned, about noon, the party embarked; but they soon found that the rafts were so small and slender that the baggage was wet, and therefore it was necessary to abandon them and go by land. They therefore crossed the plains, and at the distance of twelve miles came to the river, through a cold storm from the northeast, accompanied by showers of rain. The abruptness of the cliffs compelled them, after going a few miles, to leave the river and meet the storm in the plains. Here they directed their course too far northward, in consequence of which they did not meet the river till late at night, after having travelled twenty-three miles since noon, and halted at a little below the entrance of Lark creek. They had the good fortune to kill two buffaloe, which supplied them with supper; but spent a very uncomfortable night, without any shelter from the rain, which continued till morning,

Friday, 7th, when, at an early hour, they continued down

the river. The route was extremely unpleasant, as the wind was high from the N. E. accompanied with rain, which made the ground so slippery that they were unable to walk over the bluffs which they had passed on ascending the river. The land is the most thirsty we have ever seen; notwithstanding all the rain which has fallen, the earth is not wet for more than two inches deep, and resembles thawed ground; but if it requires more water to saturate it than the common soils, on the other hand it yields its moisture with equal difficulty. In passing along the side of one of these bluffs, at a narrow pass thirty yards in length, Captain Lewis slipped, and but for a fortunate recovery, by means of his espontoon, would have been precipitated into the river over a precipice of about ninety feet. He had just reached a spot where, by the assistance of his espontoon, he could stand with tolerable safety, when he heard a voice behind him cry out, "Good God! Captain, what shall I do?" He turned instantly and found it was Windsor, who had lost his foothold about the middle of the narrow pass, and had slipped down to the very verge of the precipice, where he lay on his belly, with his right arm and leg over the precipice, while with the other leg and arm he was with difficulty holding on to keep himself from being dashed to pieces below. His dreadful situation was instantly perceived by Captain Lewis, who, stifling his alarm, calmly told him that he was in no danger; that he should take his knife out of his belt with the right hand, and dig a hole in the side of the bluff to receive his right foot. With great presence of mind he did this, and then raised himself on his knees. Captain Lewis then told him to take off his moccasins and come forward on his hands and knees, holding the knife in one hand and his rifle in the other. He immediately crawled in this way till he came to a secure spot. The men who had not attempted this passage were ordered to return and wade the river at the foot of the bluff, where they found the water breast-high. This adventure taught them the danger of crossing the slippery heights of the river; but as the plains were intersected by deep ravines almost as difficult to pass, they continued down the river, sometimes in the mud of the low grounds, sometimes up to their arms in the water, and when it became too deep to wade, they cut footholds with their knives in the sides of the banks. In this way they travelled through the rain, mud, and water, and having made only eighteen miles during the whole day, encamped in an old Indian lodge of sticks, which afforded them a dry shelter. Here they cooked part of six deer they had killed in the course of their walk, and having eaten the only morsel they had tasted during the whole day, slept comfortably on some willow boughs.

## CHAPTER X.

Return of Captain Lewis—Account of Captain Clark's researches with his exploring party—Perilous situation of one of his party—Tansy river described—The party still believing the southern fork the Missouri, Captain Lewis resolves to ascend it—Mode of making a place to deposit provisions, called *cache*—Captain Lewis explores the southern fork—Falls of the Missouri discovered, which ascertains the question—Romantic scenery of the surrounding country—Narrow escape of Captain Lewis—The main body, under Captain Clark, approach within five miles of the falls, and prepare for making a portage over the rapids.

SATURDAY, 8th. It continued to rain moderately all last night, and the morning was cloudy till about ten o'clock, when it cleared off, and became a fine day. They breakfasted about sunrise, and then proceeded down the river in the same way as they had done yesterday, except that the travelling was somewhat better, as they had not so often to wade, though they passed some very dangerous bluffs. The only timber to be found is in the low grounds which are occasionally on the river, and these are the haunts of innumerable birds, who, when the sun began to shine, sang very delightfully. Among these birds they distinguished the brown thrush, robin, turtledove, linnet, goldfinch, the large and small blackbird, the wren, and some others. As they came along, the whole of the party were of opinion that this river was the true Missouri; but Captain Lewis, being fully persuaded that it was neither the main stream nor that which it would be advisable to ascend, gave it the name of Maria's river. After travelling all day they reached the camp at five o'clock in the afternoon, and found Captain Clark and the party very anxious for their safety, as they had staid two days longer than had been expected, and as Captain Clark had returned at the appointed time, it was feared that they had met with some accident.

Captain Clark, on setting out with five men on the 4th, went seven miles on a course S. 25° W. to a spring; thence he went S. 20° W. for eight miles to the river, where was an island, from which he proceeded in a course N. 45° W. and approached the river at the distance of three, five, and thirteen miles, at which place they encamped in an old Indian lodge made of sticks and bark. In crossing the plains they observed several herds of buffaloe, some mule-deer, antelopes, and wolves. The river is rapid and closely hemmed in by high bluffs, crowded with bars of gravel, with little timber on the low grounds and none on the highlands. Near the camp this evening a white bear attacked one of the men, whose gun, happening to be wet, would not go off; he instantly made towards a tree, but was so closely pursued that as he ascended the tree he struck the bear with his foot. The bear, not being able to climb, waited till he should be forced to come down; and as the rest of the party were separated from him by a perpendicular cliff of rocks, which they could not descend, it was not in their power to give him any assistance; fortunately, however, at last the bear became frighted at their cries and firing, and released the man. In the afternoon it rained, and during the night there fell both rain and snow, and in the morning,

June 5, the hills to the S. E. were covered with snow, and the rain continued. They proceeded on in a course N. 20° W. near the river several miles, till at the distance of eleven miles they reached a ridge, from the top of which, on the north side, they could plainly discern a mountain to the S. and W. at a great distance, covered with snow; a high ridge projecting from the mountains to the southeast approaches the river on the southeast side, forming some cliffs of dark hard stone. They also saw that the river ran for a great distance west of south, with a rapid current, from which, as well as its continuing of the same width and depth, Captain Clark thought it useless to advance any farther, and therefore returned across the level plain in a direction north 30° east, and reached, at the distance of twenty miles, the little river which is already mentioned as falling into the north fork, and to which they gave the name of Tansy river, from the great quantity of that herb growing on its banks. Here they dined, and then proceeded on a few miles by a place where the Tansy breaks through a high ridge on its north side, and encamped.

The next day, 6th, the weather was cold, raw, and cloudy, with a high northeast wind. They set out early, down the Tansy, whose low grounds resemble precisely, except as to extent, those of the Missouri before it branches, containing a great proportion of a species of cottonwood, with a leaf like that of the wild cherry. After halting at twelve o'clock for dinner, they ascended the plain, and at five o'clock reached the camp through the rain, which had fallen without intermission since noon. During his absence the party had been occupied in dressing skins, and being able to rest themselves were nearly freed from their lameness and swollen feet. All this night and the whole of the following day, 7th, it rained, the wind being from the southwest off the mountains; yet the rivers are falling, and the thermometer 40° above zero. The rain continued till the next day, 8th, at ten o'clock, when it cleared off and the weather became fine, the wind high from the southwest. The rivers at the point have now fallen six inches since our arrival, and this morning the water of the south fork became of a reddish brown colour, while the north branch continued of its usual whitish appearance. The mountains to the south are covered with snow.

Sunday, 9th. We now consulted upon the course to be pursued. On comparing our observations we were more than ever convinced of what we already suspected, that Mr. Arrowsmith is incorrect in laying down in the chain of Rocky mountains one remarkable mountain, called the Tooth, nearly as far south as 45°, and said to be so marked from the discoveries of a Mr. Fidler. We are now within one hundred miles of the Rocky mountains and in the latitude of 47° 24′ 12″ 8, and therefore it is highly improbable that the Missouri should make such a bend to the south before it reaches the Rocky mountains as to have suffered Mr. Fidler to come as low as 45° along the eastern borders without touching that river; yet the general course of Maria's river from this place for fifty-nine miles, as far as Captain Lewis ascended, was north 69° west, and the south branch, or what we consider the Missouri, which Captain Clark had examined as far as forty-five miles in a straight line, ran in a course south 29° west, and as far as it could be seen went considerably west of south, whence we conclude that the Missouri itself enters the Rocky mountains to the north of 45°. In writing to the President from our winter quarters, we had already taken the liberty of advancing the southern extremity of Mr. Fidler's discoveries about a degree to the northward, and this from Indian information as to the bearing of the point at which the Missouri enters the mountain; but we think actual observation will place it one degree still farther to the northward. This information of Mr. Fidler, however, incorrect as it is, affords an additional reason for not pursuing Maria's river; for if he came as low even as 47° and saw only small streams coming down from the mountains, it is to be presumed that these rivulets do not penetrate the Rocky mountains so far as to approach any navigable branch of the Columbia, and they are most probably the remote waters of some northern branch of the Missouri. In short, being already in latitude 47° 24′ we cannot reasonably hope, by going farther to the northward, to find between this place and the Saskashawan any stream which can, as the Indians assure us the Missouri does, possess a navigable current for some distance in the Rocky mountains; the Indians had assured us, also, that the water of the Missouri was nearly transparent at the falls; this is the case with the southern branch; that the falls lay a little to the south of sunset from them; this, too, is in favour of the southern fork, for it bears considerably south of this place, which is only a few minutes to the northward of fort Mandan; that the falls are below the Rocky mountains, and near the northern termination of one range of those mountains; now, there is a ridge of mountains which appear behind the South mountains and terminates to the southwest of us, at a sufficient distance from the unbroken chain of the Rocky mountains to allow space for several falls, indeed, we fear for too many of them. If, too, the Indians had ever passed any stream as large as this southern fork on their way up the Missouri, they would have mentioned it; so that their silence seems to prove that this branch must be the Missouri. The body of water, also, which it discharges must have been acquired from a considerable distance in the mountains, for it could not have been collected in the parched plains between the Yellowstone and the Rocky mountains, since that country could not supply nourishment for the dry channels which we passed on the south, and the travels of Mr. Fidler forbid us to believe that it could have been obtained from the mountains towards the northwest.

These observations, which satisfied our minds completely,

we communicated to the party; but every one of them were of a contrary opinion; and much of their belief depended on Crusatte, an experienced waterman on the Missouri, who gave it as his decided judgment that the north fork was the genuine Missouri. The men, therefore, mentioned that although they would most cheerfully follow us wherever we should direct, yet they were afraid that the south fork would soon terminate in the Rocky mountains and leave us at a great distance from the Columbia. In order that nothing might be omitted which could prevent our falling into an error, it was agreed that one of us should ascend the southern branch by land until we reached either the falls or the mountains. In the meantime, in order to lighten our burdens as much as possible, we determined to deposit here one of the periogues and all the heavy baggage which we could possibly spare, as well as some provision, salt, powder, and tools; this would at once lighten the other boats, and give them the crew which had been employed on board the periogue.

Monday, 10th. The weather being fair and pleasant, we dried all our baggage and merchandize and made our deposit. These holes, or *caches* as they are called by the Missouri traders, are very common, particularly among those who deal with the Sioux, as the skins and merchandize will keep perfectly sound for years, and are protected from robbery. Our cache is built in this manner: In the high plain on the north side of the Missouri, and forty yards from a steep bluff, we chose a dry situation, and then describing a small circle of about twenty inches diameter, removed the sod as gently and carefully as possible; the hole is then sunk perpendicularly for a foot deep, or more if the ground be not firm. It is now worked gradually wider as they descend, till at length it becomes six or seven feet deep, shaped nearly like a kettle or the lower part of a large still, with the bottom somewhat

## UP THE MISSOURI 275

sunk at the centre. As the earth is dug it is handed up in a vessel and carefully laid on a skin or cloth, in which it is carried away and usually thrown into the river or concealed so as to leave no trace of it. A floor of three or four inches in thickness is then made of dry sticks, on which is thrown hay or a hide perfectly dry. The goods, being well aired and dried, are laid on this floor, and prevented from touching the wall by other dried sticks in proportion as the merchandize is stowed away; when the hole is nearly full, a skin is laid over the goods, and on this earth is thrown and beaten down until, with the addition of the sod first removed, the whole is on a level with the ground, and there remains not the slightest appearance of an excavation. In addition to this we made another, of smaller dimensions, in which we placed all the baggage, some powder, and our blacksmith's tools, having previously repaired such of the tools we carry with us as require mending. To guard against accident, we hid two parcels of lead and powder in the two distinct places. The red periogue was drawn up on the middle of a small island at the entrance of Maria's river, and secured by being fastened to the trees from the effect of any floods. In the evening there was a high wind from the southwest, accompanied with thunder and rain. We now made another observation of the meridian altitude of the sun, and found that the mean latitude of the entrance of Maria's river, as deduced from three observations, is 47° 25′ 17″ 2 north. We saw a small bird like the blue thrush, or catbird, which we had not before met, and also observed that the beemartin, or kingbird, is common to this country although there are no bees here, and in fact we have not met with the honey-bee since leaving the Osage river.

Tuesday, 11th. This morning Captain Lewis, with four men, set out on their expedition up the south branch. They soon reached the point where the Tansy river approaches the Missouri, and observing a large herd of elk before them, descended and killed several, which they hung up along the river so that the party in the boats might see them as they came along. They then halted for dinner; but Captain Lewis, who had been for some days afflicted with the dysentery, was now attacked with violent pains attended by a high fever, and was unable to go on. He therefore encamped for the night under some willow boughs; having brought no medicine, he determined to try an experiment with the small twigs of the chokecherry, which, being stripped of their leaves and cut into pieces about two inches long, were boiled in pure water, till they produced a strong black decoction of an astringent bitter taste; a pint of this he took at sunset, and repeated the dose an hour afterwards. By ten o'clock he was perfectly relieved from pain, a gentle perspiration ensued, his fever abated, and in the morning he was quite recovered. One of the men caught several dozen fish of two species: the first is about nine inches long, of a white colour, round in shape; the mouth is beset both above and below with a rim of fine sharp teeth, the eye moderately large, the pupil dark, and the iris narrow and of a yellowish brown colour; in form and size it resembles the white chub of the Potomac, though its head is proportionably smaller; they readily bite at meat or grasshoppers; but the flesh, though soft and of a fine white colour, is not highly flavoured. The second species is precisely of the form and about the size of the fish known by the name of the hickory shad or old wife, though it differs from it in having the outer edge of both the upper and lower jaw set with a rim of teeth, and the tongue and palate also are defended by long sharp teeth bending inwards; the eye is very large, the iris wide and of a silvery colour; they do not inhabit muddy water, and the flavour is much superior to

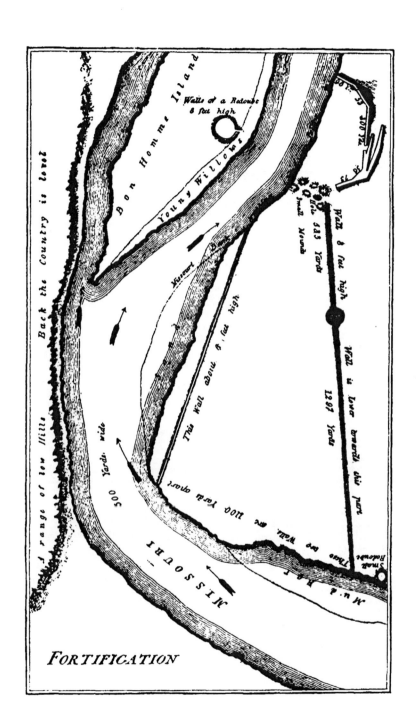

that of the former species. Of the first kind we had seen a few before we reached Maria's river; but had found none of the last before we caught them in the Missouri above its junction with that river. The white cat continues as high as Maria's river, but they are scarce in this part of the river, nor have we caught any of them since leaving the Mandans which weighed more than six pounds.

Of other game they saw a great abundance, even in their short march of nine miles.

Wednesday, 12th. This morning Captain Lewis left the bank of the river, in order to avoid the steep ravines which generally run from the shore to the distance of one or two miles in the plain; having reached the opened country, he went for twelve miles in a course a little to the west of southwest, when, the sun becoming warm by nine o'clock, he returned to the river in quest of water and to kill something for breakfast, there being no water in the plain, and the buffaloe, discovering them before they came within gunshot, took to flight. They reached the banks, in a handsome open low ground with cottonwood, after three miles walk. Here they saw two large brown bears, and killed them both at the first fire, a circumstance which has never before occurred since we have seen that animal. Having made a meal of a part and hung the remainder on a tree, with a note for Captain Clark, they again ascended the bluffs into the open plains. Here they saw great numbers of the burrowing squirrel, also some wolves, antelopes, mule-deer, and vast herds of buffaloe. They soon crossed a ridge considerably higher than the surrounding plains, and from its top had a beautiful view of the Rocky mountains, which are now completely covered with snow; their general course is from southeast to the north of northwest, and they seem to consist of several ranges which successively rise above each other till the most distant mingles with the clouds. After travelling twelve miles they again met the river, where there was a handsome plain of cottonwood; and although it was not sunset, and they had only come twenty-seven miles, yet Captain Lewis felt weak from his late disorder, and therefore determined to go no farther that night. In the course of the day they killed a quantity of game, and saw some signs of otter, as well as beaver, and many tracks of the brown bear; they also caught great quantities of the white fish mentioned yesterday. With the broad-leafed cottonwood, which has formed the principal timber of the Missouri, is here mixed another species, differing from the first only in the narrowness of its leaf and the greater thickness of its bark. The leaf is long, oval, acutely pointed, about two and a half or three inches long and from three-quarters of an inch to an inch in width; it is smooth and thick, sometimes slightly grooved or channeled, with the margin a little serrate, the upper disk of a common, the lower of a whitish, green. This species seems to be preferred by the beaver to the broad-leaved, probably because the former affords a deeper and softer bark.

Thursday, 13th. They left their encampment at sunrise, and ascending the river hills went for six miles in a course generally southwest, over a country which, though more waving than that of yesterday, may still be considered level. At the extremity of this course they overlooked a most beautiful plain, where were infinitely more buffaloe than we had ever before seen at a single view. To the southwest arose from the plain two mountains of a singular appearance, and more like ramparts of high fortifications than works of nature. They are square figures with sides rising perpendicularly to the height of two hundred and fifty feet, formed of yellow clay, and the tops seemed to be level plains. Finding that the river here bore considerably to the south, and fearful of

## UP THE MISSOURI 279

passing the falls before reaching the Rocky mountains, they now changed their course to the south, and leaving those insulated hills to the right proceeded across the plain. In this direction Captain Lewis had gone about two miles when his ears were saluted with the agreeable sound of a fall of water, and as he advanced a spray, which seemed driven by the high southwest wind, arose above the plain like a column of smoke and vanished in an instant. Towards this point he directed his steps, and the noise, increasing as he approached, soon became too tremendous to be mistaken for anything but the great falls of the Missouri. Having travelled seven miles after first hearing the sound, he reached the falls about twelve o'clock; the hills as he approached were difficult of access and two hundred feet high; down these he hurried with impatience, and seating himself on some rocks under the centre of the falls, enjoyed the sublime spectacle of this stupendous object which since the creation had been lavishing its magnificence upon the desert, unknown to civilization.

The river immediately at its cascade is three hundred yards wide, and is pressed in by a perpendicular cliff on the left, which rises to about one hundred feet and extends up the stream for a mile; on the right the bluff is also perpendicular for three hundred yards above the falls. For ninety or a hundred yards from the left cliff, the water falls in one smooth even sheet, over a precipice of at least eighty feet. The remaining part of the river precipitates itself with a more rapid current, but being received as it falls by the irregular and somewhat projecting rocks below, forms a splendid prospect of perfectly white foam, two hundred yards in length and eighty in perpendicular elevation. This spray is dissipated into a thousand shapes, sometimes flying up in columns of fifteen or twenty feet, which are then oppressed by larger masses of the white foam, on all which the sun impresses the brightest colours of the rainbow. As it rises from the fall, it beats with fury against a ledge of rocks which extend across the river at one hundred and fifty yards from the precipice. From the perpendicular cliff on the north, to the distance of one hundred and twenty yards, the rocks rise only a few feet above the water, and when the river is high the stream finds a channel across them forty yards wide, and near the higher parts of the ledge, which then rise about twenty feet, and terminate abruptly within eighty or ninety yards of the southern side. Between them and the perpendicular cliff on the south, the whole body of water runs with great swiftness. A few small cedars grow near this ridge of rocks, which serves as a barrier to defend a small plain of about three acres, shaded with cottonwood, at the lower extremity of which is a grove of the same tree, where are several Indian cabins of sticks; below the point of them the river is divided by a large rock, several feet above the surface of the water, and extending down the stream for twenty yards. At the distance of three hundred yards from the same ridge is a second abutment, of solid perpendicular rock about sixty feet high, projecting at right angles from the small plain on the north for one hundred and thirty-four yards into the river. After leaving this, the Missouri again spreads itself to its usual distance of three hundred yards, though with more than its ordinary rapidity.

The hunters who had been sent out now returned loaded with buffaloe meat, and Captain Lewis encamped for the night under a tree near the falls. The men were again despatched to hunt for food against the arrival of the party, and Captain Lewis walked down the river to discover if possible some place where the canoes might be safely drawn on shore, in order to be transported beyond the falls. He returned, however, without discovering any such spot, the river for three miles below being one continued succession of rapids and

## UP THE MISSOURI 281

cascades, overhung with perpendicular bluffs from one hundred and fifty to two hundred feet high; in short, it seems to have worn itself a channel through the solid rock. In the afternoon they caught in the falls some of both kinds of the white fish, and half a dozen trout from sixteen to twenty-three inches long, precisely resembling in form and the position of its fins the mountain or speckled trout of the United States, except that the specks of the former are of a deep black, while those of the latter are of a red or gold colour; they have long sharp teeth on the palate and tongue, and generally a small speck of red on each side behind the front ventral fins; the flesh is of a pale yellowish red, or when in good order of a rose-coloured red.

Friday, 14th. This morning one of the men was sent to Captain Clark with an account of the discovery of the falls, and after employing the rest in preserving the meat which had been killed yesterday, Captain Lewis proceeded to examine the rapids above. From the falls he directed his course southwest up the river; after passing one continued rapid and three small cascades, each three or four feet high, he reached, at the distance of five miles, a second fall. The river is about four hundred yards wide, and for the distance of three hundred throws itself over to the depth of nineteen feet, and so irregularly that he gave it the name of the Crooked falls. From the southern shore it extends obliquely upwards about one hundred and fifty yards, and then forms an acute angle downwards nearly to the commencement of four small islands close to the northern side. From the perpendicular pitch to these islands, a distance of more than one hundred yards, the water glides down a sloping rock with a velocity almost equal to that of its fall. Above this fall the river bends suddenly to the northward; while viewing this place Captain Lewis heard a loud roar above him, and crossing the point of a hill for a few hundred yards, he saw one of the most beautiful objects in nature: the whole Missouri is suddenly stopped by one shelving rock, which, without a single niche, and with an edge as straight and regular as if formed by art, stretches itself from one side of the river to the other for at least a quarter of a mile. Over this it precipitates itself in an even uninterrupted sheet to the perpendicular depth of fifty feet, whence, dashing against the rocky bottom, it rushes rapidly down, leaving behind it a spray of the purest foam across the river. The scene which it presented was indeed singularly beautiful, since, without any of the wild irregular sublimity of the lower falls, it combined all the regular elegances which the fancy of a painter would select to form a beautiful waterfall. The eye had scarcely been regaled with this charming prospect, when, at the distance of half a mile, Captain Lewis observed another of a similar kind; to this he immediately hastened, and found a cascade stretching across the whole river for a quarter of a mile, with a descent of fourteen feet, though the perpendicular pitch was only six feet. This, too, in any other neighbourhood would have been an object of great magnificence, but after what he had just seen it became of secondary interest; his curiosity being, however, awakened, he determined to go on, even should night overtake him, to the head of the falls. He therefore pursued the southwest course of the river, which was one constant succession of rapids and small cascades, at every one of which the bluffs grew lower, or the bed of the river became more on a level with the plains. At the distance of two and a half miles he arrived at another cataract, of twenty-six feet. The river is here six hundred yards wide, but the descent is not immediately perpendicular, though the river falls generally with a regular and smooth sheet; for about one-third of the descent a rock protrudes to a small distance, receives the water in its passage and gives it a curve. On the south side is a beautiful plain a few feet above the level of the falls; on the north the country is more broken, and there is a hill not far from the river. Just below the falls is a little island in the middle of the river, well covered with timber. Here, on a cottonwood tree, an eagle had fixed its nest, and seemed the undisputed mistress of a spot to contest whose dominion neither man nor beast would venture across the gulfs that surround it, and which is further secured by the mist rising from the falls. This solitary bird could not escape the observation of the Indians, who made the eagle's nest a part of their description of the falls, which now proves to be correct in almost every particular, except that they did not do justice to their height. Just above this is a cascade of about five feet, beyond which, as far as could be discerned, the velocity of the water seemed to abate. Captain Lewis now ascended the hill which was behind him, and saw from its top a delightful plain extending from the river to the base of the Snow mountains to the south and southwest. Along this wide level country the Missouri pursued its winding course, filled with water to its even and grassy banks, while about four miles above it was joined by a large river flowing from the northwest through a valley three miles in width, and distinguished by the timber which adorned its shores; the Missouri itself stretches to the south in one unruffled stream of water as if unconscious of the roughness it must soon encounter, and bearing on its bosom vast flocks of geese, while numerous herds of buffaloe are feeding on the plains which surround it.

Captain Lewis then descended the hill, and directed his course towards the river falling in from the west. He soon met a herd of at least a thousand buffaloe, and being desirous of providing for supper, shot one of them; the animal immediately began to bleed, and Captain Lewis, who had forgotten to reload his rifle, was intently watching to see him fall, when he beheld a large brown bear who was stealing on him unperceived, and was already within twenty steps. In the first moment of surprise he lifted his rifle, but remembering instantly that it was not charged, and that he had not time to reload, he felt that there was no safety but in flight. It was in the open level plain, not a bush nor a tree within three hundred yards, the bank of the river sloping and not more than three feet high, so that there was no possible mode of concealment. Captain Lewis, therefore, thought of retreating in a quick walk, as fast as the bear advanced, towards the nearest tree; but as soon as he turned the bear ran, open mouth and at full speed, upon him. Captain Lewis ran about eighty yards, but finding that the animal gained on him fast, it flashed on his mind that by getting into the water to such a depth that the bear would be obliged to attack him swimming, there was still some chance of his life; he, therefore, turned short, plunged into the river about waist deep, and facing about, presented the point of his espontoon. The bear arrived at the water's edge within twenty feet of him, but as soon as he put himself in this posture of defence he seemed frightened, and wheeling about, retreated with as much precipitation as he had pursued. Very glad to be released from this danger, Captain Lewis returned to the shore, and observed him run with great speed, sometimes looking back as if he expected to be pursued, till he reached the woods. He could not conceive the cause of the sudden alarm of the bear, but congratulated himself on his escape when he saw his own track torn to pieces by the furious animal, and learnt from the whole adventure never to suffer his rifle to be a moment unloaded. He now resumed his progress in the direction which the bear had taken towards the western river, and found it a handsome stream about two hundred yards wide, apparently deep, with a gentle current; its waters clear, and its banks, which were formed principally of dark brown and blue clay, are about the same height as those of the Missouri, that is, from three to five feet. What was singular was that the river does not seem to overflow its banks at any season, while it might be presumed, from its vicinity to the mountains, that the torrents arising from the melting of the snows would sometimes cause it to swell beyond its limits. The contrary fact would induce a belief that the Rocky mountains yield their snows very reluctantly and equally to the sun, and are not often drenched by very heavy rains. This river is no doubt that which the Indians call Medicine river, which they mentioned as emptying into the Missouri just above the falls. After examining Medicine river, Captain Lewis set out at half after six o'clock in the evening on his return towards the camp, which he estimated at the distance of twelve miles. In going through the low grounds on Medicine river he met an animal which at a distance he thought was a wolf, but on coming within sixty paces it proved to be some brownish yellow animal standing near its burrow, which, when he came nigh, crouched and seemed as if about to spring on him. Captain Lewis fired and the beast disappeared in its burrow. From the track and the general appearance of the animal he supposed it to be of the tiger kind. He then went on, but as if the beasts of the forests had conspired against him, three buffaloe bulls, which were feeding with a large herd at the distance of half a mile, left their companions and ran at full speed towards him. He turned round, and, unwilling to give up the field, advanced towards them; when they came within a hundred yards, they stopped, looked at him for some time, and then retreated as they came. He now pursued his route in the dark, reflecting on the strange adventures and sights of the day, which crowded on his mind so rapidly that he should have been inclined to believe it all enchantment, if the thorns of the prickly pear piercing his feet did not dispel at every moment the illusion. He at last reached the party, who had been very anxious for his safety, and who had already decided on the route which each should take in the morning to look for him. Being much fatigued, he supped and slept well during the night.

Saturday, 15th. The men were again sent out to bring in the game killed yesterday and to procure more; they also obtained a number of fine trout, and several small catfish weighing about four pounds, and differing from the white catfish lower down the Missouri. On awaking this morning Captain Lewis found a large rattlesnake coiled on the trunk of a tree under which he had been sleeping. He killed it, and found it like those we had seen before, differing from those of the Atlantic states, not in its colours but in the form and arrangement of them; it had one hundred and seventy-six scuta on the abdomen, and seventeen half-formed scuta on the tail. There is a heavy dew on the grass about the camp every morning, which no doubt proceeds from the mist of the falls, as it takes place nowhere in the plains nor on the river except here. The messenger sent to Captain Clark returned with information of his having arrived five miles below at a rapid, which he did not think it prudent to ascend, and would wait till Captain Lewis and his party rejoined him.

On Tuesday, 11th, the day when Captain Lewis left us, we remained at the entrance of Maria's river and completed the deposits of all the articles with which we could dispense. The morning had been fair, with a high wind from the southwest, which shifted in the evening to northwest, when the weather became cold and the wind high. The next morning,

Wednesday, 12th, we left our encampment, with a fair day and a southwest wind. The river was now so crowded with islands that within the distance of ten miles and a half we passed eleven, of different dimensions, before reaching a high black bluff in a bend on the left, where we saw a great number of swallows. Within one mile and a half farther we passed four small islands, two on each side, and at fifteen miles from our encampment reached a spring which the men called Grog spring; it is on the northern shore, and at the point where Tansy river approaches within one hundred yards of the Missouri. From this place we proceeded three miles to a low bluff on the north, opposite to an island, and spent the night in an old Indian encampment. The bluffs under which we passed were composed of a blackish clay and coal for about eighty feet, above which for thirty or forty feet is a brownish yellow earth. The river is very rapid, and obstructed by bars of gravel and stone, of different shapes and sizes, so that three of our canoes were in great danger in the course of the day. We had a few drops of rain about two o'clock in the afternoon. The only animals we killed were elk and deer, but we saw great numbers of rattlesnakes.

Thursday, 13th. The morning was fair, and there was some dew on the ground. After passing two islands, we reached, at the distance of a mile and a half, a small rapid stream fifty yards wide, emptying itself on the south, rising in a mountain to the southeast about twelve or fifteen miles distant, and at this time covered with snow. As it is the channel for the melted snow of that mountain we called it Snow river; opposite to its entrance is another island; at one mile and three-quarters is a black bluff of slate on the south, nine miles beyond which, after passing ten islands, we came to on the southern shore near an old Indian fortified camp, opposite the lower point of an island, having made thirteen miles. The number of islands and shoals, the rapidity of the river, and the quantity of large stones, rendered the navigation very disagreeable; along the banks we distinguished several low bluffs or cliffs of slate. There were great numbers of geese and goslings, the geese not being able to fly at this season. Gooseberries are ripe and in great abundance; the yellow currant is also common, but not yet ripe. Our game consisted of buffaloe and goats.

Friday, 14th. Again the day is fine. We made two miles, to a small island in the southern bend, after passing several bad rapids. The current becomes indeed swifter as we ascend, and the canoes frequently receive water, as we drag them with difficulty along. At the distance of six miles we reached Captain Clark's camp on the fourth, which is on the north side and opposite to a large gravelly bar. Here the man sent by Captain Lewis joined us, with the pleasing intelligence that he had discovered the falls, and was convinced that the course we were pursuing was that of the true Missouri. At a mile and a half we reached the upper point of an island, three-quarters of a mile beyond which we encamped on the south, after making only ten and a quarter miles. Along the river was but little timber, but much hard slate in the bluffs.

Saturday, 15th. The morning being warm and fair we set out at the usual hour, but proceeded with great difficulty in consequence of the increased rapidity of the current. The channel is constantly obstructed by rocks and dangerous rapids. During the whole progress the men are in the water hauling the canoes, and walking on sharp rocks and round stones which cut their feet or cause them to fall. The rattlesnakes, too, are so numerous that the men are constantly on their guard against being bitten by them; yet they bear the fatigues with the most undiminished cheerfulness. We hear the roar of the falls very distinctly this morning. At three and three-quarter miles we came to a rock, in a bend to the south, resembling a tower. At six and three-quarter miles we reached a large creek on the south, which, after one of our men, we called Shields's creek. It is rapid in its course, about thirty yards wide, and, on sending a person five miles up, it proved to have a fall of fifteen feet, and some timber on its low ground. Above this river the bluffs of the Missouri are of red earth mixed with stratas of black stone; below it we passed some white clay in the banks, which mixes with water in every respect like flour. At three and three-quarter miles we reached a point on the north, opposite an island and a bluff; and one mile and a quarter farther, after passing some red bluffs, came to on the north side, having made twelve miles. Here we found a rapid so difficult that we did not think proper to attempt the passage this evening, and therefore sent to Captain Lewis to apprise him of our arrival. We saw a number of geese, ducks, crows, and blackbirds to-day, the two former with their young. The river rose a little this evening, but the timber is still so scarce that we could not procure enough for our use during the night.

Sunday, June 16. Some rain fell last night, and this morning the weather was cloudy and the wind high from the southwest. We passed the rapid by doubly manning the periogue and canoes, and halted at the distance of a mile and a quarter to examine the rapids above, which we found to be a continued succession of cascades as far as the view extended, which was about two miles. About a mile above where we halted was a large creek falling in on the south, opposite to which is a large sulphur spring falling over the rocks on the north. Captain Lewis arrived at two from the falls about five miles above us, and after consulting upon the subject of the portage, we crossed the river and formed a camp on the north, having come three-quarters of a mile to-day. From our own observation we had deemed the south side to be the most favourable for a portage, but two men sent out for the purpose of examining it reported that the creek and the ravines intersected the plain so deeply that it was impossible to cross it. Captain Clark therefore resolved to examine more minutely what was the best route; the four canoes were unloaded at the camp and then sent across the river, where, by means of strong cords, they were hauled over the first rapid, whence they may be easily drawn into the creek. Finding, too, that the portage would be at all events too long to enable us to carry the boats on our shoulders, six men were set to work to make wheels for carriages to transport them. Since leaving Maria's river the wife of Chaboneau, our interpreter, has been dangerously ill, but she now found great relief from the mineral water of the sulphur spring. It is situated about two hundred yards from the Missouri, into which it empties over a precipice of rock about twenty-five feet high. The water is perfectly transparent, strongly impregnated with sulphur, and we suspect iron also, as the colour of the hills and bluffs in the neighbourhood indicates the presence of that metal. In short, the water to all appearance is precisely similar to that of Bowyer's sulphur spring in Virginia.

Monday, 17th. Captain Clark set out with five men to explore the country; the rest were employed in hunting, making wheels, and in drawing the five canoes and all the baggage up the creek, which we now called Portage creek. From this creek there is a gradual ascent to the top of the high plain, while the bluffs of the creek lower down, and of the Missouri, both above and below its entrance, were so steep as to have rendered it almost impracticable to drag them up from the Missouri. We found great difficulty and some danger in even ascending the creek thus far, in consequence of the rapids and rocks of the channel of the creek, which just above where we brought the canoes has a fall of five feet, and high and steep bluffs beyond it. We were very fortunate in finding, just below Portage creek, a cottonwood tree about twenty-two inches in diameter, and large enough to make the carriage wheels; it was perhaps the only one of the same size within twenty miles; and the cottonwood, which we are obliged to employ in the other parts of the work, is extremely soft and brittle. The mast of the white periogue, which we mean to leave behind, supplied us with two axletrees. There are vast quantities of buffaloe feeding in the plains or watering in the river, which is also strewed with the floating carcases and limbs of these animals. They go in large herds to water about the falls, and as all the passages to the river near that place are narrow and steep, the foremost are pressed into the river by the impatience of those behind. In this way we have seen ten or a dozen disappear over the falls in a few minutes. They afford excellent food for the wolves, bears, and birds of prey; and this circumstance may account for the reluctance of the bears to yield their dominion over the neighbourhood.

Tuesday, 18th. The periogue was drawn up a little below our camp and secured in a thick copse of willow bushes. We now began to form a cache or place of deposit, and to dry our goods and other articles which required inspection. The wagons, too, are completed. Our hunters brought us ten deer, and we shot two out of a herd of buffaloe that came to water at the sulphur spring. There is a species of gooseberry growing abundantly among the rocks on the sides of the cliffs; it is now ripe, of a pale red colour, about the size of a common gooseberry, and like it is an ovate pericarp of soft pulp enveloping a number of small whitish-coloured seeds, and consisting of a yellowish slimy mucilaginous substance, with a sweet taste; the surface of the berry is covered with a glutinous adhesive matter, and its fruit, though ripe, retains its withered corolla. The shrub itself seldom rises more than two feet high, is much branched, and has no thorns. The leaves resemble those of the common gooseberry except in being smaller, and the berry is supported by separate peduncles or footstalks half an inch long. There are also immense quantities of grasshoppers, of a brown colour, in the plains, and they no doubt contribute to the lowness of the grass, which is not generally more than three inches high, though it is soft, narrow-leafed, and affords a fine pasture for the buffaloe.

Wednesday, 19th. The wind blew violently to-day, as it did yesterday, and as it does frequently in this open country, where there is not a tree to break or oppose its force. Some men were sent for the meat killed yesterday, which, fortunately, had not been discovered by the wolves. Another party went to Medicine river in quest of elk, which we hope may be induced to resort there, from there being more wood in that neighbourhood than on the Missouri. All the rest were occupied in packing the baggage and mending their moccasins, in order to prepare for the portage. We caught a number of the white fish, but no catfish or trout. Our poor Indian woman, who had recovered so far as to walk out, imprudently ate a quantity of the white apple, which, with some dried fish, occasioned a return of her fever.

The meridian altitude of the sun's lower limb, as observed with octant by back observation, was 53° 15', giving as the latitude of our camp, 47° 8' 59" 5'''.

Thursday, 20th. As we were desirous of getting meat enough to last us during the portage, so that the men might not be diverted from their labour to look for food, we sent out four hunters to-day; they killed eleven buffaloe. This was indeed an easy labour, for there are vast herds coming constantly to the opposite bank of the river to water; they seem also to make much use of the mineral water of the sulphur spring, but whether from choice, or because it is more convenient than the river, we cannot determine, as they sometimes pass near the spring and go on to the river. Besides this spring, brackish water, or that of a dark colour, impregnated with mineral salts, such as we have frequently met on the Missouri, may be found in small quantities in some of the steep ravines on the north side of the river opposite to us, and at the falls.

Captain Clark returned this evening, having examined the whole course of the river and fixed the route most practicable for the portage. The first day, 17th, he was occupied in measuring the heights and distances along the banks of the river, and slept near a ravine at the foot of the crooked falls, having very narrowly escaped falling into the river, where he would have perished inevitably, in descending the cliffs near the grand cataract. The next day, 18th, he continued the same occupation, and arrived in the afternoon at the junction of Medicine and Missouri rivers; up the latter he ascended, and passed, at the distance of a mile, an island and a little timber in an eastwardly bend of the river. One mile beyond this he came to the lower point of a large island; another small island in the middle of the river, and one near the left shore at the distance of three miles, opposite to the head of which he encamped, near the mouth of a creek which appeared to rise in the South mountain. These three islands are opposite to each other, and we gave them the name of the Whitebear islands, from observing some of those animals on them. He killed a beaver, an elk, and eight buffaloe. One of the men, who was sent a short distance from the camp to bring home some meat, was attacked by a white bear, and closely pursued within forty paces of the camp, and narrowly escaped being caught. Captain Clark immediately went with three men in quest of the bear, which he was afraid might surprise another of the hunters who was out collecting the game. The bear was, however, too quick, for before Captain Clark could reach the man, the bear had attacked him and compelled him to take refuge in the water. He now ran off as they approached, and, it being late, they deferred pursuing him till the next morning.

## CHAPTER XI.

Description and romantic appearance of the Missouri at the junction of the Medicine river—The difficulty of transporting the baggage at the falls—The party employed in the construction of a boat of skins—The embarrassments they had to encounter for want of proper materials—During the work the party much troubled by white bears—Violent hail-storm, and providential escape of Captain Clark and his party—Description of a remarkable fountain—Singular explosion heard from the Black mountains—The boat found to be insufficient, and the serious disappointment of the party—Captain Clark undertakes to repair the damage by building canoes, and accomplishes the task.

ON the 19th, Captain Clark, not being able to find the bear mentioned in the last chapter, spent the day in examining the country both above and below the Whitebear islands, and concluded that the place of his encampment would be the best point for the extremity of the portage. The men were therefore occupied in drying the meat to be left here. Immense numbers of buffaloe are everywhere round, and they saw a summer duck, which is now sitting. The next morning, 20th, he crossed the level plain, fixed stakes to mark the route of the portage, till he passed a large ravine which would oblige us to make the portage farther from the river; after this, there being no other obstacle, he went to the river where he had first struck it, and took its courses and distances down to the camp. From the draught and survey of Captain Clark, we had now a clear and connected view of the falls, cascades, and rapids of the Missouri.

This river is three hundred yards wide at the point where it receives the waters of Medicine river, which is one hundred and thirty-seven yards in width. The united current continues three hundred and twenty-eight poles to a small rapid on the north side, from which it gradually widens to one thousand four hundred yards, and at the distance of five hundred and forty-eight poles reaches the head of the rapids, narrowing as it approaches them. Here the hills on the north, which had withdrawn from the bank, closely border the river, which, for the space of three hundred and twenty poles, makes its way over the rocks with a descent of thirty feet; in this course the current is contracted to five hundred and eighty yards, and after throwing itself over a small pitch of five feet, forms a beautiful cascade of twenty-six feet five inches; this does not, however, fall immediately perpendicular, being stopped by a part of the rock which projects at about one-third of the distance. After descending this fall, and passing the cottonwood island on which the eagle has fixed its nest, the river goes on for five hundred and thirty-two poles over rapids and little falls, the estimated descent of which is thirteen feet six inches, till it is joined by a large fountain boiling up underneath the rocks near the edge of the river, into which it falls with a cascade of eight feet. It is of the most perfect clearness and rather of a bluish cast, and even after falling into the Missouri it preserves its colour for half a mile. From this fountain the river descends with increased rapidity for the distance of two hundred and fourteen poles, during which the estimated descent is five feet; from this, for a distance of one hundred and thirty-five poles, the river descends fourteen feet seven inches, including a perpendicular fall of six feet seven inches. The river has now become pressed into a space of four hundred and seventy-three yards, and here forms a grand cataract by falling over a plain rock the whole distance across the river, to the depth of forty-seven feet eight inches; after recovering itself the Missouri then proceeds with an estimated descent of three feet, till at the distance of one hundred and two poles it

## UP THE MISSOURI

again is precipitated down the Crooked falls of nineteen feet perpendicular; below this, at the mouth of a deep ravine, is a fall of five feet, after which, for the distance of nine hundred and seventy poles, the descent is much more gradual, not being more than ten feet, and then succeeds a handsome level plain for the space of one hundred and seventy-eight poles, with a computed descent of three feet, making a bend towards the north. Thence it descends, during four hundred and eighty poles, about eighteen feet and a half, when it makes a perpendicular fall of two feet, which is ninety poles beyond the great cataract, in approaching which it descends thirteen feet within two hundred yards, and gathering strength from its confined channel, which is only two hundred and eighty yards wide, rushes over the fall to the depth of eighty-seven feet and three-quarters of an inch. After raging among the rocks and losing itself in foam, it is compressed immediately into a bed of ninety-three yards in width; it continues for three hundred and forty poles to the entrance of a run or deep ravine, where there is a fall of three feet, which, joined to the decline of the river during that course, makes the descent six feet. As it goes on the descent within the next two hundred and forty poles is only four feet; from this, passing a run or deep ravine, the descent for four hundred poles is thirteen feet; within two hundred and forty poles, a second descent of eighteen feet; thence one hundred and sixty poles, a descent of six feet; after which to the mouth of Portage creek, a distance of two hundred and eighty poles, the descent is ten feet. From this survey and estimate it results that the river experiences a descent of three hundred and fifty-two feet in the course of two and three-quarter miles, from the commencement of the rapids to the mouth of Portage creek, exclusive of the almost impassable rapids which extend for a mile below its entrance.

The latitude of our camp below the entrance of Portage creek was found to be 47° 7' 10" 3, as deduced from a meridian altitude of the sun's lower limb, taken with octant by back observation, giving 53° 10'.

Friday, June 21. Having made the necessary preparations for continuing our route, a part of the baggage was carried across the creek into the high plain, three miles in advance, and placed on one of the carriages with truck wheels; the rest of the party was employed in drying meat and dressing elk skins. We killed several mule-deer and an elk, and observed as usual vast quantities of buffaloe who came to drink at the river. For the first time on the Missouri we have seen near the falls a species of fishing duck, the body of which is brown and white, the wings white, and the head and upper part of the neck of a brick-red, with a narrow beak, which seems to be of the same kind common in the Susquehanna, Potomac, and James' river. The little wood which this neighbourhood affords consists of the broad and narrow leafed cottonwood, the box alder, the narrow and broad leafed willow, the large or sweet willow, which was not common below Maria's river, but which here attains the same size and has the same appearance as in the Atlantic states. The undergrowth consists of roses, gooseberries, currants, small honeysuckles, and the redwood, the inner part of which the *engages*, or watermen, are fond of smoking when mixed with tobacco.

Saturday, 22d. We now set out to pass the portage, and halted for dinner at eight miles distance near a little stream. The axletrees of our carriage, which had been made of an old mast, and the cottonwood tongues broke before we came there; but we renewed them with the timber of the sweet willow, which lasted till within half a mile of our intended camp, when the tongues gave way, and we were obliged to take as much baggage as we could carry on our backs down to the river, where we formed an encampment in a small grove of timber opposite to the Whitebear islands. Here the banks on both sides of the river are handsome, level, and extensive; that near our camp is not more than two feet above the surface of the water. The river is about eight hundred yards wide just above these islands, ten feet deep in most places, and with a very gentle current. The plains, however, on this part of the river are not so fertile as those from the mouth of the Muscleshell and thence downwards; there is much more stone on the sides of the hills and on the broken lands than is to be found lower down. We saw in the plains vast quantities of buffaloe, a number of small birds, and the large brown curlew, which is now sitting, and lays its eggs, which are of a pale blue with black specks, on the ground without any nest. There is also a species of lark much resembling the bird called the oldfield lark, with a yellow breast and a black spot on the croup, though it differs from the latter in having its tail formed of feathers of an uneqal length and pointed; the beak, too, is somewhat longer and more curved, and the note differs considerably. The prickly pear annoyed us very much to-day by sticking through our moccasins. As soon as we had kindled our fires we examined the meat which Captain Clark had left here, but found that the greater part of it had been taken by the wolves.

Sunday, 23d. After we had brought up the canoe and baggage, Captain Clark went down to the camp at Portage creek, where four of the men had been left with the Indian woman. Captain Lewis, during the morning, prepared the camp, and in the afternoon went down in a canoe to Medicine river to look after the three men who had been sent thither to hunt on the 19th, and from whom nothing had as yet been heard. He went up the river about half a mile, and then walked along on the right bank, hallooing as he went, till at the distance of five miles he found one of them who had fixed his camp on the opposite bank, where he had killed seven deer and dried about six hundred pounds of buffaloe meat, but had killed no elk, the animal chiefly wanted. He knew nothing of his companions except that on the day of their departure from camp he had left them at the falls and come on to Medicine river, not having seen them since. As it was too late to return, Captain Lewis passed over on a raft which he made for the purpose, and spent the night at Shannon's camp, and the next morning,

Monday, 24th, sent J. Fields up the river with orders to go four miles and return, whether he found the two absent hunters or not; then descending the southwest side of Medicine river, he crossed the Missouri in the canoe, and sent Shannon back to his camp to join Fields and bring the meat which they had killed; this they did, and arrived in the evening at the camp on Whitebear islands. A part of the men from Portage creek also arrived with two canoes and baggage. On going down yesterday Captain Clark cut off several angles of the former route so as to shorten the Portage considerably, and marked it with stakes; he arrived there in time to have two of the canoes carried up in the high plain about a mile in advance. Here they all repaired their moccasins, and put on double soals to protect them from the prickly pear and from the sharp points of earth which have been formed by the trampling of the buffaloe during the late rains; this of itself is sufficient to render the portage disagreeable to one who had no burden, but as the men are loaded as heavily as their strength will permit, the crossing is really painful; some are limping with the soreness of their feet, others are scarcely able to stand for more than a few minutes from the heat and fatigue; they are all obliged to halt and rest frequently, and at almost every stopping place they fall, and many of them are asleep in an instant; yet no one complains, and they go on with great cheerfulness. At their camp Drewyer and Fields joined them, and while Captain Lewis was looking for them at Medicine river, they returned to report the absence of Shannon, about whom they had been very uneasy. They had killed several buffaloe at the bend of the Missouri above the falls, and dried about eight hundred pounds of meat and got one hundred pounds of tallow; they had also killed some deer, but had seen no elk. After getting the party in motion with the canoes, Captain Clark returned to his camp at Portage creek.

We were now occupied in fitting up a boat of skins, the frame of which had been prepared for the purpose at Harper's ferry. It was made of iron, thirty-six feet long, four feet and a half in the beam, and twenty-six inches wide in the bottom. Two men had been sent this morning for timber to complete it, but they could find scarcely any even tolerably straight sticks four and a half feet long, and as the cottonwood is too soft and brittle, we were obliged to use the willow and box-alder.

Tuesday, 25th. The party returned to the lower camp. Two men were sent on the large island to look for timber. J. Fields was sent up the Missouri to hunt elk; but he returned about noon, and informed us that a few miles above he saw two white bear near the river, and in attempting to fire at them came suddenly on a third, who, being only a few steps off, immediately attacked him; that in running to escape from the monster he leaped down a steep bank of the river, where, falling on a bar of stone, he cut his hand and knee and bent his gun; but fortunately for him the bank concealed him from his antagonist, or he would have been most probably lost. The other two returned with a small quantity of bark and timber, which was all they could find on the island, but they had killed two elk; these were valuable, as we are desirous of procuring the skins of that animal in order to cover the boat, as they are more strong and durable than those of the buffaloe, and do not shrink so much in drying. The party that went to the lower camp had one canoe and the baggage carried into the high plain to be ready in the morning, and then all who could make use of their feet had a dance on the green to the music of a violin. We have been unsuccessful in our attempt to catch fish, nor does there seem to be any in this part of the river. We observe a number of water terrapins. There are great quantities of young blackbirds in these islands, just beginning to fly. Among the vegetable productions we observe a species of wild rye, which is now heading; it rises to the height of eighteen or twenty inches, the beard remarkably fine and soft; the culen is jointed, and in every respect except in height it resembles the wild rye. Great quantities of mint, too, like the peppermint, are found here.

The winds are sometimes violent in these plains. The men inform us that as they were bringing one of the canoes along on truck-wheels, they hoisted the sail and the wind carried her along for some distance.

Wednesday, 26th. Two men were sent on the opposite side of the river for bark and timber, of which they procured some, but by no means enough for our purposes. The bark of the cottonwood is too soft, and our only dependence is on the sweet willow, which has a tough, strong bark; the two hunters killed seven buffaloe. A party arrived from below with two canoes and baggage, and, the wind being from the southeast, they had made considerable progress with the sails. On their arrival one of the men, who had been considerably heated and fatigued, swallowed a very hearty draught of water, and was immediately taken ill; Captain Lewis bled him with a penknife, having no other instrument at hand, and succeeded in restoring him to health the next day. Captain Clark formed a second cache or deposit near the camp, and placed the swivel under the rocks near the river. The antelopes are still scattered through the plains, the females with their young, which are generally two in number, and the males by themselves.

Thursday, 27th. The party were employed in preparing timber for the boat, except two who were sent to hunt. About one in the afternoon a cloud arose from the southwest, and brought with it violent thunder, lightning, and hail; soon after it passed the hunters came in from about four miles above us. They had killed nine elk and three bear. As they were hunting on the river they saw a low ground covered with thick brushwood, where from the tracks along shore they thought a bear had probably taken refuge; they therefore landed, without making a noise, and climbed a tree about twenty feet above the ground. Having fixed themselves securely, they raised a loud shout, and a bear instantly rushed towards them. These animals never climb, and therefore when he came to the tree and stopped to look at them Drewyer shot him in the head; he proved to be the largest we have yet seen: his noise appeared to be like that of a common ox, his fore feet measured nine inches across, and the hind feet were seven inches wide and eleven and three-quarters long, exclusive of the talons. One of these animals came within thirty yards of the camp last night, and carried off some buffaloe meat which we had placed on a pole. In the evening, after the storm, the water on this side of the river became of a deep crimson colour, probably caused by some stream above washing down a kind of soft red stone, which we observe in the neighbouring bluffs and gullies. At the camp below, the men who left us in the morning were busy in preparing their load for to-morrow, which were impeded by the rain, hail, and the hard wind from the northwest.

Friday, 28th. The party all occupied in making the boat; they obtained a sufficient quantity of willow bark to line her, and over these were placed the elk skins, and when they failed they were obliged to use the buffaloe hide. The white bear have now become exceedingly troublesome; they constantly infest our camp during the night, and though they have not attacked us, as our dog who patroles all night gives us notice of their approach, yet we are obliged to sleep with our arms by our sides for fear of accident, and we cannot send one man alone to any distance, particularly if he has to pass through brushwood. We saw two of them to-day on the large island opposite to us, but as we are all so much occupied now, we mean to reserve ourselves for some leisure moment, and then make a party to drive them from the islands. The river has risen nine inches since our arrival here.

At Portage creek Captain Clark completed the cache, in which we deposited whatever we could spare from our baggage: some ammunition, provisions, books, the specimens of plants and minerals, and a draught of the river from its entrance to fort Mandan. After closing it he broke up the encampment, and took on all the remaining baggage to the high plain, about three miles. Portage creek has risen considerably in consequence of the rain, and the water had become of a deep crimson colour, and ill tasted. On overtaking the canoe he found that there was more baggage than could be carried on the two carriages, and therefore left some of the heavy articles which could not be injured, and proceeded on to Willow run, where he encamped for the night. Here they made a supper on two buffaloe which they killed on the way; but passed the night in the rain, with a high wind from the southwest. In the morning,

Saturday, 29th, finding it impossible to reach the end of the portage with their present load, in consequence of the state of the road after the rain, he sent back nearly all his party to bring on the articles which had been left yesterday. Having lost some notes and remarks which he had made on first ascending the river, he determined to go up to the Whitebear islands along its banks, in order to supply the deficiency. He there left one man to guard the baggage, and went on to the falls, accompanied by his servant York, Chaboneau and his wife with her young child. On his arrival there he observed a very dark cloud rising in the west which threatened rain, and looked around for some shelter, but could find no place where they would be secure from being blown into the river if the wind should prove as violent as it sometimes does in the plains. At length, about a quarter of a mile above the falls, he found a deep ravine where there were some shelving rocks, under which he took refuge. They were on the upper side of the ravine near the river, perfectly safe from the rain, and therefore laid down their guns, compass, and other articles which they carried with them. The shower was at first moderate, it then increased to a heavy rain, the effects of which they did not feel; soon after a torrent of rain and hail descended; the rain seemed to fall in a solid mass, and instantly collecting in the ravine came rolling down in a dreadful current, carrying the mud and rocks, and every thing that opposed it. Captain Clark fortunately saw it a moment before it reached them, and springing up with his gun and shotpouch in his left hand, with his right clambered up the steep bluff, pushing on the Indian woman with her child in her arms; her husband, too, had seized her hand, and was pulling her up the hill, but he was so terrified at the danger that, but for Captain Clark, himself and his wife and child would have been lost. So instantaneous was the rise of the water that before Captain Clark had reached his gun and began to ascend the bank the water was up to his waist, and he could scarce get up faster than it rose, till it reached the height of fifteen feet with a furious current, which, had they waited a moment longer, would have swept them into the river just above the great falls, down which they must inevitably have been precipitated. They reached the plain in safety, and found York, who had separated from them just before the storm to hunt some buffaloe, and was now returning to find his master. They had been obliged to escape so rapidly that Captain Clark lost his compass and umbrella, Chaboneau left his gun, shotpouch, and tomahawk, and the Indian woman had just time to grasp her child, before the net in which it lay at her feet was carried down the current. He now relinquished his intention of going up the river, and returned to the camp at Willow run. Here he found that the party sent this morning for the baggage had all returned to camp in great confusion, leaving their loads in the plain. On account of the heat they generally go nearly naked, and with no covering on their heads. The hail was so large, and driven so furiously against them by the high wind, that it knocked several of them down; one of them particularly was thrown on the ground three times, and most of them bleeding freely and complained of being much bruised. Willow run had risen six feet since the rain, and as the plains were so wet that they could not proceed, they passed the night at their camp.

At the Whitebear camp, also, we had not been insensible to the hail-storm, though less exposed. In the morning there had been a heavy shower of rain, after which it became fair. After assigning to the men their respective employments,

Captain Lewis took one of them and went to see the large fountain near the falls. For about six miles he passed through a beautiful, level plain, and then, on reaching the break of the river hills, was overtaken by the gust of wind from the southwest, attended by lightning, thunder, and rain; fearing a renewal of the scene on the 27th, they took shelter in a little gully where there were some broad stones with which they meant to protect themselves against the hail; but fortunately there was not much, and that of a small size, so that they felt no inconvenience except that of being exposed without shelter for an hour, and being drenched by the rain. After it was over they proceeded to the fountain, which is perhaps the largest in America. It is situated in a pleasant, level plain, about twenty-five yards from the river, into which it falls over some steep irregular rocks, with a sudden ascent of about six feet in one part of its course. The water boils up from among the rocks, and with such force near the centre that the surface seems higher there than the earth on the sides of the fountain, which is a handsome turf of fine green grass. The water is extremely pure, cold, and pleasant to the taste, not being impregnated with lime or any foreign substance. It is perfectly transparent, and continues its bluish cast for half a mile down the Missouri, notwithstanding the rapidity of the river. After examining it for some time, Captain Lewis returned to the camp.

Sunday, 30th. In the morning Captain Clark sent the men to bring up the baggage left in the plains yesterday. On their return the axletrees and carriages were repaired, and the baggage conveyed on the shoulders of the party across Willow run, which had fallen as low as three feet. The carriages being then taken over, a load of baggage was carried to the six-mile stake, deposited there, and the carriages brought back. Such is the state of the plains that this operation consumed the day. Two men were sent to the falls to look for the articles lost yesterday, but they found nothing but the compass, covered with mud and sand, at the mouth of the ravine; the place at which Captain Clark had been caught by the storm was filled with large rocks. The men complain much of the bruises received yesterday from the hail. A more than usual number of buffaloe appeared about the camp to-day, and furnished plenty of meat; Captain Clark thought that at one view he must have seen at least ten thousand. In the course of the day there was a heavy gust of wind from the southwest, after which the evening was fair.

At the Whitebear camp we had a heavy dew this morning, which is quite a remarkable ocurrence. The party continues to be occupied with the boat, the crossbars for which are now finished, and there remain only the strips to complete the woodwork; the skins necessary to cover it have already been prepared, and they amount to twenty-eight elk skins and four buffaloe skins. Among our game were two beaver, which we have had occasion to observe always are found wherever there is timber. We also killed a large bat, or goatsucker, of which there are many in this neighbourhood, resembling in every respect those of the same species in the United States. We have not seen the leather-winged bat for some time, nor are there any of the small goatsucker in this part of the Missouri. We have not seen either that species of goatsucker or nighthawk called the whippoorwill, which is commonly confounded in the United States with the large goatsucker which we observe here; this last prepares no nest, but lays its eggs in the open plains; they generally begin to sit on two eggs, and we believe raise only one brood in a season; at the present moment they are just hatching their young.

Monday, July 1. After a severe day's work Captain Clark

reached our camp in the evening, accompanied by his party and all the baggage except that left at the six-mile stake, for which they were too much fatigued to return. The route from the lower camp on Portage creek to that near Whitebear island, having been now measured and examined by Captain Clark, was as follows:

From our camp opposite the last considerable rapid to the entrance of Portage creek south 9° east for three-quarters of a mile; thence on a course south 10° east for two miles, though for the canoes the best route is to the left of this course, and strikes Portage one mile and three-quarters from its entrance, avoiding in this way a very steep hill which lies above Portage creek; from this south 18° west for four miles, passing the head of a drain or ravine which falls into the Missouri below the great falls, and to the Willow run, which has always a plentiful supply of good water and some timber; here the course turns to south 45° west for four miles farther; then south 66° west three miles, crossing at the beginning of the course the head of a drain which falls into the Missouri at the Crooked falls, and reaching an elevated point of the plain, from which south 42° west. On approaching the river on this course there is a long and gentle descent from the high plain, after which the road turns a little to the right of the course up the river to our camp. The whole portage is seventeen and three-quarter miles.

At the Whitebear camp we were occupied with the boat and digging a pit for the purpose of making some tar. The day has been warm, and the musquitoes troublesome. We were fortunate enough to observe equal altitudes of the sun with sextant, which, since our arrival here, we have been prevented from doing by flying clouds and storms in the evening.

Tuesday, July 2. A shower of rain fell very early this

morning. We then despatched some men for the baggage left behind yesterday, and the rest were engaged in putting the boat together. This was accomplished in about three hours, and then we began to sew on the leather over the crossbars of iron on the inner side of the boat which form the ends of the sections. By two o'clock the last of the baggage arrived, to the great delight of the party, who were anxious to proceed. The musquitoes we find very troublesome.

Having completed our celestial observations, we went over to the large island to make an attack upon its inhabitants, the bears, who have annoyed us very much of late, and who were prowling about our camp all last night. We found that the part of the island frequented by the bear forms an almost impenetrable thicket of the broad-leafed willow; into this we forced our way in parties of three, but could see only one bear, who instantly attacked Drewyer. Fortunately, as he was rushing on, the hunter shot him through the heart within twenty paces, and he fell, which enabled Drewyer to get out of his way; we then followed him one hundred yards, and found that the wound had been mortal. Not being able to discover any more of these animals, we returned to camp; here, in turning over some of the baggage, we caught a rat somewhat larger than the common European rat, and of a lighter colour; the body and outer parts of the legs and head of a light lead colour; the inner side of the legs, as well as the belly, feet, and ears, are white; the ears are not covered with hair, and are much larger than those of the common rat; the toes also are longer, the eyes black and prominent, the whiskers very long and full; the tail rather longer than the body, and covered with fine fur and hair of the same size with that on the back, which is very close, short, and silky in its texture. This was the first we had met, although its nests are very frequent among the cliffs of rocks and hollow trees, where we also found large quantities of the shells and seed of the prickly pear, on which we conclude they chiefly subsist. The musquitoes are uncommonly troublesome. The wind was again high from the southwest; these winds are in fact always the coldest and most violent which we experience, and the hypothesis which we have formed on that subject is, that the air, coming in contact with the Snowy mountains, immediately becomes chilled and condensed, and being thus rendered heavier than the air below, it descends into the rarified air below, or into the vacuum formed by the constant action of the sun on the open unsheltered plains. The clouds rise suddenly near these mountains, and distribute their contents partially over the neighbouring plains. The same cloud will discharge hail alone in one part, hail and rain in another, and rain only in a third, and all within the space of a few miles, while at the same time there is snow falling on the mountains to the southeast of us. There is at present no snow on those mountains, that which covered them on our arrival, as well as that which has since fallen, having disappeared. The mountains to the north and northwest of us are still entirely covered with snow, and indeed there has been no perceptible diminution of it since we first saw them, which induces a belief either that the clouds prevailing at this season do not reach their summits, or that they deposit their snow only. They glisten with great beauty when the sun shines on them in a particular direction, and most probably from this glittering appearance have derived the name of the Shining mountains.

Wednesday, 3d. Nearly the whole party were employed in different labours connected with the boat, which is now almost completed; but we have not as yet been able to obtain tar from our kiln, a circumstance that will occasion us not a little embarrassment. Having been told by the Indians that on leaving the falls we should soon pass the buffaloe country, we have before us the prospect of fasting occasionally; but in order to provide a supply, we sent out the hunters, who killed only a buffaloe and two antelopes, which, added to six beaver and two otter, have been all our game for two or three days. At ten in the morning we had a slight shower, which scarcely wet the grass.

Thursday, July 4. The boat was now completed, except what is in fact the most difficult part, the making her seams secure. We had intended to despatch a canoe with part of our men to the United States early this spring; but not having yet seen the Snake Indians, or knowing whether to calculate on their friendship or enmity, we have decided not to weaken our party, which is already scarcely sufficient to repel any hostility. We were afraid, too, that such a measure might dishearten those who remain; and as we have never suggested it to them, they are all perfectly and enthusiastically attached to the enterprise, and willing to encounter any danger to ensure its success. We had a heavy dew this morning.

Since our arrival at the falls we have repeatedly heard a strange noise coming from the mountains in a direction a little to the north of west. It is heard at different periods of the day and night, sometimes when the air is perfectly still and without a cloud, and consists of one stroke only, or of five or six discharges in quick succession. It is loud, and resembles precisely the sound of a six-pound piece of ordnance at the distance of three miles. The Minnetarees frequently mentioned this noise like thunder, which they said the mountains made; but we had paid no attention to it, believing it to have been some superstition, or perhaps a falsehood. The watermen also of the party say that the Pawnees and Ricaras give the same account of a noise heard in the Black mountains to the westward of them. The solution of the mystery given by the philosophy of the watermen is, that it is occasioned by the bursting of the rich mines of silver confined within the bosom of the mountain. An elk and a beaver are all that were killed to-day; the buffaloe seemed to have withdrawn from our neighbourhood, though several of the men who went to-day to visit the falls for the first time mention that they are still abundant at that place. We contrived, however, to spread, not a very sumptuous, but a comfortable table in honour of the day, and in the evening gave the men a drink of spirits, which was the last of our stock. Some of them appeared sensible to the effects of even so small a quantity, and, as is usual among them on all festivals, the fiddle was produced and a dance begun, which lasted till nine o'clock, when it was interrupted by a heavy shower of rain. They continued, however, their merriment till a late hour.

Friday, 5th. The boat was brought up into a high situation and fires kindled under her, in order to dry her more expeditiously. Despairing now of procuring any tar, we formed a composition of pounded charcoal with beeswax and buffaloe tallow to supply its place; should this resource fail us it will be very unfortunate, as in every other respect the boat answers our purposes completely. Although not quite dry, she can be carried with ease by five men; her form is as complete as could be wished, very strong, and will carry at least eight thousand pounds with her complement of hands. Besides our want of tar, we have been unlucky in sewing the skins with a needle which had sharp edges instead of a point merely, although a large thong was used in order to fill the hole; yet it shrinks in drying and leaves the hole open, so that we fear the boat will leak.

A large herd of buffaloe came near us and we procured

three of them, besides which were killed two wolves and three antelopes. In the course of the day other herds of buffaloe came near our camp on their way down the river. These herds move with great method and regularity. Although ten or twelve herds are seen scattered from each other over a space of many miles, yet if they are undisturbed by pursuit they will be uniformly travelling in the same direction.

Saturday, 6th. Last night there were several showers of rain and hail, attended with thunder and lightning, and about daybreak a heavy storm came on from the southwest, with one continued roar of thunder, and rain, and hail. The hail, which was as large as musket balls, covered the ground completely, and on collecting some of it, it lasted during the day, and served to cool the water. The red and yellow currant is abundant and now ripe, although still a little acid. We have seen in this neighbourhood what we have not met before, a remarkably small fox, which associates in bands and burrows in the prairie, like the small wolf, but have not yet been able to obtain any of them, as they are extremely vigilant, and betake themselves on the slightest alarm to their burrows, which are very deep.

Sunday, 7th. The weather is warm but cloudy, so that the moisture retained by the bark after the rain leaves it slowly, though we have small fires constantly under the boat. We have no tents, and therefore are obliged to use the sails to keep off the bad weather. Our buffaloe skins, too, are scarcely sufficient to cover our baggage, but the men are now dressing others to replace their present leather clothing, which soon rots by being so constantly exposed to water. In the evening the hunters returned with the skins of only three buffaloe, two antelope, four deer, and three wolf skins, and reported that the buffaloe had gone farther down the river; two other hunters who left us this morning could find nothing except one elk; in addition to this we caught a beaver. The musquitoes still disturb us very much, and the blowing-flies swarm in vast numbers round the boat. At four in the afternoon we had a light shower of rain, attended with some thunder and lightning.

Monday, 8th. In order more fully to replace the notes of the river which he had lost, and which he was prevented from supplying by the storm of the twenty-ninth ult., Captain Clark set out after breakfast, taking with him nearly the whole party, with a view of shooting buffaloe if there should be any near the falls. After getting some distance in the plains the men were divided into squads, and he with two others struck the Missouri at the entrance of Medicine river, and thence proceeded down to the great cataract. He found that the immense herds of buffaloe have entirely disappeared, and he thought had gone below the falls. Having made the necessary measurements, he returned through the plains and reached camp late in the evening; the whole party had killed only three buffaloe, three antelopes, and a deer; they had also shot a small fox, and brought a living ground-squirrel, somewhat larger than those of the United States.

The day was warm and fair, but a slight rain fell in the afternoon. The boat having now become sufficiently dry we gave it a coat of the composition, which after a proper interval was repeated, and the next morning,

Tuesday, 9th, she was lanched into the water, and swam perfectly well; the seats were then fixed and the oars fitted; but after we had loaded her, as well as the canoes, and were on the point of setting out, a violent wind caused the waves to wet the baggage, so that we were forced to unload them. The wind continued high till evening, when to our great disappointment we discovered that nearly all the composition had separated from the skins, and left the seams perfectly exposed, so that the boat now leaked very much. To repair this misfortune without pitch is impossible, and as none of that article is to be procured, we therefore, however reluctantly, are obliged to abandon her, after having had so much labour in the construction. We now saw that the section of the boat covered with buffaloe skins on which hair had been left answered better than the elk skins and leaked but little; while that part which was covered with hair about one-eighth of an inch retained the composition perfectly, and remained sound and dry. From this we perceived that had we employed buffaloe instead of elk skins, and not singed them so closely as we have done, carefully avoiding to cut the leather in sewing, the boat would have been sufficient even with the present composition, or had we singed instead of shaving the elk skins we might have succeeded. But we discovered our error too late; the buffaloe had deserted us, the travelling season was so fast advancing that we had no time to spare for experiments, and therefore, finding that she could be no longer useful, she was sunk in the water, so as to soften the skins and enable us the more easily to take her to pieces. It now became necessary to provide other means for transporting the baggage which we had intended to stow in her. For this purpose we shall want two canoes, but for many miles below the mouth of the Muscleshell river to this place we have not seen a single tree fit to be used in that way. The hunters, however, who had hitherto been sent after timber mention that there is a low ground on the opposite side of the river, about eight miles above us by land, and more than twice that distance by water, in which we may probably find trees large enough for our purposes. Captain Clark, therefore, determined to set out by land for that place with ten of the best workmen, who would be occupied in building the canoes till the rest of the party, after taking the boat to pieces and making the necessary deposits, should transport the baggage and join them with the other six canoes.

Wednesday, 10th. He accordingly passed over to the opposite side of the river with his party, and proceeded on eight miles by land, the distance by water being twenty-three and three-quarter miles. Here he found two cottonwood trees, but on cutting them down, one proved to be hollow, split at the top in falling, and both were much damaged at the bottom. He searched the neighbourhood but could find none which would suit better, and therefore was obliged to make use of those which he had felled, shortening them in order to avoid the cracks, and supplying the deficiency by making them as wide as possible. They were equally at a loss for wood of which they might make handles for their axes, the eyes of which not being round they were obliged to split the timber in such a manner that thirteen of the handles broke in the course of the day, though made of the best wood they could find for the purpose, which was the chokecherry.

The rest of the party took the frame of the boat to pieces, deposited it in a cache or hole, with a draught of the country from fort Mandan to this place, and also some other papers and small articles of less importance. After this we amused ourselves with fishing, and although we had thought on our arrival that there were none in this part of the river, we caught some of a species of white chub below the falls, but few in number and small in size.

Serjeant Ordway, with four canoes and eight men, had set sail in the morning, with part of the baggage, to the place where Captain Clark had fixed his camp, but the wind was so high that he only reached within three miles of that place, and encamped for the night.

Thursday, July 11. In the morning one of the canoes

joined Captain Clark; the other three, having on board more valuable articles which would have been injured by the water, went on more cautiously, and did not reach the camp till the evening. Captain Clark then had the canoes unloaded and sent back, but the high wind prevented their floating down nearer than about eight miles above us. His party were busily engaged with the canoes, and their hunters supplied them with three fat deer and a buffaloe, in addition to two deer and an antelope killed yesterday. The few men who were with Captain Lewis were occupied in hunting, but with not much success, having killed only one buffaloe. They heard about sunset two discharges of the tremendous mountain artillery; they also saw several very large gray eagles, much larger than those of the United States, and most probably a distinct species, though the bald-eagle of this country is not quite so large as that of the United States. The men have been much afflicted with painful whitlows, and one of them disabled from working by this complaint in his hand.

Friday, 12th. In consequence of the wind the canoes did not reach the lower camp till late in the afternoon, before which time Captain Lewis sent all the men he could spare up the river to assist in building the boats, and the day was too far advanced to reload and send them up before morning. The musquitoes are very troublesome, and they have a companion not less so, a large black gnat which does not sting, but attacks the eyes in swarms. The party with Captain Clark are employed on the canoes; in the course of the work Serjeant Pryor dislocated his shoulder yesterday, but it was replaced immediately, and, though painful, does not threaten much injury. The hunters brought in three deer and two otter. This last animal has been numerous since the water has become sufficiently clear for them to take fish. The blue-crested fisher, or, as it is sometimes called, the kingfisher, is

## UP THE MISSOURI

an inhabitant of this part of the river; it is a bird rare on the Missouri, indeed we had not seen more than three or four of them from its entrance to Maria's river, and even those did not seem to reside on the Missouri but on some of the clearer streams which empty into it, as they were seen near the mouths of those streams.

Saturday, 13th. The morning being fair and calm, Captain Lewis had all the remaining baggage embarked on board the six canoes, which sailed with two men in each for the upper camp. Then, with a sick man and the Indian woman, he left the encampment, and crossing over the river went on by land to join Captain Clark. From the head of the Whitebear islands he proceeded in a southwest direction, at the distance of three miles, till he struck the Missouri, which he then followed till he reached the place where all the party were occupied in boat-building. On his way he passed a very large Indian lodge, which was probably designed as a great council-house, but it differs in its construction from all that we have seen lower down the Missouri or elsewhere. The form of it was a circle two hundred and sixteen feet in circumference at the base, and composed of sixteen large cottonwood poles about fifty feet long, and at their thicker ends, which touched the ground, about the size of a man's body; they were distributed at equal distances, except that one was omitted to the east, probably for the entrance. From the circumference of this circle the poles converged towards the centre, where they were united and secured by large withes of willow brush. There was no covering over this fabric, in the centre of which were the remains of a large fire, and round it the marks of about eighty leathern lodges. He also saw a number of turtledoves, and some pigeons, of which he shot one differing in no respect from the wild pigeon of the United States. The country exhibits its usual appearances, the timber confined to the river, the country on both sides as far as the eye can reach being entirely destitute of trees or brush. In the low ground in which we are building the canoes the timber is larger and more abundant than we have seen it on the Missouri for several hundred miles. The soil, too, is good, for the grass and weeds reach about two feet high, being the tallest we have observed this season, though on the high plains and prairies the grass is at no season above three inches in height. Among these weeds are the sandrush, and nettle in small quantities; the plains are still infested by great numbers of the small birds already mentioned, among whom is the brown curlew. The current of the river is here extremely gentle; the buffaloe have not yet quite gone, for the hunters brought in three in very good order. It requires some diligence to supply us plentifully, for as we reserve our parched meal for the Rocky mountains, where we do not expect to find much game, our principal article of food is meat, and the consumption of the whole, thirty-two persons belonging to the party, amounts to four deer, an elk and a deer, or one buffaloe every twenty-four hours. The musquitoes and gnats persecute us as violently as below, so that we can get no sleep unless defended by biers, with which we are all provided. We here found several plants hitherto unknown to us, and of which we preserved specimens.

Serjeant Ordway proceeded with the six canoes five miles up the river, but the wind becoming so high as to wet the baggage, he was obliged to unload and dry it. The wind abated at five o'clock in the evening, when he again proceeded eight miles and encamped. The next morning,

Sunday, July 14, he joined us about noon. On leaving the Whitebear camp he passed at a short distance a little creek or run coming in on the left. This had been already examined, and called Flattery run; it contains back water only, with very extensive low grounds, which, rising into large plains, reach the mountains on the east; then passed a willow island on the left within one mile and a half, and reached, two miles farther, a cliff of rocks in a bend on the same side. In the course of another mile and a half he passed two islands covered with cottonwood, box-alder, sweet-willow, and the usual undergrowth, like that of the Whitebear islands. At thirteen and three-quarter miles he came to the mouth of a small creek on the left; within the following nine miles he passed three timbered islands, and after making twenty-three and a quarter miles from the lower camp, arrived at the point of woodland on the north where the canoes were constructed.

The day was fair and warm; the men worked very industriously, and were enabled by the evening to lanch the boats, which now want only seats and oars to be complete. One of them is twenty-five, the other thirty-three, feet in length and three feet wide. Captain Lewis walked out between three and four miles over the rocky bluffs to a high situation, two miles from the river, a little below Fort mountain creek. The country which he saw was in most parts level, but occasionally became varied by gentle rises and descents, but with no timber except along the water. From this position, the point at which the Missouri enters the first chain of the Rocky mountains bore south 28° west about twenty-five miles, according to our estimate.

The northern extremity of that chain north 73° west at the distance of eighty miles.

To the same extremity of the second chain north 65° west one hundred and fifty miles.

To the most remote point of a third and continued chain of these mountains north 50° west about two hundred miles.

The direction of the first chain was from south 20° east

to north 20° west; of the second, from south 45° east to north 45° west; but the eye could not reach their southern extremities, which most probably may be traced to Mexico. In a course south 75° west, and at the distance of eight miles, is a mountain, which from its appearance we shall call Fort mountain. It is situated in the level plain, and forms nearly a square, each side of which is a mile in extent. These sides, which are composed of a yellow clay with no mixture of rock or stone whatever, rise perpendicularly to the height of three hundred feet, where the top becomes a level plain, covered, as Captain Lewis now observed, with a tolerably fertile mould two feet thick, on which was a coat of grass similar to that of the plain below; it has the appearance of being perfectly inaccessible, and although the mounds near the falls somewhat resemble it, yet none of them are so large.

## CHAPTER XII.

The party embark on board the canoes—Description of Smith's river—Character of the country, etc.—Dearborne's river described—Captain Clark precedes the party for the purpose of discovering the Indians of the Rocky mountains—Magnificent rocky appearances on the borders of the river denominated the Gates of the Rocky mountains—Captain Clark arrives at the three forks of the Missouri without overtaking the Indians—The party arrive at the three forks, of which a particular and interesting description is given.

MONDAY, July 15. We rose early, embarked all our baggage on board the canoes, which, though light in number, are still heavily loaded, and at ten o'clock set out on our journey. At the distance of three miles we passed an island, just above which is a small creek coming in from the left, which we called Fort mountain creek, the channel of which is ten yards wide, but now perfectly dry. At six miles we came to an island opposite to a bend towards the north side, and reached, at seven and a half miles, the lower point of a woodland at the entrance of a beautiful river, which, in honour of the secretary of the navy, we called Smith's river. This stream falls into a bend on the south side of the Missouri, and is eighty yards wide. As far as we could discern its course it wound through a charming valley towards the southeast, in which many herds of buffaloe were feeding, till, at the distance of twenty-five miles, it entered the Rocky mountains, and was lost from our view. After dining near this place, we proceeded on four and three-quarter miles to the head of an island, four and a quarter miles beyond which is a second island, on the left; three and a quarter miles farther, in a bend of the river towards the north, is a wood where we encamped for the night, after making nineteen and three-quarter miles.

We find the prickly pear, one of the greatest beauties as well as the greatest inconveniences of the plains, now in full bloom. The sunflower, too, a plant common on every part of the Missouri from its entrance to this place, is here very abundant and in bloom. The lambs-quarter, wild cucumber, sandrush, and narrowdock are also common. Two elk, a deer, and an otter were our game to-day.

The river has now become so much more crooked than below that we omit taking all its short meanders, but note only its general course, and lay down the small bends on our daily chart by the eye. The general width is from one hundred to one hundred and fifty yards. Along the banks are large beds of sand raised above the plains, and as they always appear on the sides of the river opposite to the southwest exposure, seem obviously brought there from the channel of the river by the incessant winds from that quarter; we find also more timber than for a great distance below the falls.

Tuesday, 16th. There was a heavy dew last night. We soon passed about forty little booths, formed of willow bushes as a shelter against the sun. These seemed to have been deserted about ten days, and as we supposed by the Snake Indians, or Shoshonees, whom we hope soon to meet, as they appeared from the tracks to have a number of horses with them. At three and three-quarter miles we passed a creek or run in a bend on the left side, and four miles farther another run or small rivulet on the right. After breakfasting on a buffaloe shot by one of the hunters, Captain Lewis resolved to go on ahead of the party to the point where the river enters the Rocky mountains and make the necessary observations before our arrival. He therefore set out with Drewyer and two of the sick men, to whom he supposed the walk would be useful; he travelled on the north side of the river through a handsome level plain, which continued on the opposite side also, and at the distance of eight miles passed a small stream on which he observed a considerable quantity of the aspen tree. A little before twelve o'clock he halted on a bend to the north, in a low ground well covered with timber, about four and a half miles below the mountains, and obtained a meridian altitude, by which he found the latitude was N. 46° 46' 50" 2'". His route then lay through a high waving plain to a rapid where the Missouri first leaves the Rocky mountains, and here he encamped for the night.

In the mean time we had proceeded, after breakfast, one mile to a bend in the left, opposite to which was the frame of a large lodge situated in the prairie, constructed like that already mentioned above the Whitebear islands, but only sixty feet in diameter; round it were the remains of about eighty leathern lodges, all which seemed to have been built during the last autumn; within the next fifteen and a quarter miles we passed ten islands, on the last of which we encamped, near the right shore, having made twenty-three miles. The next morning,

Wednesday, 17th, we set out early, and at four miles distance joined Captain Lewis at the foot of the rapids, and after breakfast began the passage of them; some of the articles most liable to be injured by the water were carried round. We then double-manned the canoes, and with the aid of the towing-line got them up without accident. For several miles below the rapids the current of the Missouri becomes stronger as you approach, and the spurs of the mountain advance towards the river, which is deep and not more than seventy yards wide; at the rapids the river is closely hemmed in on both sides by the hills, and foams for half a mile over the rocks which obstruct its channel. The low grounds are now not more than a few yards in width, but they furnish room for an Indian road which winds under the hills on the north side of the river. The general range of these hills is from southeast to northwest, and the cliffs themselves are about eight hundred feet above the water, formed almost entirely of a hard black granite, on which are scattered a few dwarf pine and cedar trees. Immediately in the gap is a large rock four hundred feet high, which on one side is washed by the Missouri, while on its other sides a handsome little plain separates it from the neighbouring mountains. It may be ascended with some difficulty nearly to its summit, and affords a beautiful prospect of the plains below, in which we could observe large herds of buffaloe. After ascending the rapids for half a mile we came to a small island at the head of them, which we called Pine island, from a large pine tree at the lower end of it, which is the first we have seen near the river for a great distance, A mile beyond Captain Lewis's camp we had a meridian altitude, which gave us the latitude of 46° 42′ 14″ 7‴. As the canoes were still heavily loaded all those not employed in working them walked on shore. The navigation is now very laborious. The river is deep but with little current, and from seventy to one hundred yards wide; the low grounds are very narrow, with but little timber, and that chiefly the aspen tree. The cliffs are steep, and hang over the river so much that often we could not cross them, but were obliged to pass and repass from one side of the river to the other in order to make our way. In some places the banks are formed of rocks, of dark black granite, rising perpendicularly to a great height, through which the river seems in the progress of time to have worn its channel. On these mountains we see more pine than usual, but it is still in small quantities, Along the bottoms, which have a covering of high grass, we observe the sunflower blooming in great abundance. The Indians of the Missouri, and more especially those who do not cultivate maize, make great use of the seed of this plant for bread or in thickening their soup. They first parch and then pound it between two stones until it is reduced to a fine meal. Sometimes they add a portion of water, and drink it thus diluted; at other times they add a sufficient proportion of marrow grease to reduce it to the consistency of common dough, and eat it in that manner. This last composition we preferred to all the rest, and thought it at that time a very palatable dish. There is, however, little of the broad-leafed cottonwood on this side of the falls, much the greater part of what we see being of the narrow-leafed species. There are also great quantities of red, purple, yellow, and black currants. The currants are very pleasant to the taste, and much preferable to those of our common garden. The bush rises to the height of six or eight feet, the stem simple, branching, and erect. These shrubs associate in corps, either in upper or timbered lands near the water courses. The leaf is peteolate, of a pale green, and in form resembles the red currant so common in our gardens. The perianth of the fruit is one leaved, five cleft, abbriviated, and tubular. The corolla is monopetallous, funnel-shaped, very long, and of a fine orange colour. There are five stamens and one pistillum of the first, the filaments are capillar, inserted in the corolla, equal and converging, the anther ovate and incumbent. The germ of the second species is round, smooth, inferior, and pidicclled; the style long and thicker than the stamens, simple, cylindrical, smooth, and erect; it remains with the corolla until the fruit is ripe; the stamen is simple and obtuse, and the fruit much the size and shape of our common garden currants, growing like them in clusters supported by a compound footstalk. The peduncles are longer in this species, and the berries are more scattered.

The fruit is not so acid as the common currant, and has a more agreeable flavour.

The other species differs in no respect from the yellow currant excepting in the colour and flavour of the berries.

The serviceberry differs in some points from that of the United States. The bushes are small, sometimes not more than two feet high, and rarely exceed eight inches. They are proportionably small in their stems, growing very thickly, associated in clumps. The fruit is of the same form, but for the most part larger, and of a very dark purple. They are now ripe and in great perfection. There are two species of gooseberry here, but neither of them yet ripe; nor are the chokecherry, though in great quantities. Besides, there are also at that place the box-alder, red willow, and a species of sumach. In the evening we saw some mountain rams or big-horned animals, but no other game of any sort. After leaving Pine island we passed a small run on the left, which is formed by a large spring rising at the distance of half a mile under the mountain. One mile and a half above the island is another, and two miles farther a third island, the river making small bends constantly to the north. From this last island to a point of rocks on the south side the low grounds become rather wider, and three-quarters of a mile beyond these rocks, in a bend on the north, we encamped opposite to a very high cliff, having made during the day eleven and a half miles.

Thursday, 18th. This morning early, before our departure, we saw a large herd of the big-horned animals, who were bounding among the rocks in the opposite cliff with great agility. These inaccessible spots secure them from all their enemies, and the only danger is in wandering among these precipices, where we should suppose it scarcely possible for any animal to stand; a single false step would precipitate them at least five hundred feet into the water. At one mile and a quarter we passed another single cliff on the left, at the same distance beyond which is the mouth of a large river emptying itself from the north. It is a handsome, bold, and clear stream, eighty yards wide, that is, nearly as broad as the Missouri, with a rapid current over a bed of small smooth stones of various figures. The water is extremely transparent; the low grounds are narrow, but possess as much wood as those of the Missouri; and it has every appearance of being navigable, though to what distance we cannot ascertain, as the country which it waters is broken and mountainous. In honour of the secretary at war we called it Dearborn's river. Being now very anxious to meet with the Shoshonees or Snake Indians, for the purpose of obtaining the necessary information of our route, as well as to procure horses, it was thought best for one of us to go forward with a small party and endeavour to discover them before the daily discharge of our guns, which is necessary for our subsistence, should give them notice of our approach; if by an accident they hear us, they will most probably retreat to the mountains, mistaking us for their enemies, who usually attack them on this side. Accordingly, Captain Clark set out with three men, and followed the course of the river on the north side; but the hills were so steep at first that he was not able to go much faster than ourselves. In the evening, however, he cut off many miles of the circuitous course of the river by crossing a mountain over which he found a wide Indian road, which in many places seems to have been cut or dug down in the earth. He passed also two branches of a stream which he called Ordway's creek, where he saw a number of beaver-dams extending in close succession towards the mountains as far as he could distinguish; on the cliffs were many of the big-horned animals. After crossing this mountain he encamped near a small stream of running water, having travelled twenty miles.

On leaving Dearborn's river we passed at three and a half miles a small creek, and at six beyond it an island on the north side of the river, which makes within that distance many small bends. At two and a half miles farther is another island; three-quarters of a mile beyond this is a small creek on the north side. At a mile and a half above the creek is a much larger stream thirty yards wide, and discharging itself with a bold current on the north side; the banks are low, and the bed formed of stones altogether. To this stream we gave the name of Ordway's creek, after Serjeant John Ordway. At two miles beyond this the valley widens; we passed several bends of the river, and encamped in the centre of one on the south, having made twenty-one miles. Here we found a small grove of the narrow-leafed cottonwood, there being no longer any of the broad-leafed kind since we entered the mountains. The water of these rivulets which come down from the mountains is very cold, pure, and well tasted. Along their banks, as well as on the Missouri, the aspen is very common, but of a small kind. The river is somewhat wider than we found it yesterday; the hills more distant from the river and not so high; there are some pines on the mountains, but they are principally confined to the upper regions of them; the low grounds are still narrower, and have little or no timber. The soil near the river is good, and produces a luxuriant growth of grass and weeds; among these productions the sunflower holds a very distinguished place. For several days past we have observed a species of flax in the low grounds, the leaf-stem and pericarp of which resemble those of the flax commonly cultivated in the United States; the stem rises to the height of two and a half or three feet, and spring to the number of eight or ten from the same root, with a strong thick bark apparently well calculated for use; the root seems to be perennial, and it is probable that the cutting of the stems may not at all injure it, for although the seeds are not yet ripe, there are young suckers shooting up from the root, whence we may infer that the stems, which are fully grown and in the proper stage of vegetation to produce the best flax, are not essential to the preservation or support of the root, a circumstance which would render it a most valuable plant. To-day we have met with a second species of flax smaller than the first, as it seldom obtains a greater height than nine or twelve inches; the leaf and stem resemble those of the species just mentioned, except that the latter is rarely branched, and bears a single monopetalous bell-shaped blue flower, suspended with its limb downwards, We saw several herds of the big-horn, but they were in the cliffs beyond our reach. We killed an elk this morning, and found part of a deer which had been left for us by Captain Clark. He pursued his route,

Friday, 19th, early in the morning, and soon passed the remains of several Indian camps formed of willow brush, which seemed to have been deserted this spring. At the same time he observed that the pine trees had been stripped of their bark about the same season, which our Indian woman says her countrymen do in order to obtain the sap and the soft parts of the wood and bark for food. About eleven o'clock he met a herd of elk and killed two of them, but such was the want of wood in the neighbourhood that he was unable to procure enough to make a fire, and he was therefore obliged to substitute the dung of the buffaloe, with which he cooked his breakfast. They then resumed their course along an old Indian road. In the afternoon they reached a handsome valley watered by a large creek, both of which extend a considerable distance into the mountain; this they crossed, and during the evening travelled over a mountainous country covered with sharp fragments of flint-rock;

these bruised and cut their feet very much, but were scarcely less troublesome than the prickly pear of the open plains, which have now become so abundant that it is impossible to avoid them, and the thorns are so strong that they pierce a double soal of dressed deer skin; the best resource against them is a soal of buffaloe hide in parchment. At night they reached the river, much fatigued, having passed two mountains in the course of the day and having travelled thirty miles. Captain Clark's first employment on lighting a fire was to extract from his feet the briars, which he found seventeen in number.

In the meantime we proceeded on very well, though the water appears to increase in rapidity as we advance; the current has indeed been strong during the day and obstructed by some rapids, which are not, however, much broken by rocks, and are perfectly safe; the river is deep, and its general width is from one hundred to one hundred and fifty yards wide. For more than thirteen miles we went along the numerous bends of the river, and then reached two small islands, three and three-quarter miles beyond which is a small creek in a bend to the left, above a small island on the right side of the river. We were regaled about ten o'clock P. M. with a thunder-storm of rain and hail which lasted for an hour, but during the day, in this confined valley through which we are passing, the heat is almost insupportable; yet whenever we obtain a glimpse of the lofty tops of the mountains we are tantalized with a view of the snow. These mountains have their sides and summits partially varied with little copses of pine, cedar, and balsam fir. A mile and a half beyond this creek the rocks approach the river on both sides, forming a most sublime and extraordinary spectacle. For five and three-quarter miles these rocks rise perpendicularly from the water's edge to the height of nearly twelve hundred feet.

They are composed of a black granite near its base, but from its lighter colour above and from the fragments we suppose the upper part to be flint of a yellowish brown and cream colour. Nothing can be imagined more tremendous than the frowning darkness of these rocks, which project over the river and menace us with destruction. The river, of one hundred and fifty yards in width, seems to have forced its channel down this solid mass, but so reluctantly has it given way that during the whole distance the water is very deep, even at the edges, and for the first three miles there is not a spot, except one of a few yards, in which a man could stand between the water and the towering perpendicular of the mountain; the convulsion of the passage must have been terrible, since at its outlet there are vast columns of rock torn from the mountain which are strewed on both sides of the river, the trophies, as it were, of the victory. Several fine springs burst out from the chasms of the rock, and contribute to increase the river, which has now a strong current, but very fortunately we are able to overcome it with our oars, since it would be impossible to use either the cord or the pole. We were obliged to go on some time after dark, not being able to find a spot large enough to encamp on; but at length, about two miles above a small island in the middle of the river, we met with a spot on the left side, where we procured plenty of lightwood and pitchpine. This extraordinary range of rocks we called the Gates of the Rocky mountains. We had made twenty-two miles, and four and a quarter miles from the entrance of the Gates. The mountains are higher to-day than they were yesterday. We saw some big-horns, a few antelopes and beaver, but since entering the mountains have found no buffaloe; the otter are, however, in great plenty; the musquitoes have become less troublesome than they were.

Saturday, 20th. By employing the towrope whenever the

banks permitted the use of it, the river being too deep for the pole, we were enabled to overcome the current, which is still strong. At the distance of half a mile we came to a high rock in a bend to the left in the Gates. Here the perpendicular rocks cease, the hills retire from the river, and the vallies suddenly widen to a greater extent than they have been since we entered the mountains. At this place was some scattered timber, consisting of the narrow-leafed cottonwood, the aspen, and pine. There are also vast quantities of gooseberries, serviceberries, and several species of currant, among which is one of a black colour, the flavour of which is preferable to that of the yellow, and would be deemed superior to that of any currant in the United States. We here killed an elk, which was a pleasant addition to our stock of food. At a mile from the Gates, a large creek comes down from the mountains and empties itself behind an island in the middle of a bend to the north. To this stream, which is fifteen yards wide, we gave the name of Potts's creek, after John Potts, one of our men. Up this valley about seven miles we discovered a great smoke, as if the whole country had been set on fire; but were at a loss to decide whether it had been done accidently by Captain Clark's party, or by the Indians as a signal on their observing us. We afterwards learnt that this last was the fact; for they had heard a gun fired by one of Captain Clark's men, and believing that their enemies were approaching had fled into the mountains, first setting fire to the plains as a warning to their countrymen. We continued our course along several islands, and having made in the course of the day fifteen miles, encamped just above an island, at a spring on a high bank on the left side of the river. In the latter part of the evening we had passed through a low range of mountains, and the country became more open, though still unbroken and without timber, and the lowlands not very extensive; and just above our camp the river is again closed in by the mountains. We found on the banks an elk which Captain Clark had left us, with a note mentioning that he should pass the mountains just above us and wait our arrival at some convenient place. We saw, but could not procure, some red-headed ducks and sandhill cranes along the sides of the river, and a woodpecker about the size of the lark-woodpecker, which seems to be a distinct species; it is as black as a crow, with a long tail, and flies like a jaybird. The whole country is so infested by the prickly pear that we could scarcely find room to lie down at our camp.

Captain Clark, on setting out this morning, had gone through the valley about six miles to the right of the river. He soon fell into an old Indian road, which he pursued till he reached the Missouri, at the distance of eighteen miles from his last encampment, just above the entrance of a large creek, which we afterwards called Whiteearth creek. Here he found his party so much cut and pierced with the sharp flint and the prickly pear that he proceeded only a small distance farther, and then halted to wait for us. Along his track he had taken the precaution to strew signals, such as pieces of cloth, paper, and linen, to prove to the Indians, if by accident they met his track, that we were white men. But he observed a smoke some distance ahead, and concluded that the whole country had now taken the alarm.

Sunday, 21st. On leaving our camp we passed an island at half a mile, and reached at one mile a bad rapid at the place where the river leaves the mountain; here the cliffs are high and covered with fragments of broken rocks; the current is also strong, but although more rapid, the river is wider and shallower, so that we are able to use the pole occasionally, though we principally depend on the towline. On leaving the rapid, which is about half a mile in extent, the country opens on each side; the hills become lower; at one mile is a large island on the left side, and four and a half beyond it a large and bold creek twenty-eight yards wide, coming in from the north, where it waters a handsome valley; we called it Pryor's creek, after one of the sergeants, John Pryor. At a mile above this creek, on the left side of the Missouri, we obtained a meridian altitude, which gave 46° 10' 32" 9'" as the latitude of the place. For the following four miles the country, like that through which we passed during the rest of the day, is rough and mountainous, as we found it yesterday; but at the distance of twelve miles, we came towards evening into a beautiful plain ten or twelve miles wide, and extending as far as the eye could reach. This plain, or rather valley, is bounded by two nearly parallel ranges of high mountains whose summits are partially covered with snow, below which the pine is scattered along the sides down to the plain in some places, though the greater part of their surface has no timber and exhibits only a barren soil, with no covering except dry parched grass or black rugged rocks. On entering the valley the river assumes a totally different aspect; it spreads to more than a mile in width, and though more rapid than before, is shallow enough in almost every part for the use of the pole, while its bed is formed of smooth stones and some large rocks, as it has been, indeed, since we entered the mountains; it is also divided by a number of islands, some of which are large, near the northern shore. The soil of the valley is a rich black loam, apparently very fertile, and covered with a fine green grass about eighteen inches or two feet in height, while that of the high grounds is perfectly dry and seems scorched by the sun. The timber, though still scarce, is in greater quantities in this valley than we have seen it since entering the mountains, and seems to prefer the borders of the small creeks to the banks of the river itself. We advanced three and a half miles in this valley, and encamped on the left side, having made in all fifteen and a half miles.

Our only large game to-day was one deer. We saw, however, two pheasants, of a dark brown colour, much larger than the same species of bird in the United States. In the morning, too, we saw three swans, which, like the geese, have not yet recovered the feathers of the wing, and were unable to fly; we killed two of them, and the third escaped by diving and passing down the current. These are the first we have seen on the river for a great distance, and as they had no young with them, we presume that they do not breed in this neighbourhood. Of the geese we daily see great numbers, with their young perfectly feathered except on the wings, where both young and old are deficient; the first are very fine food, but the old ones are poor and unfit for use. Several of the large brown or sandhill crane are feeding in the low grounds on the grass, which forms their principal food. The young crane cannot fly at this season; they are as large as a turkey, of a bright reddish bay colour. Since the river has become shallow we have caught a number of trout to-day, and a fish, white on the belly and sides, but of a bluish cast on the back, and a long pointed mouth opening somewhat like that of the shad.

This morning Captain Clark, wishing to hunt, but fearful of alarming the Indians, went up the river for three miles, when, finding neither any of them nor of their recent tracks, returned, and then his little party separated to look for game. They killed two bucks and a doe, and a young curlew nearly feathered; in the evening they found the musquitoes as troublesome as we did; these animals attack us as soon as the labours and fatigues of the day require some rest, and annoy us till several hours after dark, when the coldness of the air obliges them to disappear; but such is their persecution that were it not for our biers we should obtain no repose.

Monday, 22d. We set out at an early hour. The river being divided into so many channels by both large and small islands that it was impossible to lay it down accurately by following in a canoe any single channel, Captain Lewis walked on shore, took the general courses of the river, and from the rising grounds laid down the situation of the islands and channels, which he was enabled to do with perfect accuracy, the view not being obstructed by much timber. At one mile and a quarter we passed an island somewhat larger than the rest, and four miles farther reached the upper end of another, on which we breakfasted. This is a large island, forming in the middle of a bend to the north a level fertile plain, ten feet above the surface of the water and never overflowed. Here we found great quantities of a small onion about the size of a musket ball, though some were larger; it is white, crisp, and as well flavoured as any of our garden onions; the seed is just ripening, and as the plant bears a large quantity to the square foot, and stands the rigours of the climate, it will no doubt be an acquisition to settlers. From this production we called it Onion island. During the next seven and three-quarter miles we passed several long circular bends, and a number of large and small islands which divide the river into many channels, and then reached the mouth of a creek on the north side. It is composed of three creeks, which unite in a handsome valley about four miles before they discharge themselves into the Missouri, where it is about fifteen feet wide and eight feet deep, with clear, transparent water. Here we halted for dinner, but as the canoes took different channels in ascending, it was some time before they all joined. Here we were delighted to find that the Indian woman recognizes the country; she tells us that to this creek her countrymen make excursions to procure a white paint on its banks, and we therefore call it Whiteearth creek. She says also that the three forks of the Missouri are at no great distance, a piece of intelligence which has cheered the spirits of us all, as we hope soon to reach the head of that river. This is the warmest day except one we have experienced this summer. In the shade the mercury stood at 80° above zero, which is the second time it has reached that height during this season. We encamped on an island, after making nineteen and three-quarter miles.

In the course of the day we saw many geese, cranes, small birds common to the plains, and a few pheasants; we also observed a small plover or curlew, of a brown colour, about the size of the yellow-legged plover or jack curlew, but of a different species. It first appeared near the mouth of Smith's river, but is so shy and vigilant that we were unable to shoot it. Both the broad and narrow leafed willow continue, though the sweet willow has become very scarce. The rosebush, small honeysuckle, the pulpy-leafed thorn, southern wood, sage, and box-alder, narrow-leafed cotonwood, redwood, and a species of sumach, are all abundant. So, too, are the red and black gooseberries, serviceberries, chokecherry, and the black, red, yellow, and purple currant, which last seems to be a favourite food of the bear. Before encamping we landed and took on board Captain Clark, with the meat he had collected during this day's hunt, which consisted of one deer and an elk; we had ourselves shot a deer and an antelope. The musquitoes and gnats were unusually fierce this evening.

Tuesday, 23d. Captain Clark again proceeded, with four men, along the right bank. During the whole day the river is divided by a number of islands, which spread it out sometimes to the distance of three miles; the current is very rapid and has many ripples, and the bed formed of gravel and smooth stones. The banks along the low grounds are of a rich loam, followed occasionally by low bluffs of yellow and red clay, with a hard red slatestone intermixed. The low grounds are wide, and have very little timber but a thick underbrush of willow, and rose and currant bushes; these are succeeded by high plains, extending on each side to the base of the mountains, which lie parallel to the river about eight or twelve miles apart, and are high and rocky, with some small pine and cedar interspersed on them. At the distance of seven miles a creek twenty yards wide, after meandering through a beautiful low ground on the left for several miles, parallel to the river, empties itself near a cluster of small islands; the stream we called Whitehouse creek, after Joseph Whitehouse, one of the party, and the islands, from their number, received the name of the "Ten islands." About ten o'clock we came up with Drewyer, who had gone out to hunt yesterday, and not being able to find our encampment had staid out all night; he now supplied us with five deer. Three and a quarter miles beyond Whitehouse creek we came to the lower point of an island where the river is three hundred yards wide, and continued along it for one mile and a quarter, and then passed a second island just above it. We halted rather early for dinner, in order to dry some part of the baggage which had been wet in the canoes; we then proceeded, and at five and a half miles had passed two small islands. Within the next three miles we came to a large island, which from its figure we called Broad island. From that place we made three and a half miles, and encamped on an island to the left, opposite to a much larger one on the right. Our journey to-day was twenty-two and a quarter miles, the greater part of which was made by means of our poles and cords, the use of which the banks much favoured. During the whole time we had the small flags hoisted in the canoes to apprise the Indians, if there were any in the neighbourhood, of our being white men and their friends, but we were not so fortunate as to discover any of them. Along the shores we saw great quantities of the common thistle, and procured a further supply of wild onions and a species of garlic growing on the highlands, which is now green and in bloom; it has a flat leaf, and is strong, tough, and disagreeable. There was also much of the wild flax, of which we now obtained some ripe seed, as well as some bullrush and cat-tail flag. Among the animals we met with a black snake about two feet long, with the belly as dark as any other part of the body, which was perfectly black, and which had one hundred and twenty-eight scuta on the belly and sixty-three on the tail; we also saw antelopes, crane, geese, ducks, beaver, and otter, and took up four deer which had been left on the water side by Captain Clark. He had pursued all day an Indian road on the right side of the river, and encamped late in the evening at the distance of twenty-five miles from our camp of last night. In the course of his walk he met, besides deer, a number of antelopes and a herd of elk, but all the tracks of Indians, though numerous, were of an old date.

Wednesday, 24th. We proceeded for four and a quarter miles along several islands to a small run, just above which the low bluffs touch the river. Within three and a half miles farther we came to a small island on the north, and a remarkable bluff composed of earth of a crimson colour, intermixed with stratas of slate, either black or of a red resembling brick. The following six and three-quarter miles brought us to an assemblage of islands, having passed four at different distances; and within the next five miles we met the same number of islands, and encamped on the north, after making nineteen and a half miles. The current of the river was strong, and obstructed, as indeed it has been for some days, by small rapids or ripples which descend from one to three feet in the course of one hundred and fifty yards, but they are rarely incommoded by any fixed rocks, and therefore, though the water is rapid, the passage is not attended with danger. The valley through which the river passes is like that of yesterday, the nearest hills generally concealing the most distant from us; but when we obtain a view of them they present themselves in amphitheatre, rising above each other as they recede from the river, till the most remote are covered with snow. We saw many otter and beaver to-day; the latter seem to contribute very much to the number of islands and the widening of the river. They begin by damming up the small channels of about twenty yards between the islands; this obliges the river to seek another outlet, and as soon as this is effected the channel stopped by the beaver becomes filled with mud and sand. The industrious animal is then driven to another channel, which soon shares the same fate, till the river spreads on all sides, and cuts the projecting points of the land into islands. We killed a deer, and saw great numbers of antelopes, cranes, some geese, and a few red-headed ducks. The small birds of the plains and the curlew are still abundant; we saw, but could not come within gunshot of, a large bear. There is much of the track of elk but none of the animals themselves, and from the appearance of bones and old excrement we suppose that buffaloe have sometimes strayed into the valley, though we have as yet seen no recent sign of them. Along the water are a number of snakes, some of a brown uniform colour, others black, and a third speckled on the abdomen, and striped with black and a brownish yellow on the back and sides. The first, which are the largest, are about four feet long; the second is of the kind mentioned yesterday, and the third resembles in size and appearance the garter-snake of the United States. On examining the teeth of all these several kinds we found them free from poison; they are fond of the water, in which they take shelter on being pursued. The musquitoes, gnats, and prickly pear, our three persecutors, still continue with us, and, joined with the labour of working the canoes, have fatigued us all excessively. Captain Clark continued along the Indian road, which led him up a creek. About ten o'clock he saw at the distance of six miles a horse feeding in the plains. He went towards him, but the animal was so wild that he could not get within several hundred paces of him; he then turned obliquely to the river, where he killed a deer and dined, having passed in this valley five handsome streams, only one of which had any timber; another had some willows, and was very much dammed up by the beaver. After dinner he continued his route along the river, and encamped at the distance of thirty miles. As he went along he saw many tracks of Indians, but none of recent date. The next morning,

Thursday, 25th, at the distance of a few miles he arrived at the three forks of the Missouri. Here he found that the plains had been recently burnt on the north side, and saw the track of a horse which seemed to have passed about four or five days since. After breakfast he examined the rivers, and finding that the north branch, although not larger, contained more water than the middle branch, and bore more to the westward, he determined to ascend it. He therefore left a note informing Captain Lewis of his intention, and then went up that stream on the north side for about twenty-five miles. Here Chaboneau was unable to proceed any farther, and the party therefore encamped, all of them much fatigued, their feet blistered and wounded by the prickly pear.

In the meantime we left our camp and proceeded on very well, though the water is still rapid and has some occasional ripples. The country is much like that of yesterday; there are, however, fewer islands, for we passed only two. Behind one of them is a large creek twenty-five yards wide, to which we gave the name of Gass's creek, from one of our serjeants, Patrick Gass; it is formed by the union of five streams, which descend from the mountains and join in the plain near the river. On this island we saw a large brown bear, but he retreated to the shore and ran off before we could approach him. These animals seem more shy than they were below the mountains. The antelopes have again collected in small herds, composed of several females with their young, attended by one or two males, though some of the males are still solitary or wander in parties of two over the plains, which the antelope invariably prefers to the woodlands, and to which it always retreats if by accident it is found straggling in the hills, confiding no doubt in its wonderful fleetness. We also killed a few young geese, but as this game is small and very incompetent to the subsistence of the party, we have forbidden the men any longer to waste their ammunition on them. About four and a half miles above Gass's creek the valley in which we have been travelling ceases; the high craggy cliffs again approach the river, which now enters, or rather leaves, what appears to be a second great chain of the Rocky mountains. About a mile after entering these hills, or low mountains, we passed a number of fine bold springs, which burst out near the edge of the river under the cliffs on the left, and furnished a fine freestone water; near these we met with two of the worst rapids we have seen since entering the mountains, a ridge of sharp pointed rocks stretching across the river, leaving but small and dangerous channels for the navigation. The cliffs are of a lighter colour than those we have already passed, and in the bed of the river is some limestone, which is small and worn smooth, and seems to have been brought down by the current. We went about a mile farther and encamped under a high bluff on the right, opposite to a cliff of rocks, having made sixteen miles.

All these cliffs appeared to have been undermined by the water at some period and fallen down from the hills on their sides, the stratas of rock sometimes lying with their edges upwards; others not detached from the hills are depressed obliquely on the side next the river, as if they had sunk to fill up the cavity formed by the washing of the river.

In the open places among the rocky cliffs are two kinds of gooseberry, one yellow and the other red. The former species was observed for the first time near the falls; the latter differs from it in no respect except in colour and in being of a larger size; both have a sweet flavour, and are rather indifferent fruit.

Friday, 26th. We again found the current strong and the ripples frequent; these we were obliged to overcome by means of the cord and the pole, the oar being scarcely ever used except in crossing to take advantage of the shore. Within three and three-quarter miles we passed seven small islands, and reached the mouth of a large creek which empties itself in the centre of a bend on the left side; it is a bold running stream fifteen yards wide, and received the name of Howard creek, after John P. Howard, one of the party. One mile beyond it is a small run which falls in on the same side just above a rocky cliff. Here the mountains recede from the river, and the valley widens to the extent of several miles. The river now becomes crowded with islands, of which we passed ten in the next thirteen and three-quarter miles; then, at the distance of eighteen miles, we encamped on the left shore near a rock in the centre of a bend towards the left, and opposite to two more islands. This valley has wide low grounds covered with high grass, and in many with a fine turf of green sward. The soil of the highlands is thin and meagre, without any covering except a low sedge and a dry kind of grass, which is almost as inconvenient as the prickly pear. The seeds of it are armed with a long twisted hard beard at their upper extremity, while the lower part is a sharp firm point, beset at its base with little stiff bristles, with the points in a direction contrary to the subulate point to which they answer as a barb. We see also another species of prickly pear. It is of a globular form, composed of an assemblage of little conic leaves springing from a common root to which their small points are attached as a common centre, and the base of the cone forms the apex of the leaf, which is garnished with a circular range of sharp thorns like the cochineal plant, and quite as stiff and even more keen than those of the common flat-leafed species. Between the hills the river had been confined within one hundred and fifty or two hundred yards, but in the valley it widens to two hundred or two hundred and fifty yards, and sometimes is spread by its numerous islands to the distance of three-quarters of a mile. The banks are low, but the river never overflows them. On entering the valley we again saw the snow-clad mountains before us, but the appearance of the hills, as well as of the timber near us, is much as heretofore.

Finding Chaboneau unable to proceed, Captain Clark left him with one of the men, and, accompanied by the other, went up the river about twelve miles to the top of a mountain, Here he had an extensive view of the river valley upwards, and saw a large creek which flowed in on the right side. He however discovered no fresh sign of the Indians, and therefore determined to examine the middle branch, and join us by the time we reached the forks. He descended the mountain by an Indian path which wound through a deep valley, and at length reached a fine cold spring. The day had been very warm, the path unshaded by timber, and his thirst was excessive; he was therefore tempted to drink; but although he took the precaution of previously wetting his head, feet and hands, he soon found himself very unwell; he continued his route, and after resting with Chaboneau at his camp, resumed his march across the north fork near a large island. The first part was knee-deep, but on the other side of the island the water came to their waists, and was so rapid that Chaboneau was on the point of being swept away, and not being able to swim would have perished if Captain Clark had not rescued him. While crossing the island they killed two brown bear and saw great quantities of beaver. He then went on to a small river which falls into the north fork some miles above its junction with the two others; here, finding himself grow more unwell, he halted for the night at the distance of four miles from his last encampment.

Saturday, 27th. We proceeded on but slowly, the current being still so rapid as to require the utmost exertions of us all to advance, and the men are losing their strength fast in consequence of their constant efforts. At half a mile we passed an island, and a mile and a quarter farther again entered a ridge of hills, which now approach the river, with cliffs apparently sinking like those of yesterday. They are composed of a solid limestone, of a light lead colour when exposed to the air, though when freshly broken it is of a deep blue, and of an excellent quality and very fine grain. On these cliffs were numbers of the big-horn. At two and a half miles we reached the centre of a bend towards the south, passing a small island, and at one mile and a quarter beyond this reached, about nine in the morning, the mouth of a river seventy yards wide, which falls in from the southeast. Here the country suddenly opens into extensive and beautiful meadows and plains, surrounded on every side with distant and lofty mountains. Captain Lewis went up this stream for about half a mile, and from the height of a limestone cliff could observe its course about seven miles, and the three forks of the Missouri, of which this river is one. Its extreme point bore S. 65° E., and during the seven miles it passes through a green extensive meadow of fine grass, dividing itself into several streams, the largest passing near the ridge of hills on which he stood. On the right side of the Missouri a high, wide, and extensive plain succeeds to this low meadow which reaches the hills. In the meadow a large spring rises about a quarter of a mile from this southeast fork, into which it discharges itself on the right side about four hundred paces from where he stood. Between the southeast and middle forks a distant range of snow-topped mountains spread from east to south above the irregular broken hills nearer to this spot; the middle and southwest forks unite at half a mile above the entrance of the southeast fork. The extreme point at which the former can be seen bears S. 15° E., and at the distance of fourteen miles, where it turns to the right round the point of a high plain and disappears from the view. Its low grounds are several miles in width, forming a smooth and beautiful green meadow, and like the southeast fork it divides itself into several streams. Between these two forks, and near their junction with that from the southwest, is a position admirably well calculated for a fort. It is a limestone rock of an oblong form, rising from the plain perpendicularly to the height of twenty-five feet on three of its sides; the fourth, towards the middle fork, being a gradual ascent and covered with a fine green sward, as is also the top, which is level and contains about two acres. An extensive plain lies between the middle and southwest forks, the last of which, after watering a country like that of the other two branches, disappears about twelve miles off, at a point bearing south 30° west. It is also more divided and serpentine in its course than the other two, and possesses more timber in its meadows. This timber consists almost exclusively of the narrow-leafed cottonwood, with an intermixture of box-alder and sweet-willow, the underbrush being thick and like that of the Missouri lower down. A range of high mountains partially covered with snow is seen at a considerable distance, running from south to west, and nearly all around us are broken ridges of country like that below, through which those united streams appear to have forced their passage. After observing the country, Captain Lewis descended to breakfast. We then left the mouth of the southeast fork, which, in honour of the secretary of the treasury we called Gallatin's river, and at the distance of half a mile reached the confluence of the southwest and middle branches of the Missouri. Here we found the letter from Captain Clark, and as we agreed with him that the direction of the southwest fork gave it a decided preference over the others, we ascended that branch of the river for a mile, and encamped in a level handsome plain on the left, having advanced only seven miles. Here we resolved to wait the return of Captain Clark, and in the meantime make the necessary celestial observations, as this seems an essential point in the geography of the western world, and also to recruit the men and air the baggage. It was accordingly all unloaded and stowed away on shore. Near the three forks we saw many collections of the mud-nests of the small martin attached to the smooth faces of the limestone rock, where they were sheltered by projections of the rock above it; and in the meadows were numbers of the duck or mallard, with their young, who are now nearly grown. The hunters returned towards evening with six deer, three otter, and a muskrat, and had seen great numbers of antelopes, and much sign of the beaver and elk.

During all last night Captain Clark had a high fever and chills, accompanied with great pain. He, however, pursued his route eight miles to the middle branch, where, not finding any fresh Indian track, he came down it and joined us about three o'clock, very much exhausted with fatigue and the violence of his fever. Believing himself bilious, he took a dose of Rush's pills, which we have always found sovereign in such cases, and bathing the lower extremities in warm water.

We are now very anxious to see the Snake Indians. After advancing for several hundred miles into this wild and mountainous country, we may soon expect that the game will abandon us. With no information of the route, we may be unable to find a passage across the mountains when we reach the head of the river, at least such a one as will lead us to the Columbia, and even were we so fortunate as to find a branch of that river, the timber which we have hitherto seen in these mountains does not promise us any fit to make canoes, so that our chief dependence is on meeting some tribe from whom we may procure horses. Our consolation is, that this southwest branch can scarcely head with any other river than the Columbia, and that if any nation of Indians can live in the mountains we are able to endure as much as they, and have even better means of procuring subsistence.

## CHAPTER XIII.

The name of the Missouri changed, as the river now divides itself into three forks, one of which is called after Jefferson, the other after Madison, and the other after Gallatin—Their general character—The party ascend the Jefferson branch—Description of the river Philosophy, which enters into the Jefferson—Captain Lewis and a small party go in advance in search of the Shoshonees—Description of the country, etc., bordering on the river—Captain Lewis still preceding the main party in quest of the Shoshonees—A singular accident which prevented Captain Clark from following Captain Lewis's advice, and ascending the middle fork of the river—Description of Philanthropy river, another stream running into the Jefferson—Captain Lewis and a small party, having been unsuccessful in their first attempt, set off a second time in quest of the Shoshonees.

SUNDAY, July 28. Captain Clark continued very unwell during the night, but was somewhat relieved this morning. On examining the two streams, it became difficult to decide which was the larger or the real Missouri; they are each ninety yards wide, and so perfectly similar in character and appearance that they seem to have been formed in the same mould, We were therefore induced to discontinue the name of Missouri, and gave to the southwest branch the name of Jefferson, in honour of the President of the United States, and the projector of the enterprise, and called the middle branch Madison, after James Madison, secretary of state. These two, as well as Gallatin river, run with great velocity and throw out large bodies of water. Gallatin river is, however, the most rapid of the three, and though not quite as deep, yet navigable for a considerable distance. Madison river, though much less rapid than the Gallatin, is somewhat more rapid than the Jefferson; the beds of all of them are formed of smooth pebble and gravel, and the waters are perfectly transparent.

The timber in the neighbourhood would be sufficient for the ordinary uses of an establishment, which, however, it would be adviseable to build of brick, as the earth appears calculated for that purpose, and along the shores are some bars of fine pure sand. The greater part of the men, having yesterday put their deer skins in water, were this day engaged in dressing them, for the purpose of making clothing. The weather was very warm, the thermometer in the afternoon was at 90° above zero, and the musquitoes more than usually inconvenient; we were, however, relieved from them by a high wind from the southwest which came on at four o'clock, bringing a storm of thunder and lightning, attended by refreshing showers, which continued till after dark. In the evening the hunters returned with eight deer and two elk; and the party who had been sent up the Gallatin reported that after passing the point, where it escaped from Captain Lewis's view yesterday, it turned more towards the east, as far as they could discern the opening of the mountains, formed by the valley which bordered it. The low grounds were still wide but not so extensive as near its mouth, and though the stream is rapid and much divided by islands, it is still sufficiently deep for navigation with canoes. The low grounds, although not more than eight or nine feet above the water, seem never to be overflowed, except a part on the west side of the middle fork, which is stony and seems occasionally inundated, are furnished with great quantities of small fruit, such as currants and gooseberries; among the last of which is a black species, which we observe not only in the meadows but along the mountain rivulets. From the same root rise a number of stems to the height of five or six feet, some of them particularly branched and all reclining. The berry is attached by a long peduncle to the stem from which they hang, of a smooth ovate form, as large as the common garden gooseberry, and as black as jet, though the pulp is of a bright crimson colour. It is extremely acid; the form of the leaf resembles that of the common gooseberry, though larger. The stem is covered with very sharp thorns or briars. The grass, too, is very luxuriant, and would yield fine hay in parcels of several acres. The sand-rushes will grow in many places as high as a man's breast, and as thick as stalks of wheat; it would supply the best food during the winter to cattle of any trading or military post.

Sacajawea, our Indian woman, informs us that we are encamped on the precise spot where her countrymen, the Snake Indians, had their huts five years ago, when the Minnetarees of Knife river first came in sight of them, and from which they hastily retreated three miles up the Jefferson, and concealed themselves in the woods. The Minnetarees, however, pursued and attacked them, killed four men, as many women, and a number of boys, and made prisoners of four other boys, and all the females, of whom Sacajawea was one; she does not, however, show any distress at these recollections, nor any joy at the prospect of being restored to her country; for she seems to possess the folly or the philosophy of not suffering her feelings to extend beyond the anxiety of having plenty to eat and a few trinkets to wear.

Monday, 29th. This morning the hunters brought in some fat deer of the long-tailed red kind, which are quite as large as those of the United States, and are, indeed, the only kind we have found at this place. There are numbers of the sand-hill cranes feeding in the meadows; we caught a young one of the same colour as the red deer, which, though it had nearly attained its full growth, could not fly; it is very fierce and strikes a severe blow with its beak. The kingfisher has become quite common on this side of the falls, but we have seen none of the summer duck since leaving that place. The mallard duck, which we saw for the first time on the 20th instant, with their young, are now abundant, though they do not breed on the Missouri, below the mountains. The small birds already described are also abundant in the plains; here, too, are great quantities of grasshoppers or crickets; and among other animals, a large ant with a reddish brown body and legs, and a black head and abdomen, who build little cones of gravel ten or twelve inches high, without a mixture of sticks, and but little earth. In the river we see a great abundance of fish, but we cannot tempt them to bite by anything on our hooks. The whole party have been engaged in dressing skins, and making them into moccasins and leggings. Captain Clark's fever has almost left him, but he still remains very languid and has a general soreness in his limbs. The latitude of our camp, as the mean of two observations of the meridian altitude of the sun's lower limb with octant by back observation, is N. 45° 24′ 8″ 5‴.

Tuesday, 30th. Captain Clark was this morning much restored, and therefore, having made all the observations necessary to fix the longitude, we reloaded our canoes and began to ascend Jefferson river. The river now becomes very crooked, and forms bends on each side; the current, too, is rapid, and cut into a great number of channels, and sometimes shoals, the beds of which consist of coarse gravel. The islands are unusually numerous; on the right are high plains, occasionally forming cliffs of rocks and hills; while the left was an extensive low ground and prairie, intersected by a number of bayous or channels falling into the river. Captain Lewis, who had walked through it with Chaboneau, his wife, and two invalids, joined us at dinner, a few miles above our camp. Here the Indian woman said was the place where she had been made prisoner. The men, being too few to contend with the Minnetarees, mounted their horses and fled as soon as the attack began. The women and children dispersed, and Sacajawea, as she was crossing at a shoal place, was overtaken in the middle of the river by her pursuers. As we proceeded, the low grounds were covered with cottonwood and a thick underbrush, and on both sides of the river, except where the high hills prevented it, the ground was divided by bayous, which are dammed up by the beaver, which are very numerous here. We made twelve and a quarter miles, and encamped on the north side. Captain Lewis proceeded, after dinner, through an extensive low ground of timber and meadow land intermixed; but the bayous were so obstructed by beaver dams that, in order to avoid them, he directed his course towards the high plain on the right. This he gained with some difficulty, after wading up to his waist through the mud and water of a number of beaver dams. When he desired to rejoin the canoes he found the underbrush so thick and the river so crooked that this, joined to the difficulty of passing the beaver dams, induced him to go on and endeavour to intercept the river at some point where it might be more collected into one channel and approach nearer to the high plain. He arrived at the bank about sunset, having gone only six miles in a direct course from the canoes; but he saw no traces of the men, nor did he receive any answer to his shouts nor the firing of his gun. It was now nearly dark; a duck lighted near him, and he shot it. He then went on the head of a small island, where he found some driftwood, which enabled him to cook his duck for supper, and he laid down to sleep on some willow brush. The night was cool, but the driftwood gave him a good fire, and he suffered no inconvenience except from the musquitoes.

Wednesday, 31st. The next morning he waited till after seven o'clock, when he became uneasy lest he should have gone beyond his camp last evening, and determined to follow us. Just as he had set out with this intention, he saw one of the party in advance of the canoes; although our camp was only two miles below him, in a straight line, we could not reach him sooner, in consequence of the rapidity of the water and the circuitous course of the river. We halted for breakfast, after which Captain Lewis continued his route. At the distance of one mile from our encampment we passed the principal entrance of a stream on the left, which rises in the snowy mountains to the southwest, between Jefferson and Madison rivers, and discharges itself by seven mouths, five below, and one three miles above this, which is the largest, and about thirty yards wide; we called it Philosophy river. The water of it is abundant and perfectly clear, and the bed, like that of the Jefferson, consists of pebble and gravel. There is some timber in the bottoms of the river, and vast numbers of otter and beaver, which build on its smaller mouths and the bayous of its neighbourhood. The Jefferson continues as yesterday, shoaly and rapid, but as the islands, though numerous, are small, it is however more collected into one current than it was below, and is from ninety to one hundred and twenty yards in width. The low ground has a fertile soil of rich black loam, and contains a considerable quantity of timber, with the bullrush and cat-tail flag very abundant in the moist parts, while the drier situations are covered with fine grass, tansy, thistles, onions, and flax. The uplands are barren, and without timber; the soil is a light yellow clay intermixed with small smooth pebble and gravel, and the only produce is the prickly pear, the sedge, and the bearded grass, which is as dry and inflammable as tinder. As we proceeded, the low grounds became narrower and the timber more scarce, till, at the distance of ten miles, the high hills approach and overhang the river on both sides, forming cliffs of a hard black granite, like almost all those below the limestone cliffs at the three forks of the Missouri; they continue so for a mile and three-quarters, where we came to a point of rock on the right side, at which place the hills again retire, and the valley widens to the distance of a mile and a half. Within the next five miles we passed four islands, and reached the foot of a mountain in a bend of the river to the left; from this place we went a mile and a quarter to the entrance of a small run discharging itself on the left, and encamped on an island just above it, after making seventeen and three-quarter miles. We observe some pine on the hills on both sides of our encampment, which are very lofty. The only game which we have seen are one big-horn, a few antelopes, deer, and one brown bear, which escaped from our pursuit. Nothing was, however, killed to-day, nor have we had any fresh meat except one beaver for the last two days, so that we are now reduced to an unusual situation, for we have hitherto always had a great abundance of flesh.

Thursday, August 1. We left our encampment early, and at the distance of a mile, reached a point of rocks on the left side, where the river passes through perpendicular cliffs. Two and three-quarter miles farther we halted for breakfast, under a cedar tree in a bend to the right; here, as had been previously arranged, Captain Lewis left us, with Sergeant Gass, Chaboneau, and Drewyer, intending to go on in advance in search of the Shoshonees. He began his route along the north side of the river over a high range of mountains, as Captain Clark, who ascended them on the 26th, had observed from them a large valley spreading to the north of west, and concluded that on leaving the mountain the river took that direction; but when he reached that valley Captain Lewis found it to be the passage of a large creek falling just above the mountain into the Jefferson, which bears to the southwest. On discovering his error he bent his course towards that river, which he reached about two in the afternoon, very much exhausted with heat and thirst. The mountains were very bare of timber, and the route lay along the steep and narrow hollows of the mountain, exposed to the mid-day sun, without air, or shade, or water. Just as he arrived there a flock of elk passed, and they killed two of them, on which they made their dinner, and left the rest on the shore for the party in the canoes. After dinner they resumed their march, and encamped on the north side of the river, after making seventeen miles; in crossing the mountains Captain Lewis saw a flock of the black or dark brown pheasant, of which he killed one. This bird is one-third larger than the common pheasant of the Atlantic States; its form is much the same. The male has not, however, the tufts of long black feathers on the sides of the neck so conspicuous in the Atlantic pheasant, and both sexes are booted nearly to the toes. The colour is a uniform dark brown, with a small mixture of yellow or yellowish brown specks on some of the feathers, particularly those of the tail, though the extremities of these are perfectly black for about an inch. The eye is nearly black, and the iris has a small dash of yellowish brown; the feathers of the tail are somewhat longer than those of our pheasant, but the same in number, eighteen, and nearly equal in size, except that those of the middle are somewhat the longest; their flesh is white, and agreeably flavoured.

He also saw among the scattered pine near the top of the mountain a blue bird about the size of a robin, but in action and form something like a jay; it is constantly in motion, hopping from spray to spray, and its note, which is loud and frequent, is, as far as letters can represent it, char ah! char ah! char ah!

After breakfast, we proceeded on. At the distance of two and a quarter miles the river enters a high mountain, which forms rugged cliffs of nearly perpendicular rocks. These are of a black granite at the lower part, and the upper consists of a light coloured freestone; they continue from the point of rocks close to the river for nine miles, which we passed before breakfast, during which the current is very strong. At nine and a quarter miles we passed an island, and a rapid with a fall of six feet, and reached the entrance of a large creek on the left side. In passing this place the towline of one of the canoes broke just at the shoot of the rapids, swung on the rocks, and had nearly upset. To the creek, as well as the rapid, we gave the name of Frazier, after Robert Frazier, one of the party. Here the country opens into a beautiful valley from six to eight miles in width; the river then becomes crooked and crowded with islands; its low grounds wide and fertile, but though covered with fine grass from nine inches to two feet high, possesses but a small proportion of timber, and that consists almost entirely of a few narrow-leafed cottonwood distributed along the verge of the river. The soil of the plain is tolerably fertile, and consists of a black or dark yellow loam. It gradually ascends on each side to the bases of two ranges of high mountains, which lie parallel to the river; the tops of them are yet in part covered with snow, and while in the valley we are nearly suffocated with heat during the day, and at night the air is so cold that two blankets are not more than sufficient covering. In passing through the hills we observed some large cedar trees, and some juniper also. From Frazier's creek we went three and three-quarter miles, and encamped on the left side, having come thirteen miles. Directly opposite our camp is a large creek which we call Fields's creek, from Reuben Fields, one of our men. Soon after we halted two of the hunters went out and returned with five deer, which, with one big-horn, we killed in coming through the mountain, on which we dined, and the elk left by Captain Lewis. We were again well supplied with fresh meat. In the course of the day we saw a brown bear, but were not able to shoot him.

Friday, August 2. Captain Lewis, who slept in the valley a few miles above us, resumed his journey early, and after making five miles, and finding that the river still bore to the south, determined to cross it in hopes of shortening the route; for the first time, therefore, he waded across it, although there are probably many places above the falls where it might be attempted with equal safety. The river was about ninety yards wide, the current rapid, and about waist-deep; the bottom formed of smooth pebble with a small mixture of coarse gravel. He then continued along the left bank of the river till sunset, and encamped after travelling twenty-four miles. He met no fresh tracks of Indians. Throughout the valley are scattered the bones and excrement of the buffaloe of an old date, but there seems no hope of meeting the animals themselves in the mountains; he saw an abundance of deer and antelope, and many tracks of elk and bear. Having killed two deer, they feasted sumptuously, with a desert of currants of different colours, two species of red, others yellow, deep purple, and black; to these were added black gooseberries and deep purple serviceberries, somewhat larger than ours, from which it differs also in colour, size, and the superior excellence of its flavour. In the low grounds of the river were many beaver dams formed of willow brush, mud, and gravel, so closely interwoven that they resist the water perfectly; some of them were five feet high, and overflowed several acres of land.

In the meantime we proceeded on slowly, the current being so strong as to require the utmost exertions of the men to make any advance even with the aid of the cord and pole, the wind being from the northwest. The river is full of large and small islands, and the plain cut by great numbers of bayous or channels, in which are multitudes of beaver. In the course of the day we passed some villages of barking squirrels; we saw several rattlesnakes in the plain; young ducks, both of the duckon-mallard and red-headed fishing duck species; some geese; also the black woodpecker, and a large herd of elk. The channel, current, banks, and general appearance of the river are like that of yesterday. At fourteen and three-quarter miles we reached a rapid creek or bayou about thirty yards wide, to which we gave the name of Birth creek. After making seventeen miles we halted in a smooth plain in a bend towards the left.

Saturday, 3d. Captain Lewis continued his course along the river through the valley, which continued much as it was yesterday, except that it now widens to nearly twelve miles; the plains, too, are more broken, and have some scattered pine near the mountains, where they rise higher than hitherto. In the level parts of the plains and the river bottoms there is no timber except small cottonwood near the margin, and an undergrowth of narrow-leafed willow, small honeysuckle, rosebushes, currants, serviceberry, and gooseberry, and a little of a small species of birch; it is a finely indented oval, of a small size and a deep green colour; the stem is simple, ascending and branching, and seldom rises higher than ten or twelve feet. The mountains continue high on each side of the valley, but their only covering is a small species of pitch-pine with a short leaf, growing on the lower and middle regions, while for some distance below the snowy tops there is neither timber nor herbage of any kind. About eleven o'clock Drewyer killed a doe, on which they breakfasted, and after resting two hours, continued till night, when they reached the river near a low ground more extensive than usual. From the appearance of the timber Captain Lewis supposed that the river forked above him, and therefore encamped, with an intention of examining it more particularly in the morning. He had now made twenty-three miles, the latter part of which were for eight miles through a high plain covered with prickly pears and bearded grass, which rendered the walking very inconvenient; but even this was better than the river bottoms we crossed in the evening, which, though apparently level, were formed into deep holes as if they had been rooted up by hogs, and the holes were so covered with thick grass that they were in danger of falling at every step. Some parts of these low grounds, however, contain turf or peat of an excellent quality, for many feet deep apparently, as well as the mineral salts which we have already mentioned on the Missouri. They saw many deer, antelopes, ducks, geese, some beaver, and great traces of their work, and the small birds and curlews as usual. The only fish which they observed in this part of the river is the trout, and a species of white fish with a remarkably long small mouth, which one of our men recognize as the fish called in the eastern states the bottlenose.

On setting out with the canoes, we found the river, as usual, much crowded with islands, the current more rapid as well as shallower, so that in many places they were obliged to man the canoes double, and drag them over the stone and gravel of the channel. Soon after we set off, Captain Clark, who was walking on shore, observed a fresh track which he knew to be that of an Indian from the large toes being turned inwards, and on following it found that it led to the point of a hill from which our camp of last night could be seen. This circumstance strengthened the belief that some Indian had strayed thither, and had run off alarmed at the sight of us. At two and a quarter miles is a small creek in a bend towards the right, which runs down from the mountains at a little distance; we called it Panther creek, from an animal of that kind killed by Reuben Fields at its mouth. It is precisely the same animal common to the western parts of the United States, and measured seven and a half feet from the nose to the extremity of the tail. Six and three-quarter miles beyond this stream is another, on the left, formed by the drains which convey the melted snows from a mountain near it, under which the river passes, leaving the low grounds on the right side, and making several bends in its course. On this stream are many large beaver dams. One mile above it is a small run on the left, and after leaving which begins a very bad rapid, where the bed of the river is formed of solid rock; this we passed in the course of a mile, and encamped on the lower point of an island. Our journey had been only thirteen miles, but the badness of the river made it very laborious, as the men were compelled to be in the water during the greater part of the day. We saw only deer, antelopes, and the common birds of the country.

Sunday, 4th. This morning Captain Lewis proceeded early, and after going southeast by east for four miles reached a bold running creek, twelve yards wide, with clear cold water, furnished apparently by four drains from the snowy mountains on the left; after passing this creek he changed his direction to southeast, and leaving the valley in which he had travelled for the two last days, entered another, which bore east. At the distance of three miles on this course he passed a handsome little river, about thirty yards wide, which winds through the valley; the current is not rapid nor the water very clear, but it affords a considerable quantity of water, and appears as if it might be navigable for some miles. The banks are low, and the bed formed of stone and gravel. He now changed his route to southwest, and passing a high plain which separates the vallies, returned to the more southern, or that which he had left; in passing this he found a river about forty-five yards wide, the water of which has a whitish blue tinge, with a gentle current and a gravelly bottom. This he waded and found it waist-deep. He then continued down it, till, at the distance of three-quarters of a mile, he saw the entrance of the small river he had just passed; as he went on two miles lower down, he found the mouth of the creek he had seen in the morning. Proceeding farther on three miles, he arrived at the junction of this river with another which rises from the southwest, runs through the south valley about twelve miles before it forms its junction, where it is fifty yards wide. We now found that our camp of last night was about a mile and a half above the entrance of this large river, on the right side. This is a bold, rapid, clear stream, but its bed is so much obstructed by gravelly bars, and subdivided by islands, that the navigation must be very insecure, if not impracticable. The other, or middle stream, has about two-thirds its quantity of water, and is more gentle, and may be safely navigated. As far as it could be observed, its course was about southwest, but the opening of the valley induced him to believe that farther above it turned more towards the west. Its water is more turbid and warmer than that of the other branch, whence it may be presumed to have its sources at a greater distance in the mountains, and to pass through a more open country. Under this impression he left a note recommending to Captain Clark the middle fork, and then continued his course along the right side of the other, or more rapid, branch. After travelling twenty-three miles, he arrived near a place where the river leaves the valley and enters the mountains. Here he encamped for the night. The country he passed is like that of the rest of this valley, though there is more timber in this part on the rapid fork than there has been on the river in the same extent since we entered it; for on some parts of the valley the Indians seem to have destroyed a great proportion of the little timber there was, by setting fire to the bottoms. He saw some antelopes, deer, cranes, geese, and ducks of the two species common to this country, though the summer duck has ceased to appear, nor does it seem to be an inhabitant of this part of the river.

We proceeded soon after sunrise. The first five miles we passed four bends on the left, and several bayous on both sides. At eight o'clock we stopped to breakfast, and found the note Captain Lewis had written on the 2d instant. During the next four miles, we passed three small bends of the river to the right, two small islands, and two bayous on the same side. Here we reached a bluff on the left; our next course was six miles to our encampment. In this course we met six circular bends on the right, and several small bayous, and halted for the night in a low ground of cottonwood on the right. Our day's journey, though only fifteen miles in length, was very fatiguing. The river is still rapid, and the water, though clear, is very much obstructed by shoals or ripples at every two or three hundred yards; at all these places we are obliged to drag the canoes over the stones, as there is not a sufficient depth of water to float them, and in the other parts the current obliges us to have recourse to the cord. But as the brushwood on the banks will not permit us to walk on shore, we are under the necessity of wading through the river as we drag the boats. This soon makes our feet tender, and sometimes occasions severe falls over the slippery stones, and the men, by being constantly wet, are becoming more feeble. In the course of the day the hunters killed two deer, some geese and ducks, and the party saw antelopes, cranes, beaver, and otter.

Monday, 5th. This morning Chaboneau complained of being unable to march far to-day, and Captain Lewis therefore ordered him and Serjeant Gass to pass the rapid river and proceed through the level low ground, to a point of high timber on the middle fork, seven miles distant, and wait his return. He then went along the north side of the rapid river about four miles, where he waded it, and found it so rapid and shallow that it would be impossible to navigate it. He continued along the left side for a mile and a half, when the mountains came close on the river, and rise to a considerable height, with a partial covering of snow. From this place the course of the river was to the east of north. After ascending with some difficulty a high point of the mountain, he had a pleasing view of the valley he had passed, and which continued for about twenty miles farther on each side of the middle fork, which then seemed to enter the mountains and was lost to the view. In that direction, however, the hills which terminate the valley are much lower than those along either of the other forks, particularly the rapid one, where they continue rising in ranges above each other as far as the eye could reach. The general course, too, of the middle fork, as well as that of the gap which it forms on entering the mountains, is considerably to the south of west—circumstances which gave a decided preference to this branch as our future route. Captain Lewis now descended the mountain and crossed over to the middle fork, about five miles distant, and found it still perfectly navigable. There is a very large and plain Indian road leading up it, but it has at present no tracks except those of horses, which seem to have used it last spring. The river here made a great bend to the southeast, and he therefore directed his course, as well as he could, to the spot where he had directed Chaboneau and Gass to repair, and struck the river about three miles above their camp. It was now dark, and he therefore was obliged to make his way through the thick brush of the pulpy-leafed thorn and the prickly pear for two hours before he reached their camp. Here he

## UP THE MISSOURI

was fortunate enough to find the remains of some meat, which was his only food during the march of twenty-five miles to-day. He had seen no game of any sort except a few antelopes, who were very shy. The soil of the plains is a meagre clay, of a light yellow colour, intermixed with a large proportion of gravel, and producing nothing but twisted or bearded grass, sedge, and prickly pears. The drier parts of the low grounds are also more indifferent in point of soil than those farther down the river, and although they have but little grass, are covered with southern wood, pulpy-leafed thorn, and prickly pears, while the moist parts are fertile, and supplied with fine grass and sandrushes.

We passed within the first four and a quarter miles three small islands and the same number of bad rapids. At the distance of three-quarters of a mile is another rapid, of difficult passage; three miles and three-quarters beyond this are the forks of the river, in reaching which we had two islands and several bayous on different sides to pass. Here we had come nine miles and a quarter. The river was straighter and more rapid than yesterday, the labour of the navigation proportionally increased, and we therefore proceeded very slowly, as the feet of several of the men were swollen, and all were languid with fatigue. We arrived at the forks about four o'clock, but unluckily Captain Lewis's note had been left on a green pole which the beaver had cut down and carried off with the note, an accident which deprived us of all information as to the character of the two branches of the river. Observing, therefore, that the northwest fork was most in our direction, and contained as much water as the other, we ascended it; we found it extremely rapid, and its waters were scattered in such a manner that for a quarter of a mile we were forced to cut a passage through the willowbrush that leaned over the little channels and united at the top. After going up it for a mile we encamped on an island which had been overflowed, and was still so wet that we were compelled to make beds of brush to keep ourselves out of the mud. Our provision consisted of two deer which had been killed in the morning.

Tuesday, 6th. We proceeded up the northwest fork, which we found still very rapid and divided by several islands, while the plains near it were intersected by bayous. After passing with much difficulty over stones and rapids, we reached a bluff on the right at the distance of nine miles, our general course south 30° west, and halted for breakfast. Here we were joined by Drewyer, who informed us of the state of the two rivers and of Captain Lewis's note, and we immediately began to descend the river in order to take the other branch. On going down, one of the canoes upset and two others filled with water, by which all the baggage was wet and several articles irrecoverably lost. As one of them swung round in a rapid current, Whitehouse was thrown out of her, and whilst down the canoe passed over him, and had the water been two inches shallower would have crushed him to pieces; but he escaped with a severe bruise of his leg. In order to repair these misfortunes we hastened to the forks, where we were joined by Captain Lewis, and then passed over to the left side opposite to the entrance of the rapid fork, and encamped on a large gravelly bar, near which there was plenty of wood. Here we opened and exposed to dry all the articles which had suffered from the water; none of them were completely spoiled except a small keg of powder; the rest of the powder, which was distributed in the different canoes, was quite safe, although it had been under the water upwards of an hour. The air is indeed so pure and dry that any wood-work immediately shrinks, unless it is kept filled with water; but we had placed our powder in small canisters of lead, each containing powder enough for the canister when melted into bullets, and secured with cork and wax, which answered our purpose perfectly.

Captain Lewis had risen very early, and having nothing to eat, sent out Drewyer to the woodland on the left in search of a deer, and directed Serjeant Gass to keep along the middle branch to meet us if we were ascending it. He then set off with Chaboneau towards the forks, but five miles above them, hearing us on the left, struck the river as we were descending, and came on board at the forks.

In the evening we killed three deer and four elk, which furnished us once more with a plentiful supply of meat. Shannon, the same man who was lost before for fifteen days, was sent out this morning to hunt, up the northwest fork; when we decided on returning, Drewyer was directed to go in quest of him, but he returned with information that he had gone several miles up the river without being able to find Shannon. We now had the trumpet sounded, and fired several guns, but he did not return, and we fear he is again lost.

Wednesday, 7th. We remained here this morning for the purpose of making some celestial observations, and also in order to refresh the men and complete the drying of the baggage. We obtained a meridian altitude, which gave the latitude of our camp as north $4.5° \ 2' \ 43'' \ 8'''$. We were now completely satisfied that the middle branch was the most navigable, and the true continuation of the Jefferson. The northwest fork seems to be the drain of the melting snows of the mountains; its course cannot be so long as the other branch, and although it contains now as great a quantity of water, yet the water has obviously overflowed the old bed and spread into channels, which leave the low grounds covered with young grass, resembling that of the adjoining lands, which are not inundated; whence we readily infer that the supply is more precarious than that of the other branch, the waters of which, though more gentle, are more constant. This northwest fork we called Wisdom river.

As soon as the baggage was dried it was reloaded on board the boats, but we now found it so much diminished that we would be able to proceed with one canoe less. We therefore hauled up the superfluous one into a thicket of brush, where we secured her against being swept away by the high tide. At one o'clock all set out, except Captain Lewis, who remained till the evening in order to complete the observation of equal altitudes. We passed several bends of the river, both to the right and left, as well as a number of bayous on both sides, and made seven miles by water, though the distance by land is only three. We then encamped on a creek which rises in a high mountain to the northeast, and after passing through an open plain for several miles, discharges itself on the left, where it is a bold running stream twelve yards wide. We called it Turf creek, from the number of bogs and the quantity of turf on its waters. In the course of the afternoon there fell a shower of rain, attended with thunder and lightning, which lasted about forty minutes, and the weather remained so cloudy all night that we were unable to take any lunar observations. Uneasy about Shannon, we sent R. Fields in search of him this morning, but we have as yet no intelligence of either of them. Our only game to-day was one deer.

Thursday, 8th. There was a heavy dew this morning. Having left one of the canoes, there are now more men to spare for the chace, and four were sent out at an early hour, after which we proceeded. We made five miles, by water, along two islands and several bayous, but as the river formed seven different bends towards the left, the distance by land was only two miles south of our encampment. At the end of that course we reached the upper principal entrance of a stream, which we called Philanthropy river. This river empties itself into the Jefferson on the southeast side, by two channels a short distance from each other; from its size and its southeastern course, we presume that it rises in the Rocky mountains near the sources of the Madison. It is thirty yards wide at its entrance, has a very gentle current, and is navigable for some distance. One mile above this river we passed an island, a second at the distance of six miles farther, during which the river makes a considerable bend to the east. Reuben Fields returned about noon with information that he had gone up Wisdom river till its entrance into the mountains, but could find nothing of Shannon. We made seven miles beyond the last island, and after passing some small bayous, encamped under a few high trees on the left, at the distance of fourteen miles above Philanthropy river by water, though only six by land. The river has in fact become so very crooked that although by means of the pole, which we now use constantly, we make a considerable distance, yet being obliged to follow its windings, at the end of the day we find ourselves very little advanced on our general course. It forms itself into small circular bends, which are so numerous that within the last fourteen miles we passed thirty-five of them, all inclining towards the right; it is, however, much more gentle and deep than below Wisdom river, and its general width is from thirty-five to forty-five yards. The general appearance of the surrounding country is that of a valley five or six miles wide, enclosed between two high mountains. The bottom is rich, with some small timber on the islands and along the river, which consists rather of underbrush, and a few cottonwood, birch, and willow trees. The high grounds have some scattered pine, which just relieve the general nakedness of the hills and the plain, where there is nothing except grass. Along the bottoms we saw to-day a considerable quantity of the buffaloe clover, the sunflower, flax, green sward, thistle, and several species of rye grass, some of which rise to the height of three or four feet. There is also a grass with a soft smooth leaf which rises about three feet high, and bears its seed very much like the timothy, but it does not grow luxuriantly, nor would it apparently answer so well in our meadows as that plant. We preserved some of its seed, which are now ripe, in order to make the experiment. Our game consisted of deer and antelope, and we saw a number of geese and ducks just beginning to fly, and some cranes. Among the inferior animals, we have an abundance of the large biting or hare fly, of which there are two species, one black, the other smaller and brown, except the head, which is green. The green or blowing flies unite with them in swarms to attack us, and seem to have relieved the eye-gnats, who have now disappeared. The musquitoes, too, are in large quantities, but not so troublesome as they were below. Through the valley are scattered bogs and some very good turf; the earth of which the mud is composed is of a white or bluish white colour, and seems to be argilaceous. On all the three rivers, but particularly on the Philanthropy, are immense quantities of beaver, otter, and muskrat. At our camp there was an abundance of rosebushes and briars, but so little timber that we were obliged to use willow brush for fuel. The night was again cloudy, which prevented the lunar observations.

On our right is the point of a high plain, which our Indian woman recognizes as the place called the Beaver's-head, from a supposed resemblance to that object. This she says is not far from the summer retreat of her countrymen, which is on a river beyond the mountains, and running to the west. She is therefore certain that we shall meet them either on this river or on that immediately west of its source, which, judging from its present size, cannot be far distant. Persuaded of the absolute necessity of procuring horses to cross the mountains, it was determined that one of us should proceed in the morning to the head of the river, and penetrate the mountains till he found the Shoshonees or some other nation who could assist us in transporting our baggage, the greater part of which we shall be compelled to leave without the aid of horses.

Friday, 9th. The morning was fair and fine. We set off early, and proceeded on very well, though there were more rapids in the river than yesterday. At eight o'clock we halted for breakfast, part of which consisted of two fine geese killed before we stopped. Here we were joined by Shannon, for whose safety we had been so uneasy. The day on which he left us on his way up Wisdom river, after hunting for some time and not seeing the party arrive, he returned to the place where he had left us. Not finding us there he supposed we had passed him, and he therefore marched up the river during all the next day, when he was convinced that we had not gone on, as the river was no longer navigable. He now followed the course of the river down to the forks, and then took the branch which we are pursuing. During the three days of his absence he had been much wearied with his march, but had lived plentifully, and brought the skins of three deer. As far as he had ascended Wisdom river it kept its course obliquely down towards the Jefferson. Immediately after breakfast, Captain Lewis took Drewyer, Shields, and M'Neal, and slinging their knapsacks they set out with a resolution to meet some nation of Indians before they returned, however long they might be separated from the party. He directed his course across the low ground to the plain on the right, leaving the Beaver's-head about two miles to the left. After walking eight miles to the river, which they waded, they went on to a commanding point from which he saw the place at which it enters the mountain, but as the distance would not permit his reaching it this evening he descended towards the river, and after travelling eight miles farther, encamped for the evening some miles below the mountain. They passed, before reaching their camp, a handsome little stream formed by some large springs which rise in the wide bottom on the left side of the river. In their way they killed two antelopes, and took with them enough of the meat for their supper and breakfast the next morning.

In the meantime we proceeded, and in the course of eleven miles from our last encampment passed two small islands, sixteen short round bends in the river, and halted in a bend towards the right, where we dined. The river increases in rapidity as we advance, and is so crooked that the eleven miles, which have cost us so much labour, only bring us four miles in a direct line. The weather became overcast towards evening, and we experienced a slight shower attended with thunder and lightning. The three hunters who were sent out killed only two antelopes, game of every kind being scarce.

Saturday, 10th. Captain Lewis continued his route at an early hour through the wide bottom along the left bank of the river. At about five miles he passed a large creek, and then fell into an Indian road leading towards the point where the river entered the mountain. This he followed till he reached a high perpendicular cliff of rocks where the river makes its passage through the hills, and which he called the Rattlesnake cliff, from the number of that animal which he saw there; here he kindled a fire and waited the return of Drewyer, who had been sent out on the way to kill a deer; he came back about noon with the skin of three deer and the flesh of one of the best of them. After a hasty dinner, they returned to the Indian road, which they had left for a short distance to see the cliff. It led them sometimes over the hills, sometimes in the narrow bottoms of the river, till at the distance of fifteen miles from the Rattlesnake cliffs they reached a handsome open and level valley, where the river divided into two nearly equal branches. The mountains over which they passed were not very high, but are rugged and continue close to the river side. The river, which before it enters the mountain was rapid, rocky, very crooked, much divided by islands, and shallow, now becomes more direct in its course as it is hemmed in by the hills, and has not so many bends nor islands, but becomes more rapid and rocky, and continues as shallow. On examining the two branches of the river it was evident that neither of them was navigable farther. The road forked with the river, and Captain Lewis therefore sent a man up each of them for a short distance, in order that, by comparing their respective information, he might be able to take that which seemed to have been most used this spring. From their account he resolved to choose that which led along the southwest branch of the river, which was rather the smaller of the two; he accordingly wrote a note to Captain Clark informing him of the route, and recommending his staying with the party at the forks till he should return. This he fixed on a dry willow pole at the forks of the river, and then proceeded up the southwest branch; but after going a mile and a half the road became scarcely distinguishable, and the tracks of the horses which he had followed along the Jefferson were no longer seen. Captain Lewis therefore returned to examine the other road himself, and found that the horses had in fact passed along the western or right fork, which had the additional recommendation of being larger than the other.

This road he concluded to take, and therefore sent back Drewyer to the forks with a second letter to Captain Clark apprising him of the change, and then proceeded on. The valley of the west fork, through which he now passed, bears a little to the north of west, and is confined within the space of about a mile in width by rough mountains and steep cliffs of rock. At the distance of four and a half miles it opens into a beautiful and extensive plain, about ten miles long and five or six in width; this is surrounded on all sides by higher rolling or waving country, intersected by several little rivulets from the mountains, each bordered by its wide meadows. The whole prospect is bounded by these mountains, which nearly surround it, so as to form a beautiful cove about sixteen or eighteen miles in diameter. On entering this cove the river bends to the northwest, and bathes the foot of the hills to the right. At this place they halted for the night on the right side of the river, and having lighted a fire of dry willow brush, the only fuel which the country affords, supped on a deer. They had travelled to-day thirty miles by estimate; that is, ten to the Rattlesnake cliff, fifteen to the forks of Jefferson river, and five to their encampment. In this cove some parts of the low grounds are tolerably fertile, but much the greater proportion is covered with prickly pear, sedge, twisted grass, the pulpy-leafed thorn, southern-wood, and wild sage, and like the uplands have a very inferior soil. These last have little more than the prickly pear and the twisted or bearded grass, nor are there in the whole cove more than three or four cottonwood trees, and those are small. At the apparent extremity of the bottom above, and about ten miles to the westward, are two perpendicular cliffs rising to a considerable height on each side of the river, and at this distance seem like a gate. In the meantime we proceeded at sunrise, and found the river not so rapid as yesterday, though more narrow and still very crooked, and so shallow that we were obliged to drag the canoes over many ripples in the course of the day. At six and a half miles we had passed eight bends on the north and two small bayous on the left, and came to what the Indians call the Beaver's-head, a steep rocky cliff about one hundred and fifty feet high, near the right side of the river. Opposite to this, at three hundred yards from the water, is a low cliff about fifty feet in height, which forms the extremity of a spur of the mountain about four miles distant on the left. At four o'clock we were overtaken by a heavy shower of rain, attended with thunder, lightning, and hail. The party were defended from the hail by covering themselves with willow bushes, but they got completely wet, and in this situation, as soon as the rain ceased, continued till we encamped. This we did at a low bluff on the left, after passing, in the course of six and a half miles, four islands and eighteen bends on the right, and a low bluff and several bayous on the same side. We had now come thirteen miles, yet were only four on our route towards the mountains. The game seems to be declining, for our hunters procured only a single deer, though we found another for us that had been killed three days before by one of the hunters during an excursion, and left for us on the river.

## CHAPTER XIV.

Captain Lewis proceeds before the main body in search of the Shoshonees—His ill success on the first interview—The party with Captain Lewis at length discover the source of the Missouri—Captain Clark with the main body still employed in ascending the Missouri or Jefferson river—Captain Lewis's second interview with the Shoshonees attended with success—The interesting ceremonies of his first introduction to the natives detailed at large—Their hospitality—Their mode of hunting the antelope—The difficulties encountered by Captain Clark and the main body in ascending the river—The suspicions entertained of Captain Lewis by the Shoshonees, and his mode of allaying them—The ravenous appetites of the savages illustrated by a singular adventure—The Indians still jealous, and the great pains taken by Captain Lewis to preserve their confidence—Captain Clark arrives with the main body, exhausted by the difficulties which they underwent.

SUNDAY, August 11. Captain Lewis again proceeded on early, but had the mortification to find that the track which he followed yesterday soon disappeared. He determined, therefore, to go on to the narrow gate or pass of the river which he had seen from the camp, in hopes of being able to recover the Indian path. For this purpose he waded across the river, which was now about twelve yards wide and barred in several places by the dams of the beaver, and then went straight forward to the pass, sending one man along the river to his left and another on the right, with orders to search for the road, and if they found it to let him know by raising a hat on the muzzle of their guns. In this order they went along for about five miles, when Captain Lewis perceived, with the greatest delight, a man on horseback at the distance of two miles coming down the plain towards them. On examining him with the glass, Captain Lewis saw that he was of a different nation from any Indians we had hitherto met; he was armed with a bow and a quiver of arrows; mounted on an elegant horse without a saddle, and a small string attached to the under jaw answered as a bridle. Convinced that he was a Shoshonee, and knowing how much of our success depended on the friendly offices of that nation, Captain Lewis was full of anxiety to approach without alarming him, and endeavor to convince him that he was a white man. He therefore proceeded on towards the Indian at his usual pace. When they were within a mile of each other the Indian suddenly stopt; Captain Lewis immediately followed his example, took his blanket from his knapsack, and holding it with both hands at the two corners, threw it above his head and unfolded it as he brought it to the ground as if in the act of spreading it. This signal, which originates in the practice of spreading a robe or a skin, as a seat for guests to whom they wish to show a distinguished kindness, is the universal sign of friendship among the Indians on the Missouri and the Rocky mountains. As usual, Captain Lewis repeated this signal three times; still the Indian kept his position, and looked with an air of suspicion on Drewyer and Shields who were now advancing on each side. Captain Lewis was afraid to make any signal for them to halt, lest he should increase the suspicions of the Indian, who began to be uneasy, and they were too distant to hear his voice. He therefore took from his pack some beads, a looking-glass, and a few trinkets, which he had brought for the purpose, and leaving his gun, advanced unarmed towards the Indian. He remained in the same position till Captain Lewis came within two hundred yards of him, when he turned his horse, and began to move off slowly. Captain Lewis then called out to him, in as loud a voice as he could, repeating the words, tabba bone! which in the Shoshonee language means white man; but looking over his shoulder the Indian kept his eyes on Drewyer and

Shields, who were still advancing, without recollecting the impropriety of doing so at such a moment, till Captain Lewis made a signal to them to halt; this Drewyer obeyed, but Shields did not observe it, and still went forward. Seeing Drewyer halt, the Indian turned his horse about as if to wait for Captain Lewis, who now reached within one hundred and fifty paces, repeating the words, tabba bone! and holding up the trinkets in his hand, at the same time stripping up the sleeve of his shirt to show the colour of his skin. The Indian suffered him to advance within one hundred paces, then suddenly turned his horse, and giving him the whip, leaped across the creek and disappeared in an instant among the willow bushes; with him vanished all the hopes which the sight of him had inspired of a friendly introduction to his countrymen. Though sadly disappointed by the imprudence of his two men, Captain Lewis determined to make the incident of some use, and therefore calling the men to him they all set off after the track of the horse, which they hoped might lead them to the camp of the Indian who had fled, or if he had given the alarm to any small party, their track might conduct them to the body of the nation. They now fixed a small flag of the United States on a pole, which was carried by one of the men as a signal of their friendly intentions, should the Indians observe them as they were advancing. The route lay across an island formed by a nearly equal division of the creek in the bottom; after reaching the open grounds on the right side of the creek, the track turned towards some high hills about three miles distant. Presuming that the Indian camp might be among these hills, and that by advancing hastily he might be seen and alarm them, Captain Lewis sought an elevated situation near the creek, had a fire made of willow brush, and took breakfast. At the same time he prepared a small assortment of beads, trinkets, awls, some paint, and a looking-glass, and placed them on a pole near the fire, in order that if the Indians returned they might discover that the party were white men and friends. Whilst making these preparations a very heavy shower of rain and hail came on, and wet them to the skin; in about twenty minutes it was over, and Captain Lewis then renewed his pursuit, but as the rain had made the grass which the horse had trodden down rise again, his track could with difficulty be distinguished. As they went along they passed several places where the Indians seemed to have been digging roots to-day, and saw the fresh track of eight or ten horses, but they had been wandering about in so confused a manner that he could not discern any particular path, and at last, after pursuing it about four miles along the valley to the left under the foot of the hills, he lost the track of the fugitive Indian. Near the head of the valley they had passed a large bog covered with moss and tall grass, among which were several springs of pure cold water; they now turned a little to the left along the foot of the high hills, and reached a small creek, where they encamped for the night, having made about twenty miles, though not more than ten in a direct line from their camp of last evening.

The morning being rainy and wet we did not set out with the canoes till after an early breakfast. During the first three miles we passed three small islands, six bayous on different sides of the river, and the same number of bends towards the right. Here we reached the lower point of a large island which we called Three-thousand-mile island, on account of its being at that distance from the mouth of the Missouri. It is three miles and a half in length, and as we coasted along it we passed several small bends of the river towards the left, and two bayous on the same side. After leaving the upper point of Three-thousand-mile island, we followed the main channel on the left side, which led us by three small islands and several small bayous, and fifteen bends towards the right. Then, at the distance of seven miles and a half, we encamped on the upper end of a large island near the right. The river was shallow and rapid, so that we were obliged to be in the water during a great part of the day, dragging the canoes over the shoals and ripples. Its course, too, was so crooked that, notwithstanding we had made fourteen miles by water, we were only five miles from our encampment of last night. The country consists of a low ground on the river about five miles wide, and succeeded on both sides by plains of the same extent, which reach to the base of the mountains. These low grounds are very much intersected by bayous, and in those on the left side is a large proportion of bog covered with tall grass, which would yield a fine turf. There are very few trees, and those small narrow-leafed cottonwood, the principal growth being the narrow-leafed willow and currant bushes, among which were some bunches of privy near the river. We saw a number of geese, ducks, beaver, otter, deer, and antelopes, of all which one beaver was killed with a pole from the boat, three otters with a tomahawk, and the hunters brought in three deer and an antelope.

Monday, 12th. This morning, as soon as it was light, Captain Lewis sent Drewyer to reconnoitre if possible the route of the Indians; in about an hour and a half he returned, after following the tracks of the horse which we had lost yesterday to the mountains, where they ascended and were no longer visible. Captain Lewis now decided on making the circuit along the foot of the mountains which formed the cove, expecting by that means to find a road across them, and accordingly sent Drewyer on one side and Shields on the other. In this way they crossed four small rivulets near each other, on which were some bowers or conical lodges of willow brush, which seemed to have been made recently, From the manner in which the ground in the neighbourhood was torn up, the Indians appeared to have been gathering roots; but Captain Lewis could not discover what particular plant they were searching for, nor could he find any fresh track, till at the distance of four miles from his camp he met a large plain Indian road which came into the cove from the northeast, and wound along the foot of the mountains to the southwest, approaching obliquely the main stream he had left yesterday. Down this road he now went towards the southwest; at the distance of five miles it crossed a large run or creek, which is a principal branch of the main stream into which it falls, just above the high cliffs or gates observed yesterday, and which they now saw below them; here they halted and breakfasted on the last of the deer, keeping a small piece of pork in reserve against accident; they then continued through the low bottom along the main stream, near the foot of the mountains on their right. For the first five miles the valley continues towards the southwest from two to three miles in width; then the main stream, which had received two small branches from the left in the valley, turns abruptly to the west through a narrow bottom between the mountains. The road was still plain, and as it led them directly on towards the mountain the stream gradually became smaller, till after going two miles it had so greatly diminished in width that one of the men in a fit of enthusiasm, with one foot on each side of the river, thanked God that he had lived to bestride the Missouri. As they went along their hopes of soon seeing the waters of the Columbia arose almost to painful anxiety; when, after four miles from the last abrupt turn of the river, they reached a small gap formed by the high mountains which recede on each side, leaving room for the Indian road. From the foot of one of the lowest of these mountains, which rises with a gentle ascent of about half a mile, issues the remotest water of the Missouri. They had now reached the hidden sources of that river, which had never yet been seen by civilized man; and as they quenched their thirst at the chaste and icy fountain,—as they sat down by the brink of that little rivulet, which yielded its distant and modest tribute to the parent ocean,—they felt themselves rewarded for all their labours and all their difficulties. They left reluctantly this interesting spot, and pursuing the Indian road through the interval of the hills, arrived at the top of a ridge, from which they saw high mountains partially covered with snow still to the west of them. The ridge on which they stood formed the dividing line between the waters of the Atlantic and Pacific oceans. They followed a descent much steeper than that on the eastern side, and at the distance of three-quarters of a mile reached a handsome bold creek of cold clear water running to the westward. They stopped to taste for the first time the waters of the Columbia; and after a few minutes followed the road across steep hills and low hollows, till they reached a spring on the side of a mountain; here they found a sufficient quantity of dry willow brush for fuel, and therefore halted for the night; and having killed nothing in the course of the day, supped on their last piece of pork, and trusted to fortune for some other food to mix with a little flour and parched meal, which was all that now remained of their provisions. Before reaching the fountain of the Missouri they saw several large hawks, nearly black, and some of the heath cocks; these last have a long pointed tail, and are of a uniform dark brown colour, much larger than the common dunghill fowl, and similar in habits and the mode of flying to the grouse or prairie hen. Drewyer also wounded, at the distance of one hundred and thirty yards, an animal which we had not yet seen, but which after falling recovered itself and escaped. It seemed to be of the fox kind, rather larger than the small wolf of the plains, and with a skin in which black, reddish brown, and yellow were curiously intermixed. On the creek of the Columbia they found a species of currant which does not grow as high as that of the Missouri, though it is more branching, and its leaf, the under disk of which is covered with a hairy pubescence, is twice as large. The fruit is of the ordinary size and shape of the currant, and supported in the usual manner, but is of a deep purple colour, acid, and of a very inferior flavour.

We proceeded on in the boats, but as the river was very shallow and rapid, the navigation is extremely difficult, and the men, who are almost constantly in the water, are getting feeble and sore, and so much worn down by fatigue that they are very anxious to commence travelling by land. We went along the main channel, which is on the right side, and after passing nine bends in that direction, three islands and a number of bayous, reached, at the distance of five and a half miles, the upper point of a large island. At noon there was a storm of thunder, which continued about half an hour; after which we proceeded, but, as it was necessary to drag the canoes over the shoals and rapids, made but little progress. On leaving the island we passed a number of short bends, several bayous, and one run of water on the right side, and having gone by four small and two large islands, encamped on a smooth plain to the left near a few cottonwood trees; our journey by water was just twelve miles, and four in a direct line. The hunters supplied us with three deer and a fawn.

Tuesday, 13th. Very early in the morning Captain Lewis resumed the Indian road, which led him in a western direction, through an open broken country; on the left was a deep valley at the foot of a high range of mountains running from southeast to northwest, with their sides better clad with timber than the hills to which we have been for some time accustomed, and their tops covered in part with snow. At five miles distance, after following the long descent of another valley, he reached a creek about ten yards wide, and on rising the hill beyond it had a view of a handsome little valley on the left, about a mile in width, through which they judged, from the appearance of the timber, that some stream of water most probably passed. On the creek they had just left were some bushes of the white maple, the sumach of the small species with the winged rib, and a species of honeysuckle, resembling in its general appearance and the shape of its leaf the small honeysuckle of the Missouri, except that it is rather larger, and bears a globular berry about the size of a garden pea, of a white colour, and formed of a soft white mucilaginous substance, in which are several small brown seeds irregularly scattered, without any cell, and enveloped in a smooth thin pellicle.

They proceeded along a waving plain parallel to this valley for about four miles, when they discovered two women, a man, and some dogs, on an eminence at the distance of a mile before them. The strangers first viewed them apparently with much attention for a few minutes, and then two of them sat down as if to await Captain Lewis's arrival. He went on till he reached within about half a mile, then ordered his party to stop, put down his knapsack and rifle, and unfurling the flag advanced alone towards the Indians. The females soon retreated behind the hill, but the man remained till Captain Lewis came within a hundred yards from him, when he, too, went off, though Captain Lewis called out tabba bone! loud enough to be heard distinctly. He hastened to the top of the hill, but they had all disappeared. The dogs, however, were less shy, and came close to him; he therefore thought of tying a handkerchief with some beads round their necks, and then let them loose to convince the fugitives of his friendly disposition, but they would not suffer him to take hold of them, and soon left him. He now made a signal to the men, who joined him, and then all followed the track of the Indians, which led along a continuation of the same road they had been already travelling. It was dusty, and seemed to have been much used lately both by foot passengers and horsemen. They had not gone along it more than a mile when on a sudden they saw three female Indians, from whom they had been concealed by the deep ravines which intersected the road, till they were now within thirty paces of each other; one of them, a young woman, immediately took to flight; the other two, an elderly woman and a little girl, seeing we were too near for them to escape, sat on the ground, and holding down their heads seemed as if reconciled to the death which they supposed awaited them. The same habit of holding down the head and inviting the enemy to strike, when all chance of escape is gone, is preserved in Egypt to this day. Captain Lewis instantly put down his rifle, and advancing towards them, took the woman by the hand, raised her up, and repeated the words tabba bone! at the same time stripping up his shirt sleeve to prove that he was a white man, for his hands and face had become by constant exposure quite as dark as their own. She appeared immediately relieved from her alarm, and Drewyer and Shields now coming up, Captain Lewis gave them some beads, a few awls, pewter mirrors, and a little paint, and told Drewyer to request the woman to recall her companion who had escaped to some distance, and by alarming the Indians might cause them to attack him without any time for explanation. She did as she was desired, and the young woman returned, almost out of breath; Captain Lewis gave her an equal portion of trinkets, and painted the tawny cheeks of all three of them with vermillion, a ceremony which among the Shoshonees is emblematic of peace. After they had become composed, he informed them by signs of his wish to go to their camp in order to see their chiefs and warriors; they readily obeyed, and conducted the party along the same road down the river. In this way they marched two miles, when they met a troop of nearly sixty warriors, mounted on excellent horses, riding at full speed towards them. As they advanced Captain Lewis put down his gun, and went with the flag about fifty paces in advance. The chief, who, with two men, were riding in front of the main body, spoke to the women, who now explained that the party was composed of white men, and showed exultingly the presents they had received. The three men immediately leaped from their horses, came up to Captain Lewis and embraced him with great cordiality, putting their left arm over his right shoulder and clasping his back; applying at the same time their left cheek to his, and frequently vociferating ah hi e! ah hi e! " I am much pleased, I am much rejoiced." The whole body of warriors now came forward, and our men received the caresses, and no small share of the grease and paint, of their new friends. After this fraternal embrace, of which the motive was much more agreeable than the manner, Captain Lewis lighted a pipe and offered it to the Indians, who had now seated themselves in a circle around the party. But before they would receive this mark of friendship they pulled off their moccasins, a custom, as we afterwards learnt, which indicates the sacred sincerity of their professions when they smoke with a stranger, and which imprecates on themselves the misery of going barefoot forever if they are faithless to their words, a penalty by no means light to those who rove over the thorny plains of their country. It is not unworthy to remark the analogy which some of the customs of those wild children of the wilderness bear to those recorded in holy writ. Moses is admonished to pull off his shoes, for the place on which he stood was holy ground. Why this was enjoined as an act of peculiar reverence— whether it was from the circumstance that, in the arid region in which the patriarch then resided, it was deemed a test of the sincerity of devotion to walk upon the burning sands barefooted, in some measure analogous to the pains inflicted by the prickly pear,—does not appear. After smoking a few pipes, some trifling presents were distributed amongst them, with which they seemed very much pleased, particularly with the blue beads and the vermillion. Captain Lewis then informed the chief that the object of his visit was friendly, and should be explained as soon as he reached their camp; but that in the meantime, as the sun was oppressive and no water near, he wished to go there as soon as possible. They now put on their moccasins, and their chief, whose name was Cameahwait, made a short speech to the warriors. Captain Lewis then gave him the flag, which he informed him was among white men the emblem of peace and now that he had received it was to be in future the bond of union between them. The chief then moved on, our party followed him, and the rest of the warriors, in a squadron, brought up the rear. After marching a mile they were halted by the chief, who made a second harangue, on which six or eight young men rode forward to their camp, and no further regularity was observed in the order of march. At the distance of four miles from where they had first met they reached the Indian camp, which was in a handsome level meadow on the bank of the river. Here they were introduced into an old leathern lodge which the young men who had been sent from the party had fitted up for their reception. After being seated on green boughs and antelope skins, one of the warriors pulled up the grass in the centre of the lodge so as to form a vacant circle of two feet diameter, in which he kindled a fire. The chief then produced his pipe and tobacco, the warriors all pulled off their moccasins, and our party was requested to take off their own. This being done, the chief lighted his pipe at the fire within the magic circle, and then retreating from it began a speech several minutes long, at the end of which he pointed the stem towards the four cardinal points of the heavens, beginning with the east and concluding with the north. After this ceremony he presented the stem in the same way to Captain Lewis, who, supposing it an invitation to smoke, put out his hand to receive the pipe, but the chief drew it back, and continued to repeat the same offer three times, after which he pointed the stem first to the heavens, then to the centre of the little circle, took three whiffs himself, and presented it again to Captain Lewis. Finding that this last offer was in good earnest, he smoked a little; the pipe was then held to each of the white men, and after they had taken a few whiffs was given to the warriors. This pipe was made of a dense transparent green stone, very highly polished, about two and a half inches long, and of an oval figure, the bowl being in the same situation with the stem. A small piece of burnt clay is placed in the bottom of the bowl to separate the tobacco from the end of the stem, and is of an irregularly round figure, not fitting the tube perfectly close, in order that the smoke may pass with facility. The tobacco is of the same kind with that used by the Minnetarees, Mandans, and Ricaras of the Missouri. The Shoshonees do not cultivate this plant, but obtain it from the Rocky mountain Indians and some of the bands of their own nation who live farther south. The ceremony of smoking being concluded, Captain Lewis explained to the chief the purposes of his visit, and as by this time all the women and children of the camp had gathered around the lodge to indulge in a view of the first white man they had ever seen, he distributed among them the remainder of the small articles he had brought with him. It was now late in the afternoon, and our party had tasted no food since the night before. On apprising the chief of this circumstance, he said that he had nothing but berries to eat, and presented some cakes made of serviceberry and chokecherries which had been dried in the sun. On these Captain Lewis made a hearty meal, and then walked down towards the river. He found it a rapid clear stream, forty yards wide and three feet deep; the banks were low and abrupt, like those of the upper part of the Missouri, and the bed formed of loose stones and gravel. Its course, as far as he could observe it, was a little to the north of west, and was bounded on each side by a range of high mountains, of which those on the east are the lowest and most distant from the river.

The chief informed him that this stream discharged itself, at the distance of half a day's march, into another of twice its size coming from the southwest; but added, on further inquiry, that there was scarcely more timber below the junction of those rivers than in this neighbourhood, and that the river was rocky, rapid, and so closely confined between high mountains that it was impossible to pass down it, either by land or water, to the great lake, where, as he had understood, the white men lived. This information was far from being satisfactory, for there was no timber here that would answer the purpose of building canoes, indeed not more than just sufficient for fuel, and even that consisted of the narrow-leafed cottonwood, the red and the narrow-leafed willow, the chokecherry, serviceberry, and a few currant bushes such as are common on the Missouri. The prospect of going on by land is more pleasant, for there are great numbers of horses feeding in every direction round the camp, which will enable us to transport our stores if necessary over the mountains. Captain Lewis returned from the river to his lodge, and on his way an Indian invited him into his bower and gave him a small morsel of boiled antelope and a piece of fresh salmon roasted. This was the first salmon he had seen, and perfectly satisfied him that he was now on the waters of the Pacific. On reaching this lodge he resumed his conversation with the chief, after which he was entertained with a dance by the Indians. It now proved, as our party had feared, that the men whom they had first met this morning had returned to the camp and spread the alarm that their enemies, the Minnetarees of fort de Prairie, whom they call Pahkees, were advancing on them. The warriors instantly armed themselves and were coming down in expectation of an attack, when they were agreeably surprised by meeting our party. The greater part of them were armed with bows and arrows and shields, but a few had small fusils, such as are furnished by the northwest company traders, and which they had obtained from the Indians on the Yellowstone, with whom they are now at peace. They had reason to dread the approach of the Pahkees, who had attacked them in the course of this spring and totally defeated them. On this occasion twenty of their warriors were either killed or made prisoners, and they lost their whole camp except the leathern lodge which they had fitted up for us, and were now obliged to live in huts of a conical figure made with willow brush. The music and dancing, which was in no respect different from those of the Missouri Indians, continued nearly all night; but Captain Lewis retired to rest about twelve o'clock, when the fatigues of the day enabled him to sleep, though he was awaked several times by the yells of the dancers.

Whilst all these things were occurring to Captain Lewis we were slowly and laboriously ascending the river. For the first two and a half miles we went along the island opposite to which we encamped last evening, and soon reached a second island, behind which comes in a small creek on the left side of the river. It rises in the mountains to the east and forms a handsome valley for some miles from its mouth, where it is a bold running stream about seven yards wide; we called it M'Neal's creek, after Hugh M'Neal, one of our party. Just above this stream, and at the distance of four miles from our camp, is a point of limestone rock on the right, about seventy feet high, forming a cliff over the river. From the top of it the Beaver's-head bore north 24° east twelve miles distant; the course of Wisdom river, that is the direction of its valley through the mountains, is north 25° west; while the gap through which the Jefferson enters the mountains is ten miles above us on a course south 18° west. From this limestone rock we proceeded along several islands, on both sides, and after making twelve miles arrived at a cliff of high rocks on the right, opposite to which we encamped in a smooth level prairie, near a few cottonwood trees, but were obliged to use the dry willow brush for fuel. The river is still very crooked, the bends short and abrupt, and obstructed by so many shoals, over which the canoes were to be dragged, that the men were in the water three-fourths of the day. They saw numbers of otter, some beaver, antelopes, ducks, geese, and cranes, but they killed nothing except a single deer. They, however, caught some very fine trout, as they have done for several days past. The weather had been cloudy and cool during the fore part of the day, and at eight o'clock a shower of rain fell.

Wednesday, 14th. In order to give time for the boats to reach the forks of Jefferson river, Captain Lewis determined to remain here and obtain all the information he could collect with regard to the country. Having nothing to eat but a little flour and parched meal, with the berries of the Indians, he sent out Drewyer and Shields, who borrowed horses from the natives, to hunt for a few hours. About the same time the young warriors set out for the same purpose. There are but few elk or black-tailed deer in this neighbourhood, and as the common red deer secrete themselves in the bushes when alarmed, they are soon safe from the arrows, which are but feeble weapons against any animals which the huntsmen cannot previously run down with their horses. The chief game of the Shoshonees, therefore, is the antelope, which when pursued retreats to the open plains, where the horses have full room for the chase. But such is its extraordinary fleetness and wind that a single horse has no possible chance of outrunning it or tiring it down, and the hunters are therefore obliged to resort to stratagem. About twenty Indians, mounted on fine horses, and armed with bows and arrows, left the camp; in a short time they descried a herd of ten antelopes; they immediately separated into little squads of two or three, and formed a scattered circle round the herd for five or six miles, keeping at a wary distance, so as not to alarm them till they were perfectly inclosed, and usually selecting some commanding eminence as a stand. Having gained their positions, a small party rode towards the herd, and with wonderful dexterity the huntsman preserved his seat, and the horse his footing, as he ran at full speed over the hills, and down the steep ravines, and along the borders of the precipices. They were soon outstripped by the antelopes, which, on gaining the other extremity of the circle, were driven back and pursued by the fresh hunters. They turned and flew, rather than ran, in another direction; but there, too, they found new enemies. In this way they were alternately pursued backwards and forwards, till at length, notwithstanding the skill of the hunters, they all escaped, and the party after running for two hours returned without having caught anything, and their horses foaming with sweat. This chase, the greater part of which was seen from the camp, formed a beautiful scene; but to the hunters is exceedingly laborious, and so unproductive, even when they are able to worry the animal down and shoot him, that forty or fifty hunters will sometimes be engaged for half a day without obtaining more than two or three antelopes. Soon after they returned, our two huntsmen came in with no better success. Captain Lewis therefore made a little paste with the flour, and the addition of some berries formed a very palatable repast. Having now secured the good will of Cameahwait, Captain Lewis informed him of his wish that he would speak to the warriors and endeavour to engage them to accompany him to the forks of Jefferson river, where by this time another chief with a large party of white men were waiting his return; that it would be necessary to take about thirty horses to transport the merchandize; that they should be well rewarded for their trouble; and that when all the party should have reached the Shoshonee camp they would remain some time among them, and trade for horses, as well as concert plans for furnishing them in future with regular supplies of merchandize. He readily consented to do so, and after collecting the tribe together he made a long harangue, and in about an hour and a half returned, and told Captain Lewis that they would be ready to accompany him in the morning.

As the early part of the day was cold, and the men stiff and sore from the fatigues of yesterday, we did not set out till seven o'clock. At the distance of a mile we passed a bold stream on the right, which comes from a snowy mountain to the north, and at its entrance is four yards wide and three feet in depth; we called it Track creek. At six miles farther we reached another stream, which heads in some springs at the foot of the mountains on the left. After passing a number of bayous and small islands on each side, we encamped about half a mile by land below the Rattlesnake cliffs. The river was cold, shallow, and as it approached the mountains formed one continued rapid, over which we were obliged to drag the boats with great labour and difficulty. By using constant exertions we succeeded in making fourteen miles, but this distance did not carry us more than six and a half in a straight line; several of the men have received wounds and lamed themselves in hauling the boats over the stones. The hunters supplied them with five deer and an antelope.

Thursday, 15th. Captain Lewis rose early, and having eaten nothing yesterday except his scanty meal of flour and berries, felt the inconveniences of extreme hunger. On inquiry he found that his whole stock of provisions consisted of two pounds of flour. This he ordered to be divided into two equal parts, and one-half of it boiled with the berries into a sort of pudding; and after presenting a large share to the chief, he and his three men breakfasted on the remainder. Cameahwait was delighted at this new dish; he took a little of the flour in his hand, tasted and examined it very narrowly, asking if it was made of roots; Captain Lewis explained the process of preparing it, and he said it was the best thing he had eaten for a long time.

This being finished, Captain Lewis now endeavoured to hasten the departure of the Indians, who still hesitated and seemed reluctant to move, although the chief addressed them twice for the purpose of urging them; on inquiring the reason, Cameahwait told him that some foolish person had suggested that he was in league with their enemies the Pahkees, and had come only to draw them into ambuscade, but that he himself did not believe it. Captain Lewis felt uneasy at this insinuation; he knew the suspicious temper of the Indians, accustomed from their infancy to regard every stranger as an enemy, and saw that if this suggestion were not instantly checked it might hazard the total failure of the enterprise. Assuming therefore a serious air, he told the chief that he was sorry to find they placed so little confidence in him, but that he pardoned their suspicions because they were ignorant of the character of white men, among whom it was disgraceful to lie or entrap even an enemy by falsehood; that if they continued to think thus meanly of us they might be assured no white men would ever come to supply them with arms and merchandize; that there was at this moment a party of white men waiting to trade with them at the forks of the river; and that if the greater part of the tribe entertained any suspicion, he hoped there were still among them some who were men, who would go and see with their own eyes the truth of what he said, and who, even if there was any danger, were not afraid to die. To doubt the courage of an Indian is to touch the tenderest string of his mind, and the surest way to rouse him to any dangerous achievement. Cameahwait instantly replied that he was not afraid to die, and mounting his horse, for the third time harangued the warriors; he told them that he was resolved to go if he went alone, or if he were sure of perishing; that he hoped there were among those who heard him some who were not afraid to die, and who would prove it by mounting their horses and following him. This harangue produced an effect on six or eight only of the warriors, who now joined their chief. With these Captain Lewis smoked a pipe, and then, fearful of some change in their capricious temper, set out immediately. It was about twelve o'clock when his small party left the camp, attended by Cameahwait and the eight warriors; their departure seemed to spread a gloom over the village; those who would not venture to go were sullen and melancholy, and the women were crying and imploring the Great Spirit to protect their warriors as if they were going to certain destruction. Yet such is the wavering inconstancy of these savages that Captain Lewis's party had not gone far when they were joined by ten or twelve more warriors, and before reaching the creek which they had passed on the morning of the 13th, all the men of the nation and a number of women had overtaken them, and had changed from the surly ill temper in which they were two hours ago to the greatest cheerfulness and gayety. When they arrived at the spring on the side of the mountain where the party had encamped on the 12th, the chief insisted on halting to let the horses graze; to which Captain Lewis assented, and smoked with them. They are excessively fond of the pipe, in which, however, they are not able to indulge much, as they do not cultivate tobacco themselves, and their rugged country affords them but few articles to exchange for it. Here they remained for about an hour, and on setting out, by engaging to pay four of the party, Captain Lewis obtained permission for himself and each of his men to ride behind an Indian; but he soon found riding without stirrups more tiresome than walking, and therefore dismounted, making the Indian carry his pack. About sunset they reached the upper part of the level valley in the cove through which he had passed, and which they now called Shoshonee cove. The grass being burned on the north side of the river they crossed over to the south, and encamped about four miles above the narrow pass between the hills noticed as they traversed the cove before. The river was here about six yards wide, and frequently dammed up by the beaver. Drewyer had been sent forward to hunt, but he returned in the evening unsuccessful, and their only supper therefore was the remaining pound of flour, stirred in a little boiling water, and then divided between the four white men and two of the Indians.

In order not to exhaust the strength of the men, Captain Clark did not leave his camp till after breakfast. Although he was scarcely half a mile below the Rattlesnake cliffs, he was obliged to make a circuit of two miles by water before he reached them. The river now passed between low and rugged mountains and cliffs formed of a mixture of limestone and a hard black rock, with no covering except a few scattered pines. At the distance of four miles is a bold little stream which throws itself from the mountains down a steep precipice of rocks on the left. One mile farther is a second point of rocks and an island, about a mile beyond which is a creek on the right, ten yards wide and three feet three inches in depth, with a strong current; we called it Willard's creek after one of our men, Alexander Willard. Three miles beyond this creek, after passing a high cliff on the right opposite to a steep hill, we reached a small meadow on the left bank of the river. During its passage through these hills to Willard's creek the river had been less tortuous than usual, so that in the first six miles to Willard's creek we had advanced four miles on our route. We continued on for two miles, till we reached in the evening a small bottom covered with clover and a few cottonwood trees; here we passed the night, near the remains of some old Indian lodges of brush. The river is as it has been for some days, shallow and rapid, and our men, who are for hours together in the river, suffer not only from fatigue but from the extreme coldness of the water, the temperature of which is as low as that of the freshest springs in our country. In walking along the side of the river Captain Clark was very near being bitten twice by rattlesnakes, and the Indian woman narrowly escaped the same misfortune. We caught a number of fine trout; but the only game procured to-day was a buck, which had a peculiarly bitter taste, proceeding probably from its favourite food, the willow.

Friday, 16th. As neither our party nor the Indians had anything to eat, Captain Lewis sent two of his hunters ahead this morning to procure some provision; at the same time he requested Cameahwait to prevent his young men from going out, lest by their noise they might alarm the game. But this measure immediately revived their suspicions; it now began to be believed that these men were sent forward in order to apprise the enemy of their coming, and as Captain Lewis was fearful of exciting any further uneasiness, he made no objection on seeing a small party of Indians go on each side of the valley under pretence of hunting, but in reality to watch the movements of our two men; even this precaution, however, did not quiet the alarms of the Indians, a considerable part of whom returned home, leaving only twenty-eight men and three women. After the hunters had been gone about an hour, Captain Lewis again mounted with one of the Indians behind him, and the whole party set out; but just as they passed through the narrows they saw one of the spies coming back at full speed across the plain; the chief stopped and seemed uneasy, the whole band were moved with fresh suspicions, and Captain Lewis himself was much disconcerted, lest by some unfortunate accident some of their enemies might have perhaps straggled that way. The young Indian had scarcely breath to say a few words as he came up, when the whole troop dashed forward as fast as their horses could carry them; and Captain Lewis, astonished at this movement, was borne along for nearly a mile before he learnt with great satisfaction that it was all caused by the spy's having come to announce that one of the white men had killed a deer. Relieved from his anxiety, he now found the jolting very uncomfortable; for the Indian behind him, being afraid of not getting his share of the feast, had lashed the horse at every step since they set off; he therefore reined him in and ordered the

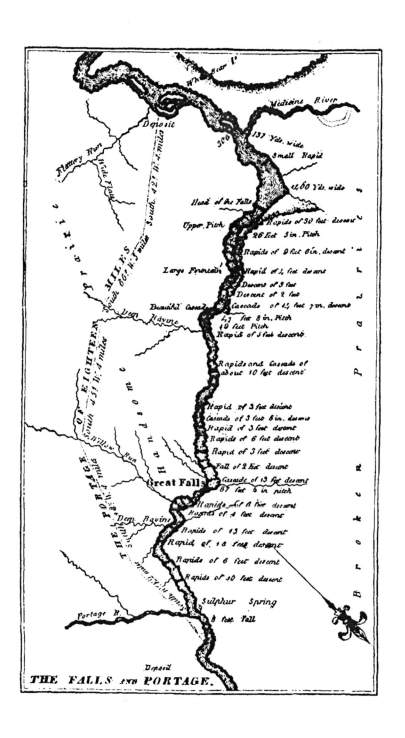

Indian to stop beating him. The fellow had no idea of losing time in disputing the point, and jumping off the horse ran for a mile at full speed. Captain Lewis slackened his pace, and followed at a sufficient distance to observe them. When they reached the place where Drewyer had thrown out the intestines, they all dismounted in confusion and ran tumbling over each other like famished dogs; each tore away whatever part he could, and instantly began to eat it; some had the liver, some the kidneys, in short no part on which we are accustomed to look with disgust escaped them; one of them who had seized about nine feet of the entrails was chewing at one end, while with his hand he was diligently clearing his way by discharging the contents at the other. It was indeed impossible to see these wretches ravenously feeding on the filth of animals, and the blood streaming from their mouths, without deploring how nearly the condition of savages approaches that of the brute creation; yet though suffering with hunger they did not attempt, as they might have done, to take by force the whole deer, but contented themselves with what had been thrown away by the hunter. Captain Lewis now had the deer skinned, and after reserving a quarter of it gave the rest of the animal to the chief to be divided among the Indians, who immediately devoured nearly the whole of it without cooking. They now went forward towards the creek where there was some brushwood to make a fire, and found Drewyer, who had killed a second deer; the same struggle for the entrails was renewed here, and on giving nearly the whole deer to the Indians, they devoured it, even to the soft part of the hoofs. A fire being made, Captain Lewis had his breakfast, during which Drewyer brought in a third deer; this, too, after reserving one quarter, was given to the Indians, who now seemed completely satisfied and in good humour. At this place they remained about two hours to let the horses graze, and then continued their journey, and towards evening reached the lower part of the cove, having on the way shot an antelope, the greater part of which was given to the Indians. As they were now approaching the place where they had been told by Captain Lewis they would see the white men, the chief insisted on halting; they therefore all dismounted, and Cameahwait, with great ceremony, and as if for ornament, put tippets or skins round the necks of our party, similar to those worn by themselves. As this was obviously intended to disguise the white men, Captain Lewis, in order to inspire them with more confidence, put his cocked hat and feather on the head of the chief, and as his own over-shirt was in the Indian form, and his skin browned by the sun, he could not have been distinguished from an Indian; the men followed his example, and the change seemed to be very agreeable to the Indians.

In order to guard, however, against any disappointment, Captain Lewis again explained the possibility of our not having reached the forks in consequence of the difficulty of the navigation, so that if they should not find us at that spot they might be assured of our not being far below. They again all mounted their horses and rode on rapidly, making one of the Indians carry their flag, so that we might recognise them as they approached us; but to the mortification and disappointment of both parties, on coming within two miles of the forks no canoes were to be seen. Uneasy lest at this moment he should be abandoned, and all his hopes of obtaining aid from the Indians be destroyed, Captain Lewis gave the chief his gun, telling him that if the enemies of his nation were in the bushes he might defend himself with it; that for his own part he was not afraid to die, and that the chief might shoot him as soon as they discovered themselves betrayed. The other three men at the same time gave their guns to the

Indians, who now seemed more easy, but still wavered in their resolutions. As they went on towards the point, Captain Lewis, perceiving how critical his situation had become, resolved to attempt a stratagem which his present difficulty seemed completely to justify. Recollecting the notes he had left at the point for us, he sent Drewyer for them with an Indian who witnessed his taking them from the pole. When they were brought, Captain Lewis told Cameahwait that on leaving his brother chief at the place where the river issues from the mountains, it was agreed that the boats should not be brought higher than the next forks we should meet; but that if the rapid water prevented the boats from coming on as fast as they expected, his brother chief was to send a note to the first forks above him to let him know where the boats were; that this note had been left this morning at the forks, and mentioned that the canoes were just below the mountains, and coming slowly up in consequence of the current. Captain Lewis added that he would stay at the forks for his brother chief, but would send a man down the river, and that if Cameahwait doubted what he said, one of their young men would go with him whilst he and the other two remained at the forks. This story satisfied the chief and the greater part of the Indians, but a few did not conceal their suspicions, observing that we told different stories, and complaining that the chief exposed them to danger by a mistaken confidence. Captain Lewis now wrote by the light of some willow brush a note to Captain Clark, which he gave to Drewyer, with an order to use all possible expedition in ascending the river, and engaged an Indian to accompany him by a promise of a knife and some beads. At bedtime the chief and five others slept round the fire of Captain Lewis, and the rest hid themselves in different parts of the willow brush to avoid the enemy, who they feared would attack them in the night.

Captain Lewis endeavoured to assume a cheerfulness he did not feel, to prevent the despondency of the savages; after conversing gayly with them he retired to his musquitoe bier, by the side of which the chief now placed himself; he lay down, yet slept but little, being in fact scarcely less uneasy than his Indian companions. He was apprehensive that, finding the ascent of the river impracticable, Captain Clark might have stopped below the Rattlesnake bluff, and the messenger would not meet him. The consequence of disappointing the Indians at this moment would most probably be, that they would retire and secrete themselves in the mountains, so as to prevent our having an opportunity of recovering their confidence; they would also spread a panic through all the neighbouring Indians, and cut us off from the supply of horses so useful and almost so essential to our success; he was at the same time consoled by remembering that his hopes of assistance rested on better foundations than their generosity—their avarice and their curiosity. He had promised liberal exchanges for their horses; but what was still more seductive, he had told them that one of their countrywomen, who had been taken with the Minnetarees, accompanied the party below; and one of the men had spread the report of our having with us a man perfectly black, whose hair was short and curled. This last account had excited a great degree of curiosity, and they seemed more desirous of seeing this monster than of obtaining the most favourable barter for their horses.

In the meantime we had set out after breakfast, and although we proceeded with more ease than we did yesterday, the river was still so rapid and shallow as to oblige us to drag the large canoes during the greater part of the day. For the first seven miles the river formed a bend to the right, so as to make our advance only three miles in a straight line; the stream is crooked, narrow, small, and shallow, with highlands occasionally on the banks, and strewed with islands, four of which are opposite to each other. Near this place we left the valley, to which we gave the name of Serviceberry valley, from the abundance of that fruit now ripe which is found in it. In the course of the four following miles we passed several more islands and bayous on each side of the river, and reached a high cliff on the right. Two and a half miles beyond this the cliffs approach on both sides and form a very considerable rapid near the entrance of a bold running stream on the left. The water was now excessively cold, and the rapids had been frequent and troublesome. On ascending an eminence Captain Clark saw the forks of the river, and sent the hunters up. They must have left it only a short time before Captain Lewis's arrival, but fortunately had not seen the note which enabled him to induce the Indians to stay with him. From the top of this eminence he could discover only three trees through the whole country, nor was there along the sides of the cliffs they had passed in the course of the day any timber except a few small pines; the low grounds were supplied with willow, currant bushes, and serviceberries. After advancing half a mile farther, we came to the lower point of an island near the middle of the river and about the centre of the valley; here we halted for the night, only four miles by land, though ten by water, below the point where Captain Lewis lay. Although we had made only fourteen miles, the labours of the men had fatigued and exhausted them very much; we therefore collected some small willow brush for a fire, and lay down to sleep.

## CHAPTER XV.

Affecting interview between the wife of Chaboneau and the chief of the Shoshonees —Council held with that nation, and favourable result—The extreme navigable point of the Missouri mentioned—General character of the river and of the country through which it passes—Captain Clark in exploring the source of the Columbia falls in company with another party of Shoshonees—The geographical information acquired from one of that party—Their manner of catching fish—The party reach Lewis river—The difficulties which Captain Clark had to encounter in his route—Friendship and hospitality of the Shoshonees—The party with Captain Lewis employed in making saddles, and preparing for the journey.

SATURDAY, August 17. Captain Lewis rose very early and despatched Drewyer and the Indian down the river in quest of the boats. Shields was sent out at the same time to hunt, while M'Neal prepared a breakfast out of the remainder of the meat. Drewyer had been gone about two hours, and the Indians were all anxiously waiting for some news, when an Indian who had straggled a short distance down the river returned with a report that he had seen the white men, who were only a short distance below, and were coming on. The Indians were all transported with joy, and the chief in the warmth of his satisfaction renewed his embrace to Captain Lewis, who was quite as much delighted as the Indians themselves. The report proved most agreeably true. On setting out at seven o'clock, Captain Clark, with Chaboneau and his wife, walked on shore; but they had not gone more than a mile before Captain Clark saw Sacajawea, who was with her husband one hundred yards ahead, begin to dance and show every mark of the most extravagant joy, turning round him and pointing to several Indians, whom he now saw advancing on horseback, sucking her fingers at the same time to indicate that they were of her native tribe. As they advanced Captain Clark discovered among them Drewyer dressed like an Indian, from whom he learnt the situation of the party. While the boats were performing the circuit he went towards the forks with the Indians, who, as they went along, sang aloud with the greatest appearance of delight. We soon drew near to the camp, and just as we approached it a woman made her way through the crowd towards Sacajawea, and recognising each other, they embraced with the most tender affection. The meeting of these two young women had in it something peculiarly touching, not only in the ardent manner in which their feelings were expressed, but from the real interest of their situation. They had been companions in childhood; in the war with the Minnetarees they had both been taken prisoners in the same battle, they had shared and softened the rigours of their captivity, till one of them had escaped from the Minnetarees, with scarce a hope of ever seeing her friend relieved from the hands of her enemies. While Sacajawea was renewing among the women the friendships of former days, Captain Clark went on, and was received by Captain Lewis and the chief, who, after the first embraces and salutations were over, conducted him to a sort of circular tent or shade of willows. Here he was seated on a white robe, and the chief immediately tied in his hair six small shells resembling pearls, an ornament highly valued by these people, who procured them in the course of trade from the seacoast. The moccasins of the whole party were then taken off, and after much ceremony the smoking began. After this the conference was to be opened, and glad of an opportunity of being able to converse more intelligibly, Sacajawea was sent for; she came into the tent, sat down, and was beginning to interpret, when in the person of Cameahwait she recognised her brother; she instantly jumped up and ran and embraced him, throwing over him her blanket and weeping profusely; the chief was himself moved, though not in the same degree. After some conversation between them she resumed her seat, and attempted to interpret for us, but her new situation seemed to overpower her, and she was frequently interrupted by her tears. After the council was finished, the unfortunate woman learnt that all her family were dead except two brothers, one of whom was absent, and a son of her eldest sister, a small boy, who was immediately adopted by her. The canoes arriving soon after, we formed a camp in a meadow on the left side, a little below the forks, took out our baggage, and by means of our sails and willow poles formed a canopy for our Indian visitors. About four o'clock the chiefs and warriors were collected, and after the customary ceremony of taking off the moccasins and smoking a pipe, we explained to them in a long harangue the purposes of our visit, making themselves one conspicuous object of the good wishes of our government, on whose strength as well as its friendly disposition we expatiated. We told them of their dependance on the will of our government for all future supplies of whatever was necessary either for their comfort or defence; that as we were sent to discover the best route by which merchandize could be conveyed to them, and no trade would be begun before our return, it was mutually advantageous that we should proceed with as little delay as possible; that we were under the necessity of requesting them to furnish us with horses to transport our baggage across the mountains, and a guide to show us the route, but that they should be amply remunerated for their horses, as well as for every other service they should render us. In the meantime our first wish was, that they should immediately collect as many horses as were necessary to transport our baggage to their village, where, at

## UP THE MISSOURI

our leisure, we would trade with them for as many horses as they could spare.

The speech made a favourable impression; the chief in reply thanked us for our expressions of friendship towards himself and his nation, and declared their willingness to render us every service. He lamented that it would be so long before they should be supplied with firearms, but that till then they could subsist as they had heretofore done. He concluded by saying that there were not horses here sufficient to transport our goods, but that he would return to the village tomorrow, and bring all his own horses, and encourage his people to come over with theirs. The conference being ended to our satisfaction, we now inquired of Cameahwait what chiefs were among the party, and he pointed out two of them. We then distributed our presents: to Cameahwait we gave a medal of the small size, with the likeness of President Jefferson, and on the reverse a figure of hands clasped with a pipe and tomahawk; to this was added a uniform coat, a shirt, a pair of scarlet leggings, a carrot of tobacco, and some small articles. Each of the other chiefs received a small medal struck during the presidency of General Washington, a shirt, handkerchief, leggings, a knife, and some tobacco. Medals of the same sort were also presented to two young warriors, who though not chiefs were promising youths and very much respected in the tribe. These honorary gifts were followed by presents of paint, moccasins, awls, knives, beads, and looking-glasses. We also gave them all a plentiful meal of Indian corn, of which the hull is taken off by being boiled in lye; and as this was the first they had ever tasted, they were very much pleased with it. They had indeed abundant sources of surprise in all they saw: the appearance of the men, their arms, their clothing, the canoes, the strange looks of the negro, and the sagacity of our dog, all in turn shared their admiration, which was raised to astonishment by a shot from the airgun; this operation was instantly considered as *a great medicine,* by which they as well as the other Indians mean something emanating directly from the Great Spirit, or produced by his invisible and incomprehensible agency. The display of all these riches had been intermixed with inquiries into the geographical situation of their country, for we had learnt by experience that to keep the savages in good temper their attention should not be wearied with too much business, but that the serious affairs should be enlivened by a mixture of what is new and entertaining. Our hunters brought in very seasonably four deer and an antelope, the last of which we gave to the Indians, who in a very short time devoured it. After the council was over, we consulted as to our future operations. The game does not promise to last here for a number of days, and this circumstance combined with many others to induce our going on as soon as possible. Our Indian information as to the state of the Columbia is of a very alarming kind, and our first object is of course to ascertain the practicability of descending it, of which the Indians discourage our expectations. It was therefore agreed that Captain Clark should set off in the morning with eleven men, furnished, besides their arms, with tools for making canoes; that he should take Chaboneau and his wife to the camp of the Shoshonees, where he was to leave them, in order to hasten the collection of horses; that he was then to lead his men down to the Columbia, and if he found it navigable, and the timber in sufficient quantity, begin to build canoes. As soon as he had decided as to the propriety of proceeding down the Columbia or across the mountains, he was to send back one of the men with information of it to Captain Lewis, who by that time would have brought up the whole party and the rest of the baggage as far as the Shoshonee village.

Preparations were accordingly made this evening for such an arrangement. The sun is excessively hot in the day time, but the nights very cold, and rendered still more unpleasant from the want of any fuel except willow brush. The appearances, too, of game for many days' subsistence are not very favourable.

Sunday, 18th. In order to relieve the men of Captain Clark's party from the heavy weight of their arms, provisions, and tools, we exposed a few articles to barter for horses, and soon obtained three very good ones, in exchange for which we gave a uniform coat, a pair of leggings, a few handkerchiefs, three knifes, and some other small articles, the whole of which did not in the United States cost more than twenty dollars; a fourth was purchased by the men for an old checkered shirt, a pair of old leggings, and a knife. The Indians seemed to be quite as well pleased as ourselves at the bargains they had made. We now found that the two inferior chiefs were somewhat displeased at not having received a present equal to that given to the great chief, who appeared in a dress so much finer than their own. To allay their discontent, we bestowed on them two old coats, and promised them that if they were active in assisting us across the mountains they should have an additional present. This treatment completely reconciled them, and the whole Indian party, except two men and two women, set out in perfect good humour to return home with Captain Clark. After going fifteen miles through a wide level valley with no wood but willows and shrubs, he encamped in the Shoshonee cove near a narrow pass where the highlands approach within two hundred yards of each other, and the river is only ten yards wide. The Indians went on farther, except the three chiefs and two young men, who assisted in eating two deer brought in by the hunters. After their departure everything was prepared for the transportation of the baggage, which was now exposed to the air and dried. Our game was one deer and a beaver, and we saw an abundance of trout in the river, for which we fixed a net in the evening.

We have now reached the extreme navigable point of the Missouri, which our observation places in latitude 43° 30′ 43″ north. It is difficult to comprise in any general description the characteristics of a river so extensive, and fed by so many streams which have their sources in a great variety of soils and climates. But the Missouri is still sufficiently powerful to give to all its waters something of a common character, which is of course decided by the nature of the country through which it passes. The bed of the river is chiefly composed of a blue mud, from which the water itself derives a deep tinge. From its junction here to the place near which it leaves the mountains, its course is embarrassed by rapids and rocks which the hills on each side have thrown into its channel. From that place, its current, with the exception of the falls, is not difficult of navigation, nor is there much variation in its appearance till the mouth of the Platte. That powerful river throws out vast quantities of coarse sand which contribute to give a new face to the Missouri, which is now much more impeded by islands. The sand, as it is drifted down, adheres in time to some of the projecting points from the shore, and forms a barrier to the mud, which at length fills to the same height with the sandbar itself; as soon as it has acquired a consistency, the willow grows there the first year, and by its roots assists the solidity of the whole; as the mud and sand accumulate, the cottonwood tree next appears; till the gradual excretion of soils raises the surface of the point above the highest freshets. Thus stopped in its course the water seeks a passage elsewhere, and as the soil on each side is light and yielding, what was only a peninsula becomes

## UP THE MISSOURI 413

gradually an island, and the river indemnifies itself for the usurpation by encroaching on the adjacent shore. In this way the Missouri, like the Mississippi, is constantly cutting off the projections of the shore, and leaving its ancient channel, which is then marked by the mud it has deposited and a few stagnant ponds.

The general appearance of the country as it presents itself on ascending may be thus described: From its mouth to the two Charletons, a ridge of highlands borders the river at a small distance, leaving between them fine rich meadows. From the mouth of the two Charletons the hills recede from the river, giving greater extent to the low grounds, but they again approach the river for a short distance near Grand river, and again at Snake creek. From that point they retire, nor do they come again to the neighbourhood of the river till above the Sauk prairie, where they are comparatively low and small. Thence they diverge and reappear at the Charaton Searty, after which they are scarcely if at all discernible, till they advance to the Missouri nearly opposite to the Kanzas.

The same ridge of hills extends on the south side in almost one unbroken chain, from the mouth of the Missouri to the Kanzas, though decreasing in height beyond the Osage. As they are nearer the river than the hills on the opposite side, the intermediate low grounds are of course narrower, but the general character of the soil is common to both sides.

In the meadows and along the shore the tree most common is the cottonwood, which with the willow forms almost the exclusive growth of the Missouri. The hills, or rather high grounds, for they do not rise higher than from one hundred and fifty to two hundred feet, are composed of a good rich black soil, which is perfectly susceptible of cultivation, though it becomes richer on the hills beyond the Platte, and are in general thinly covered with timber. Beyond these hills the country extends into high open plains, which are on both sides sufficiently fertile, but the south has the advantage of better streams of water, and may therefore be considered as preferable for settlements. The lands, however, become much better and the timber more abundant between the Osage and the Kanzas. From the Kanzas to the Nadawa the hills continue at nearly an equal distance, varying from four to eight miles from each other, except that from the little Platte to nearly opposite the ancient Kanzas village the hills are more remote, and the meadows of course wider on the north side of the river. From the Nadawa the northern hills disappear, except at occasional intervals where they are seen at a distance, till they return about twenty-seven miles above the Platte near the ancient village of the Ayoways. On the south the hills continue close to the river from the ancient village of the Kanzas up to Council bluff, fifty miles beyond the Platte, forming high prairie lands. On both sides the lands are good, and perhaps this distance from the Osage to the Platte may be recommended as among the best districts on the Missouri for the purposes of settlers.

From the Ayoway village the northern hills again retire from the river, to which they do not return till three hundred and twenty miles above, at Floyd's river. The hills on the south also leave the river at Council bluffs, and reappear at the Mahar village, two hundred miles up the Missouri. The country thus abandoned by the hills is more open and the timber in smaller quantities than below the Platte, so that although the plain is rich and covered with high grass, the want of wood renders it less calculated for cultivation than below that river.

The northern hills, after remaining near the Missouri for a few miles at Floyd's river, recede from it at the Sioux river, the course of which they follow; and though they again visit the Missouri at Whitestone river, where they are low, yet they do not return to it till beyond James river. The highlands on the south, after continuing near the river at the Mahar villages, again disappear, and do not approach it till the Cobalt bluffs, about forty-four miles from the villages, and then from those bluffs to the Yellowstone river, a distance of about one thousand miles, they follow the banks of the river with scarcely any deviation.

From the James river, the lower grounds are confined within a narrow space by the hills on both sides, which now continue near each other up to the mountains. The space between them, however, varies from one to three miles as high as the Muscleshell river, from which the hills approach so high as to leave scarcely any low grounds on the river, and near the falls reach the water's edge. Beyond the falls the hills are scattered and low to the first range of mountains.

The soil during the whole length of the Missouri below the Platte is, generally speaking, very fine, and although the timber is scarce, there is still sufficient for the purposes of settlers. But beyond that river, although the soil is still rich, yet the almost total absence of timber, and particularly the want of good water, of which there is but a small quantity in the creeks, and even that brackish, oppose powerful obstacles to its settlement. The difficulty becomes still greater between the Muscleshell river and the falls, where, besides the greater scarcity of timber, the country itself is less fertile.

The elevation of these highlands varies as they pass through this extensive tract of country. From Wood river they are about one hundred and fifty feet above the water, and continue at that height till they rise near the Osage, from which place to the ancient fortification they again diminish in size. Thence they continue higher till the Mandan village, after which they are rather lower till the neighbourhood of Muscleshell river, where they are met by the Northern hills, which have advanced at a more uniform height, varying from one hundred and fifty to two hundred or three hundred feet. From this place to the mountains the height of both is nearly the same, from three hundred to five hundred feet, and the low grounds so narrow that the traveller seems passing through a range of high country. From Maria's river to the falls, the hills descend to the height of about two or three hundred feet.

Monday, 19th. The morning was cold, and the grass perfectly whitened by the frost. We were engaged in preparing packs and saddles to load the horses as soon as they should arrive. A beaver was caught in a trap, but we were disappointed in trying to catch trout in our net; we therefore made a seine of willow brush, and by hauling it procured a number of fine trout, and a species of mullet which we had not seen before; it is about sixteen inches long, the scales small; the nose long, obtusely pointed, and exceeding the under jaw; the mouth opens with folds at the sides; it has no teeth, and the tongue and palate is smooth. The colour of its back and sides is a bluish brown, while the belly is white; it has the faggot bones, whence we concluded it to be of the mullet species. It is by no means so well flavoured a fish as the trout, which are the same as those we first saw at the falls, larger than the speckled trout of the mountains in the Atlantic states, and equally well flavoured. In the evening the hunters returned with two deer.

Captain Clark, in the meantime, proceeded through a wide level valley, in which the chief pointed out a spot where many of his tribe were killed in battle a year ago. The Indians accompanied him during the day, and as they had nothing to eat, he was obliged to feed them from his own stores, the hunters not being able to kill anything. Just as he was entering the mountains he met an Indian with two mules and a Spanish saddle, who was so polite as to offer one of them to him to ride over the hills. Being on foot, Captain Clark accepted his offer and gave him a waistcoat as a reward for his civility. He encamped for the night on a small stream, and the next morning,

Tuesday, August 20, he set out at six o'clock. In passing through a continuation of the hilly broken country, he met several parties of Indians. On coming near the camp, which had been removed since we left them two miles higher up the river, Cameahwait requested that the party should halt. This was complied with; a number of Indians came out from the camp, and with great ceremony several pipes were smoked. This being over, Captain Clark was conducted to a large leathern lodge prepared for his party in the middle of the encampment, the Indians having only shelters of willow bushes. A few dried berries and one salmon, the only food the whole village could contribute, were then presented to him; after which he proceeded to repeat in council, what had been already told them, the purposes of his visit; urged them to take their horses over and assist in transporting our baggage, and expressed a wish to obtain a guide to examine the river. This was explained and enforced to the whole village by Cameahwait, and an old man was pointed out who was said to know more of their geography to the north than any other person, and whom Captain Clark engaged to accompany him. After explaining his views he distributed a few presents, the council was ended, and nearly half the village set out to hunt the antelope, but returned without success.

Captain Clark in the meantime made particular inquiries as to the situation of the country, and the possibility of soon reaching a navigable water. The chief began by drawing on the ground a delineation of the rivers, from which it appeared that his information was very limited. The river on which the camp is he divided into two branches just above us, which, as he indicated by the opening of the mountains, were in view; he next made it discharge itself into a larger river ten miles below, coming from the southwest; the joint stream continued one day's march to the northwest, and then inclined to the westward for two days' march farther. At that place he placed several heaps of sand on each side, which, as he explained them, represented vast mountains of rock always covered with snow, in passing through which the river was so completely hemmed in by the high rocks that there was no possibility of travelling along the shore; that the bed of the river was obstructed by sharp-pointed rocks, and such its rapidity that, as far as the eye could reach, it presented a perfect column of foam. The mountains he said were equally inaccessible, as neither man nor horse could cross them; that such being the state of the country, neither he nor any of his nation had ever attempted to go beyond the mountains. Cameahwait said also that he had been informed by the Chopunnish, or pierced-nose Indians, who reside on this river west of the mountains, that it ran a great way towards the setting sun, and at length lost itself in a great lake of water which was ill-tasted, and where the white men lived. An Indian belonging to a band of Shoshonees who live to the southwest, and who happened to be at camp, was then brought in, and inquiries made of him as to the situation of the country in that direction. This he described in terms scarcely less terrible than those in which Cameahwait had represented the west. He said that his relations lived at the distance of twenty days' march from this place, on a course a little to the west of south, and not far from the whites, with whom they traded for horses, mules, cloth, metal, beads, and the shells here worn as ornaments, and which are those of a species of pearl oyster. In order to reach his country we should be obliged during the first seven days to climb over steep rocky mountains where there was no game, and we should find nothing but roots for subsistence. Even for these, however, we should be obliged to contend with a fierce warlike people, whom he called the Broken-moccasin, or moccasin with holes, who lived like bears in holes, and fed on roots and the flesh of such horses as they could steal or plunder from those who passed through the mountains. So rough, indeed, was the passage that the feet of the horses would be wounded in such a manner that many of them would be unable to proceed. The next part of the route was for ten days through a dry parched desert of sand, inhabited by no animal which would supply us with subsistence, and as the sun had now scorched up the grass and dried up the small pools of water which are sometimes scattered through this desert in the spring, both ourselves and our horses would perish for want of food and water. About the middle of this plain a large river passes from southeast to northwest, which, though navigable, afforded neither timber nor salmon. Three or four days' march beyond this plain his relations lived, in a country tolerably fertile and partially covered with timber, on another large river running in the same direction as the former; that this last discharges itself into a third large river, on which resided many numerous nations, with whom his own were at war, but whether this last emptied itself into the great or stinking lake, as they called the ocean, he did not know; that from his country to the stinking lake was a great distance, and that the route to it, taken by such of his relations as had visited it, was up the river on which they lived, and over to that on which the white people lived, and which they knew discharged itself into the ocean. This route he advised us to take, but added that we had better defer the journey till spring, when he would himself conduct us. This account persuaded us that the streams of which he spoke were southern branches of the Columbia, heading with the Rio des Apostolos and Rio Colorado, and that the route which he mentioned was to the gulf of California. Captain Clark therefore told him that this road was too much towards the south for our purpose, and then requested to know if there was no route on the left of the river where we now are, by which we might intercept it below the mountains; but he knew of none except that through the barren plains, which he said joined the mountains on that side, and through which it was impossible to pass at this season, even if we were fortunate enough to escape the Brokenmoccasin Indians. Captain Clark recompensed the Indian by a present of a knife, with which he seemed much gratified, and now inquired of Cameahwait by what route the Piercednose Indians, who he said lived west of the mountains, crossed over to the Missouri. This he said was towards the north, but that the road was a very bad one; that during the passage he had been told they suffered excessively from hunger, being obliged to subsist for many days on berries alone, there being no game in that part of the mountains, which were broken and rocky, and so thickly covered with timber that they could scarcely pass. Surrounded by difficulties as all the other routes are, this seems to be the most practicable of all the passages by land, since, if the Indians can pass the mountains with their women and children, no difficulties which they could encounter could be formidable to us; and if the Indians below the mountains are so numerous as they are represented to be, they must have some means of subsistence equally within our power. They tell us, indeed, that the nations to the westward subsist principally on fish and roots, and that their only game were a few elk, deer, and antelope, there being no buffaloe west of the mountain. The first inquiry, however, was to ascertain the truth of their information relative to the difficulty of descending the river; for this purpose Captain Clark set out at three o'clock in the afternoon, accompanied by the guide and all his men, except one, whom he left with orders to purchase a horse and join him as soon as possible. At the distance of four miles he crossed the river, and eight miles from the camp halted for the night at a small stream. The road which he followed was a beaten path through a wide rich meadow, in which were several old lodges. On the route he met a number of men, women, and children, as well as horses, and one of the men, who appeared to possess some consideration, turned back with him, and observing a woman with three salmon, obtained them from her, and presented them to the party. Captain Clark shot a mountain cock or cock of the plains, a dark brown bird larger than the dunghill fowl, with a long and pointed tail, and a fleshy protuberance about the base of the upper chop, something like that of the turkey, though without the snout. In the morning,

Wednesday, 21st, he resumed his march early, and at the distance of five miles reached an Indian lodge of brush, inhabited by seven families of Shoshonees. They behaved with great civility, gave the whole party as much boiled salmon as they could eat, and added as a present several dried salmon and a considerable quantity of chokecherries. After smoking with them all, he visited the fish weir, which was about two hundred yards distant. The river was here divided by three small islands, which occasioned the water to pass along four channels. Of these, three were narrow, and stopped by means of trees which were stretched across, and supported by willow stakes, sufficiently near each other to prevent the passage of the fish. About the centre of each was placed a basket formed of willows, eighteen or twenty feet in length, of a cylindrical form, and terminating in a conic shape at its lower extremity; this was situated with its mouth upwards, opposite to an aperture in the weir. The main channel of the water was then conducted to this weir, and as the fish entered it they were so entangled with each other that they could not move, and were taken out by untying the small end of the willow basket. The weir in the main channel was formed in a manner somewhat different; there were in fact two distinct weirs, formed of poles and willow sticks, quite across the river, approaching each other obliquely with an aperture in each side near the angle. This is made by tying a number of poles together at the top, in parcels of three, which were then set up in a triangular form at the base, two of the poles being in the range desired for the weir, and the third down the stream. To these poles two ranges of other poles are next lashed horizontally, with willow bark and wythes, and willow sticks joined in with these crosswise, so as to form a kind of wicker-work from the bottom of the river to the height of three or four feet above the surface of the water. This is so thick as to prevent the fish from passing, and even in some parts, with the help of a little gravel and some stone, enables them to give any direction which they wish to the water. These two weirs being placed near to each other, one for the purpose of catching the fish as they ascend, the other as they go down the river, is provided with two baskets made in the form already described, and which are placed at the apertures of the weir. After examining these curious objects, he returned to the lodges, and soon passed the river to the left, where an Indian brought him a tomahawk which he said he had found in the grass, near the lodge where Captain Lewis had staid on his first visit to the village. This was a tomahawk which had been missed at the time, and supposed to be stolen; it was, however, the only article which had been lost in our intercourse with the nation, and as even that was returned the inference is highly honourable to the integrity of the Shoshonees. On leaving the lodges Captain Clark crossed to the left side of the river, and despatched five men to the forks of it, in search of the man left behind yesterday, who procured a horse and passed by another road, as they learnt, to the forks. At the distance of fourteen miles they killed a very large salmon, two and a half feet long, in a creek six miles below the forks; and after travelling about twenty miles through the valley, following the course of the river, which runs nearly northwest, halted in a small meadow on the right side, under a cliff of rocks. Here they were joined by the five men who had gone in quest of Crusatte. They had been to the forks of the river, where the natives resort in great numbers for the purpose of gigging fish, of which they made our men a present of five fresh salmon. In addition to this food, one deer was killed to-day. The western branch of this river is much larger than the eastern, and after we passed the junction we found the river about one hundred yards in width, rapid and shoaly, but containing only a small quantity of timber. As Captain Lewis was the first white man who visited its waters, Captain Clark gave it the name of Lewis's river. The low grounds through which he had passed to-day were rich and wide, but at his camp this evening the hills begin to assume a formidable aspect. The cliff under which he lay is of a reddish brown colour; the rocks which have fallen from it are a dark brown flintstone. Near the place are gullies of white sandstone, and quantities of a fine sand, of a snowy whiteness; the mountains on each side are high and rugged, with some pine trees scattered over them.

Thursday, 22d. He soon began to perceive that the Indian accounts had not exaggerated; at the distance of a mile he passed a small creek, and the points of four mountains, which were rocky, and so high that it seemed almost impossible to cross them with horses. The road lay over the sharp fragments of rocks which had fallen from the mountains, and were strewed in heaps for miles together, yet the horses, altogether unshod, travelled across them as fast as the men, and without detaining them a moment. They passed two bold-running streams and reached the entrance of a small river, where a few Indian families resided. They had not been previously acquainted with the arrival of the whites; the guide was behind, and the wood so thick that we came upon them unobserved, till at a very short distance. As soon as they saw us, the women and children fled in great consternation; the men offered us everything they had: the fish on the scaffolds, the dried berries, and the collars of elk's tushes worn by the children. We took only a small quantity of the food, and gave them in return some small articles, which conduced very much to pacify them. The guide, now coming up, explained to them who we were and the object of our visit, which seemed to relieve the fears; but still a number of the women and children did not recover from their fright, but cryed during our stay, which lasted about an hour. The guide, whom we found a very intelligent, friendly old man, informed us that up this river there was a road which led over the mountains to the Missouri. On resuming his route, he went along the steep side of a mountain about three miles, and then reached the river near a small island, at the lower part of which he encamped; he here attempted to gig some fish, but could only obtain one small salmon. The river is here shoal and rapid, with many rocks scattered in various directions through its bed. On the sides of the mountains are some scattered pines, and of those on the left the tops are covered with them; there are, however, but few in the low grounds through which they passed, indeed they have seen only a single tree fit to make a canoe, and even that was small. The country has an abundant growth of berries, and we met several women and children gathering them, who bestowed them upon us with great liberality. Among the woods Captain Clark observed a species of woodpecker, the beak and tail of which were white, the wings black, and every other part of the body of a dark brown; its size was that of the robin, and it fed on the seeds of the pine.

Friday, 23d. Captain Clark set off very early, but as his route lay along the steep side of a mountain, over irregular and broken masses of rocks, which wounded the horses' feet, he was obliged to proceed slowly. At the distance of four miles he reached the river, but the rocks here became so steep, and projected so far into the river, that there was no mode of passing except through the water. This he did for some distance, though the river was very rapid, and so deep that they were forced to swim their horses. After following the edge of the water for about a mile under this steep cliff he reached a small meadow, below which the whole current of the river beat against the right shore on which he was, and which was formed of a solid rock perfectly inaccessible to horses. Here, too, the little track which he had been pursuing terminated. He therefore resolved to leave the horses and the greater part of the men at this place, and examine the river still farther, in order to determine if there were any possibility of descending it in canoes. Having killed nothing except a single goose to-day, and the whole of our provision being consumed last evening, it was by no means advisable to remain any length of time where they were. He now directed the men to fish and hunt at this place till his return, and then with his guide and three men he proceeded, clambering over immense rocks and along the sides of lofty precipices which bordered the river, when at about twelve miles distance he reached a small meadow, the first he had seen on the river since he left his party. A little below this meadow, a large creek, twelve yards wide and of some depth, discharges itself from the north. Here were some recent signs of an Indian encampment, and the tracks of a number of horses, who must have come along a plain Indian path, which he now saw following the course of the creek. This stream his guide said led towards a large river running to the north, and was frequented by another nation for the purpose of catching fish. He remained here two hours, and having taken some small fish, made a dinner on them with the addition of a few berries. From the place where he had left the party to the mouth of this creek it presents one continued rapid, in which are five shoals, neither of which could be passed with loaded canoes; and the baggage must therefore be transported for a considerable distance over the steep mountains, where it would be impossible to employ horses for the relief of the men. Even the empty canoes must be let down the rapids by means of cords, and not even in that way without great risk both to the canoes as well as to the men. At one of these shoals, indeed, the rocks rise so perpendicularly from the water as to leave no hope of a passage or even a portage without great labour in removing rocks, and in some instances cutting away the earth. To surmount these difficulties would exhaust the strength of the party, and, what is equally discouraging, would waste our time and consume our provisions, of neither of which have we much to spare. The season is now far advanced, and the Indians tell us we shall shortly have snow; the salmon, too, have so far declined that the natives themselves are hastening from the country, and not an animal of any kind larger than a pheasant or a squirrel, and of even these a few only, will then be seen in this part of the mountains; after which we shall be obliged to rely on our own stock of provisions, which will not support us more than ten days. These circumstances combine to render a passage by water impracticable in our present situation. To descend the course of the river on horseback is the other alternative, and scarcely a more inviting one. The river is so deep that there are only a few places where it can be forded, and the rocks approach so near the water as to render it impossible to make a route along the water's edge. In crossing the mountains themselves we should have to encounter, besides their steepness, one barren surface of broken masses of rock, down which in certain seasons the torrents sweep vast quantities of stone into the river. These rocks are of a whitish brown, and towards the base of a gray colour, and so hard that on striking them with steel they yield a fire like flint. This sombre appearance is in some places scarcely relieved by a single tree, though near the river and on the creeks there is more timber, among which are some tall pine; several of these might be made into canoes, and by lashing two of them together one of tolerable size might be formed.

After dinner he continued his route, and at the distance of half a mile passed another creek, about five yards wide. Here his guide informed him that by ascending the creek for some distance he would have a better road, and cut off a considerable bend of the river towards the south. He, therefore, pursued a well-beaten Indian track up this creek for about six miles, when, leaving the creek to the right, he passed over a ridge, and after walking a mile, again met the river where it flows through a meadow of about eighty acres in extent. This they passed, and then ascended a high and steep point of a mountain, from which the guide now pointed out where the river broke through the mountains about twenty miles distant. Near the base of the mountains a small river falls in from the south; this view was terminated by one of the loftiest mountains Captain Clark had ever seen, which was perfectly covered with snow. Towards this formidable barrier the river went directly on, and there it was, as the guide observed, that the difficulties and dangers of which he and Cameahwait had spoken commenced. After reaching the mountain, he said, the river continues its course towards the north for many miles, between high perpendicular rocks, which were scattered through its bed; it then penetrated the mountain through a narrow gap, on each side of which arose perpendicularly a rock as high as the top of the mountain before them; that the river then made a bend which concealed its future course from view, and as it was alike impossible to descend the river or clamber over that vast mountain, eternally covered with snow, neither he nor any of his nation had ever been lower than at a place where they could see the gap made by the river on entering the mountain. To that place he said he would conduct Captain Clark if he desired it by the next evening. But he was in need of no further evidence to convince him of the utter impracticability of the route before him. He had already witnessed the difficulties of part of the road; yet, after all these dangers, his guide, whose intelligence and fidelity he could not doubt, now assured him that the difficulties were only commencing, and what he saw before him too clearly convinced him of the Indian's veracity. He therefore determined to abandon this route, and returned to the upper part of the last creek we had passed, and reaching it an hour after dark, encamped for the night; on this creek he had seen in the morning an Indian road coming in from the north. Disappointed in finding a route by water, Captain Clark now questioned his guide more particularly as to the direction of this road, which he seemed to understand perfectly. He drew a map on the sand, and represented this road, as well as that we passed yesterday on Berry creek, as both leading towards two forks of the same great river, where resided a nation called Tushepaws, who, having no salmon on their river, came by these roads to the fish weirs on Lewis's river. He had himself been among these Tushepaws, and having once accompanied them on a fishing party to another river, he had there seen Indians who had come across the Rocky mountains. After a great deal of conversation, or rather signs, and a second and more particular map from his guide, Captain Clark felt persuaded that his guide knew of a road from the Shoshonee village they had left to the great river to the north, without coming so low down as this on a route impracticable for horses. He was desirous of hastening his return, and therefore set out early,

Saturday, 24th, and after descending the creek to the river, stopped to breakfast on berries in the meadow above the second creek. He then went on, but unfortunately fell from a rock and injured his leg very much; he however walked on as rapidly as he could, and at four in the afternoon rejoined his men. During his absence they had killed one of the mountain cocks, a few pheasants, and some small fish, on which, with haws and serviceberries, they had subsisted. Captain Clark immediately sent forward a man on horseback with a note to Captain Lewis, apprising him of the result of his inquiries, and late in the afternoon set out with the rest of the party and encamped at the distance of two miles. The men were much disheartened at the bad prospect of escaping from the mountains, and having nothing to eat but a few berries, which have made several of them sick, they all passed a disagreeable night, which was rendered more uncomfortable by a heavy dew.

Sunday, 25th. The want of provisions urged Captain Clark to return as soon as possible; he therefore set out early, and halted an hour in passing the Indian camp near the fish weirs. These people treated them with great kindness, and though poor and dirty, they willingly give what little they possess; they gave the whole party boiled salmon and dried berries, which were not, however, in sufficient quantities to appease their hunger. They soon resumed their old road, but as the abstinence or strange diet had given one of the men a very severe illness, they were detained very much on his account, and it was not till late in the day they reached the cliff under which they had encamped on the twenty-first. They immediately began to fish and hunt, in order to procure a meal. We caught several small fish, and by means of our guide obtained two salmon from a small party of women and children, who, with one man, were going below to gather berries. This supplied us with about half a meal, but after dark we were regaled with a beaver which one of the hunters brought in. The other game seen in the course of the day were one deer and a party of elk among the pines on the sides of the mountains.

Monday, 26th. The morning was fine, and three men were despatched ahead to hunt, while the rest were detained until nine o'clock, in order to retake some horses which had strayed away during the night. They then proceeded along the route by the forks of the river, till they reached the lower Indian camp where they first were when we met them. The whole camp immediately flocked around him with great appearance of cordiality, but all the spare food of the village did not amount to more than two salmon, which they gave to Captain Clark, who distributed them among his men. The hunters had not been able to kill anything, nor had Captain Clark or the greater part of the men any food during the twenty-four hours, till towards evening one of them shot a salmon in the river, and a few small fish were caught, which furnished them with a scanty meal. The only animals they had seen were a few pigeons, some very wild hares, a great number of the large black grasshopper, and a quantity of ground lizards.

Tuesday, 27th. The men, who were engaged last night in mending their moccasins, all except one went out hunting, but no game was to be procured. One of the men, however, killed a small salmon, and the Indians made a present of another, on which the whole party made a very slight breakfast. These Indians, to whom this life is familiar, seem contented, although they depend for subsistence on the scanty productions of the fishery. But our men, who are used to hardships, but have been accustomed to have the first wants of nature regularly supplied, feel very sensibly their wretched situation; their strength is wasting away; they begin to express their apprehensions of being without food in a country perfectly destitute of any means of supporting life except a few fish. In the course of the day an Indian brought into the camp five salmon, two of which Captain Clark bought, and made a supper for the party.

Wednesday, 28th. There was a frost again this morning. The Indians gave the party two salmon out of several which they caught in their traps, and having purchased two more, the party was enabled to subsist on them during the day. A camp of about forty Indians from the west fork passed us to-day, on their route to the eastward. Our prospect of provisions is getting worse every day; the hunters, who had ranged through the country in every direction where game might be reasonably expected, have seen nothing. The fishery is scarcely more productive, for an Indian who was out all day with his gig killed only one salmon. Besides the four fish procured from the Indians, Captain Clark obtained some fishroe in exchange for three small fish-hooks, the use of which he taught them, and which they very readily comprehended.

All the men who are not engaged in hunting are occupied in making pack-saddles for the horses which Captain Lewis informed us he had bought.

August 20. Two hunters were despatched early in the morning, but they returned without killing anything, and the only game we procured was a beaver, who was caught last night in a trap which he carried off two miles before he was found. The fur of this animal is as good as any we have ever seen, nor does it in fact appear to be ever out of season on the upper branches of the Missouri. This beaver, with several dozen of fine trout, gave us a plentiful subsistence for the day. The party were occupied chiefly in making pack-saddles, in the manufacture of which we supply the place of nails and boards by substituting for the first thongs of raw hide, which answer very well; and for boards we use the handles of our oars, and the plank of some boxes, the contents of which we empty into sacks of raw hides made for the purpose. The Indians who visit us behave with the greatest decorum, and the women are busily engaged in making and mending the moccasins of the party. As we had still some superfluous baggage which would be too heavy to carry across the mountains, it became necessary to make a cache or deposit. For this purpose we selected a spot on the bank of the river three-quarters of a mile below the camp, and three men were set to dig it, with a sentinel in the neighbourhood, who was ordered if the natives were to straggle that way to fire a signal for the workmen to desist and separate. Towards evening the cache was completed without being perceived by the Indians, and the packages prepared for deposit.

## CHAPTER XVI.

Contest between Drewyer and a Shoshonee—The fidelity and honour of that tribe—The party set out on their journey—The conduct of Cameahwait reproved, and himself reconciled—The easy parturition of the Shoshonee women—History of this nation—Their terror of the Pawkees—Their government and family economy in their treatment of their women—Their complaints of Spanish treachery—Description of their weapons of warfare—Their curious mode of making a shield—The caparison of their horses—The dress of the men and of the women particularly described—Their mode of acquiring new names.

WEDNESDAY, August 21. The weather was very cold, the water which stood in the vessels exposed to the air being covered with ice a quarter of an inch thick; the ink freezes in the pen, and the low grounds are perfectly whitened with frost; after this the day proved excessively warm. The party were engaged in their usual occupations, and completed twenty saddles with the necessary harness, all prepared to set off as soon as the Indians should arrive. Our two hunters who were despatched early in the morning have not returned, so that we were obliged to encroach on our pork and corn, which we consider as the last resource when our casual supplies of game fail. After dark we carried our baggage to the cache, and deposited what we thought too cumbrous to carry with us; a small assortment of medicines, and all the specimens of plants, seeds, and minerals collected since leaving the falls of the Missouri. Late at night Drewyer, one of the hunters, returned with a fawn and a considerable quantity of Indian plunder, which he had taken by way of reprisal. While hunting this morning in the Shoshonee cove he came suddenly upon an Indian camp, at which were an old man, a young one, three women, and a boy; they showed no surprise at the sight of him, and he therefore rode up to them, and after turning his horse loose to graze, sat down and began to converse with them by signs. They had just finished a repast on some roots, and in about twenty minutes one of the women spoke to the rest of the party, who immediately went out, collected their horses and began to saddle them. Having rested himself, Drewyer thought that he would continue his hunt, and rising went to catch his horse who was at a short distance, forgetting at the moment to take up his rifle. He had scarcely gone more than fifty paces when the Indians mounted their horses, the young man snatched up the rifle, and leaving all their baggage, whipt their horses, and set off at full speed towards the passes of the mountains. Drewyer instantly jumped on his horse and pursued them. After running about ten miles the horses of the women nearly gave out, and the women finding Drewyer gain on them raised dreadful cries, which induced the young man to slacken his pace, and being mounted on a very fleet horse, rode round them at a short distance. Drewyer now came up with the women, and by signs persuaded them that he did not mean to hurt them; they then stopped, and as the young man came towards them Drewyer asked him for his rifle, but the only part of the answer which he understood was Pahkee, the name by which they call their enemies, the Minnetarees of fort de Prairie. While they were thus engaged in talking, Drewyer watched his opportunity, and seeing the Indian off his guard, galloped up to him and seized his rifle; the Indian struggled for some time, but finding Drewyer getting too strong for him, had the presence of mind to open the pan and let the priming fall out; he then let go his hold, and giving his horse the whip escaped at full speed, leaving the women to the mercy of the conqueror. Drewyer then returned to where he had first seen them, where he found that their baggage had been left behind, and brought it to camp with him.

Thursday, 22d. This morning early two men were sent to complete the covering of the cache, which could not be so perfectly done during the night as to elude the search of the Indians. On examining the spoils which Drewyer had obtained, they were found to consist of several dressed and undressed skins; two bags wove with the bark of the silk-grass, each containing a bushel of dried serviceberries and about the same quantity of roots; an instrument made of bone for manufacturing the flints into heads for arrows, and a number of flints themselves; these were much of the same colour and nearly as transparent as common black glass, and when cut detached itself into flakes, leaving a very sharp edge.

The roots were of three kinds, and folded separate from each in hides of buffaloe made into parchment. The first is a fusiform root six inches long, and about the size of a man's finger at the largest end, with radicles larger than is usual in roots of the fusiform sort; the rind is white and thin; the body is also white, mealy, and easily reducible, by pounding, to a substance resembling flour, like which it thickens by boiling, and is of an agreeable flavour; it is eaten frequently in its raw state, either green or dried. The second species was much mutilated, but appeared to be fibrous; it is of a cylindrical form about the size of a small quill, hard and brittle. A part of the rind which had not been detached in the preparation was hard and black, but the rest of the root was perfectly white; this the Indians informed us was always boiled before eating; and on making the experiment we found that it became perfectly soft, but had a bitter taste, which was nauseous to our taste, but which the Indians seemed to relish, for on giving the roots to them they were very heartily swallowed.

The third species was a small nut about the size of a nutmeg, of an irregularly rounded form, something like the smallest of the Jerusalem artichokes, which, on boiling, we found them to resemble also in flavour, and is certainly the best root we have seen in use among the Indians. On inquiring of the Indians from what plant these roots were procured, they informed us that none of them grew near this place.

The men were chiefly employed in dressing the skins belonging to the party who accompanied Captain Clark. About eleven o'clock Chaboneau and his wife returned with Cameahwait, accompanied by about fifty men with their women and children. After they had encamped near us and turned loose their horses, we called a council of all the chiefs and warriors and addressed them in a speech; additional presents were then distributed, particularly to the two second chiefs, who had, agreeably to their promises, exerted themselves in our favour. The council was then adjourned, and all the Indians were treated with an abundant meal of boiled Indian corn and beans. The poor wretches, who had no animal food and scarcely anything but a few fish, had been almost starved, and received this new luxury with great thankfulness. Out of compliment to the chief we gave him a few dried squashes which we had brought from the Mandans, and he declared it was the best food he had ever tasted except sugar, a small lump of which he had received from his sister; he now declared how happy they should all be to live in a country which produced so many good things, and we told him that it would not be long before the white men would put it in their power to live below the mountains, where they might themselves cultivate all these kinds of food instead of wandering in the mountains. He appeared to be much pleased with this information, and the whole party being now in excellent temper after their repast, we began our purchase of horses. We soon obtained five very good ones on very reasonable terms, that is, by giving for each merchandise which cost us originally about six dollars. We have again to admire the perfect decency and propriety of their conduct, for although so numerous, they do not attempt to crowd round our camp or take anything which they see lying about, and whenever they borrow knives or kettles or any other article from the men, they return them with great fidelity.

Towards evening we formed a drag of bushes, and in about two hours caught five hundred and twenty-eight very good fish, most of them large trout. Among them we observed for the first time ten or twelve trout of a white or silvery colour, except on the back and head, where they are of a bluish cast; in appearance and shape they resemble exactly the speckled trout, except that they are not quite so large, though the scales are much larger, and the flavour equally good. The greater part of the fish was distributed among the Indians.

Friday, 23d. Our visitors seem to depend wholly on us for food, and as the state of our provisions obliges us to be careful of our remaining stock of corn and flour, this was an additional reason for urging our departure; but Cameahwait requested us to wait till the arrival of another party of his nation who were expected to-day. Knowing that it would be in vain to oppose his wish, we consented, and two hunters were sent out with orders to go farther up the southeast fork than they had hitherto been. At the same time the chief was informed of the low state of our provisions, and advised to send out his young men to hunt. This he recommended them to do, and most of them set out; we then sunk our canoes by means of stones to the bottom of the river, a situation which better than any other secured them against the effects of the high waters, and the frequent fires of the plains, the Indians having promised not to disturb them during our absence, a promise we believe the more readily, as they are almost too lazy to take the trouble of raising them for fire-wood. We were desirous of purchasing some more horses, but they declined selling any until we reached their camp in the mountains. Soon after starting the Indian hunters discovered a mule buck, and twelve of their horsemen pursued it for four miles. We saw the chase, which was very entertaining, and at length they rode it down and killed it. This mule buck was the largest deer of any kind we have seen, being nearly as large as a doe elk. Besides this they brought in another deer and three goats; but instead of a general distribution of the meat, and such as we have hitherto seen among all tribes of Indians, we observed that some families had a large share, while others received none. On inquiring of Cameahwait the reason of this custom, he said that meat among them was scarce, that each hunter reserved what he killed for the use of himself and his own family, none of the rest having any claim on what he chose to keep. Our hunters returned soon after with two mule deer and three common deer, three of which we distributed among the families who had received none of the game of their own hunters. About three o'clock the expected party, consisting of fifty men, women, and children, arrived. We now learnt that most of the Indians were on their way down the valley towards the buffaloe country, and some anxiety to accompany them appeared to prevail among those who had promised to assist us in crossing the mountains. We ourselves were not without some apprehension that they might leave us, but as they continued to say that they would return with us nothing was said upon the subject. We were, however, resolved to move early in the morning, and therefore despatched two men to hunt in the cove and leave the game on the route we should pass to-morrow.

Saturday, 24th. As the Indians who arrived yesterday had a number of spare horses, we thought it probable they might be willing to dispose of them, and desired the chief to speak to them for that purpose. They declined giving any positive answer, but requested to see the goods which we proposed to exchange. We then produced some battle-axes which we had made at fort Mandan, and a quantity of knives, with both of which they appeared very much pleased; and we were soon able to purchase three horses by giving for each an axe, a knife, a handkerchief, and a little paint. To this we were obliged to add a second knife, a shirt, a handkerchief, and a pair of leggings; and such is the estimation in which those animals are held that even at this price, which was double that for a horse, the fellow who sold him took upon himself great merit in having given away a mule to us. They now said that they had no more horses for sale, and as we had now nine of our own, two hired horses and a mule, we began loading them as heavily as was prudent, and placing the rest on the shoulders of the Indian women, left our camp at twelve o'clock. We were all on foot, except Sacajawea, for whom her husband had purchased a horse with some articles which we gave him for that purpose; an Indian, however, had the politeness to offer Captain Lewis one of his horses to ride, which he accepted in order better to direct the march of the party. We crossed the river below the forks, directing our course towards the cove by the route already passed, and had just reached the lower part of the cove when an Indian rode up to Captain Lewis to inform him that one of his men was very sick, and unable to come on. The party was immediately halted at a run which falls into the creek on the left, and Captain Lewis rode back two miles, and found Wiser severely afflicted with the colic; by giving him some of the essence of peppermint and laudanum, he recovered sufficiently to ride the horse of Captain Lewis, who then rejoined the party on foot. When he arrived he found that the Indians, who had been impatiently expecting his return, at last unloaded their horses and turned them loose, and had now made their camp for the night. It would have been fruitless to remonstrate, and not prudent to excite any irritation, and therefore, although the sun was still high, and we had made only six miles, we thought it best to remain with them; after we had encamped there fell a slight shower of rain. One of the men caught several fine trout, but Drewyer had been sent out to hunt without having killed anything. We therefore gave a little corn to those of the Indians who were actually engaged in carrying our baggage, and who had absolutely nothing to eat. We also advised Cameahwait, as we could not supply all his people with provisions, to recommend to all who were not assisting us to go on before us to their camp. This he did; but in the morning,

Sunday, 25th, a few only followed his advice, the rest accompanying us at some distance on each side. We set out at sunrise, and after going seventeen miles halted for dinner within two miles of the narrow pass in the mountains. The Indians who were on the sides of our party had started some antelopes, but were obliged, after a pursuit of several hours, to abandon the chase; our hunters had in the meantime brought in three deer, the greater part of which was distributed among the Indians. Whilst at dinner we learnt by means of Sacajawea that the young men who left us this morning carried a request from the chief that the village would break up its encampment and meet this party to-morrow, when they would all go down the Missouri into the buffaloe country. Alarmed at this new caprice of the Indians, which, if not counteracted, threatened to leave ourselves and our baggage on the mountains, or even if we reached the waters of the Columbia, prevent our obtaining horses to go on farther, Captain Lewis immediately called the three chiefs together. After smoking a pipe he asked them if they were men of their words, and if we can rely on their promises. They readily answered in the affirmative. He then asked if they had not agreed to assist us in carrying our baggage over the mountains. To this they also answered yes; and why then, said he, have you requested your people to meet us to-morrow where it will be impossible for us to trade for horses as you promised we should. If, he continued, you had not promised to help us in transporting our goods over the mountains, we should not have attempted it, but have returned down the river, after which no white men would ever have come into your country. If you wish the whites to be your friends, and to bring you arms and protect you from your enemies, you should never promise what you do not mean to perform. When I first met you, you doubted what I said, yet you afterwards saw that I told you the truth. How, therefore, can you doubt what I now tell you; you see that I have divided amongst you the meat which my hunters kill, and I promise to give all who assist us a share of whatever we have to eat. If, therefore, you intend to keep your promise, send one of the young men immediately to order the people to remain at the village till we arrive.

The two inferior chiefs then said that they had wished to keep their words and to assist us; that they had not sent for the people, but on the contrary had disapproved of the measure, which was done wholly by the first chief. Cameahwait remained silent for some time; at last he said that he knew he had done wrong, but that seeing his people all in want of provisions, he had wished to hasten their departure for the country where their wants might be supplied. He, however, now declared that having passed his word he would never violate it, and counter orders were immediately sent to the village by a young man, to whom we gave a handkerchief in order to ensure despatch and fidelity.

This difficulty being now adjusted, our march was resumed with an unusual degree of alacrity on the part of the Indians. We passed a spot where six years ago the Shoshonees suffered a very severe defeat from the Minnetarees, and late in the evening we reached the upper part of the cove where the creek enters the mountains. The part of the cove on the northeast side of the creek has lately been burnt, most probably as a signal on some occasion. Here we were joined by our hunters with a single deer, which Captain Lewis gave, as a proof of his sincerity, to the women and children, and remained supperless himself. As we came along we observed several large hares, some ducks, and many of the cock of the plains; in the low grounds of the cove were also considerable quantities of wild onions.

Monday, 26th. The morning was excessively cold, and the ice in our vessels was nearly a quarter of an inch in thickness. We set out at sunrise, and soon reached the fountain of the Missouri, where we halted for a few minutes, and then crossing the dividing ridge reached the fine spring where Captain Lewis had slept on the 12th in his first excursion to the Shoshonee camp. The grass on the hillsides is perfectly dry and parched by the sun, but near the spring was a fine green grass, we therefore halted for dinner and turned our horses to graze. To each of the Indians who were engaged in carrying our baggage was distributed a pint of corn, which they parched, then pounded, and made a sort of soup. One of the women who had been leading two of our pack horses halted at a rivulet about a mile behind, and sent on the two horses by a female friend; on inquiring of Cameahwait the cause of her detention, he answered, with great appearance of unconcern, that she had just stopped to lie in, but would soon overtake us. In fact we were astonished to see her in about an hour's time come on with her newborn infant and pass us on her way to the camp, apparently in perfect health.

This wonderful facility with which the Indian women bring forth their children seems rather some benevolent gift of nature, in exempting them from pains which their savage state would render doubly grievous, than any result of habit. If, as has been imagined, a pure dry air or a cold and elevated country are obstacles to easy delivery, every difficulty incident to that operation might be expected in this part of the continent; nor can another reason, the habit of carrying heavy burthens during pregnancy, be at all applicable to the Shoshonee women, who rarely carry any burdens, since their nation possesses an abundance of horses. We have indeed been several times informed by those conversant with Indian manners, and who asserted their knowledge of the fact, that Indian women pregnant by white men experience more difficulty in child-birth than when the father is an Indian. If this account be true, it may contribute to strengthen the belief that the easy delivery of the Indian women is wholly constitutional.

The tops of the high irregular mountains to the westward are still entirely covered with snow, and the coolness which the air acquires in passing them is a very agreeable relief from the heat, which has dried up the herbage on the sides of the hills. While we stopped, the women were busily employed in collecting the root of a plant with which they feed their children, who, like their mothers, are nearly half starved and in a wretched condition. It is a species of fennel which grows in the moist grounds; the radix is of the knob kind, of a long ovate form, terminating in a single radicle, the whole being three or four inches long, and the thickest part about the size of a man's little finger; when fresh it is white, firm, and crisp, and when dried and pounded makes a fine white meal. Its flavour is not unlike that of aniseed, though less pungent. From one to four of these knobbed roots are attached to a single stem, which rises to the height of three or four feet, and is jointed, smooth, cylindric, and has several small peduncles, one at each joint above the sheathing leaf. Its colour is a deep green, as is also that of the leaf, which is sheathing, sessile, and *polipartite,* the divisions being long and narrow. The flowers, which are now in bloom, are small and numerous, with white and umbellifferous petals; there are no root leaves. As soon as the seeds have matured, the roots of the present year, as well as the stem, decline, and are renewed in the succeeding spring from the little knot which unites the roots. The sunflower is also abundant here, and the seeds, which are now ripe, are gathered in considerable quantities, and after being pounded and rubbed between smooth stones, form a kind of meal, which is a favourite dish among the Indians.

After dinner we continued our route and were soon met by a party of young men on horseback, who turned with us and went to the village. As soon as we were within sight of it, Cameahwait requested that we would discharge our guns; the men were therefore drawn up in a single rank, and gave a running fire of two rounds, to the great satisfaction of the Indians. We then proceeded to the encampment, where we arrived about six o'clock, and were conducted to the leathern lodge in the centre of thirty-two others made of brush. The baggage was arranged near this tent, which Captain Lewis occupied, and surrounded by those of the men so as to secure it from pillage. This camp was in a beautiful smooth meadow near the river, and about three miles above their camp when we first visited the Indians. We here found Colter, who had been sent by Captain Clark with a note apprising us that there were no hopes of a passage by water, and that the most practicable route seemed to be that mentioned by his guide, towards the north. Whatever road we meant to take, it was now necessary to provide ourselves with horses; we therefore informed Cameahwait of our intention of going to the great river beyond the mountains, and that we would wish to purchase twenty more horses. He said the Minnetarees had stolen a great number of their horses this spring, but he still hoped they could spare us that number. In order not to lose the present favourable moment, and to keep the Indians as cheerful as possible, the violins were brought out and our men danced, to the great diversion of the Indians. This mirth was the more welcome because our situation was not precisely that which would most dispose us for gayety, for we have only a little parched corn to eat, and our means of subsistence, or of success, depend on the wavering temper of the natives, who may change their minds to-morrow.

The Shoshonees are a small tribe of the nation called Snake Indians, a vague denomination, which embraces at once the inhabitants of the southern parts of the Rocky mountains and of the plains on each side. The Shoshonees, with whom we now are, amount to about one hundred warriors, and three times that number of women and children. Within their own recollection they formerly lived in the plains, but they have been driven into the mountains by the Pawkees, or the roving Indians of the Sascatchawain, and are now obliged to visit occasionally, and by stealth, the country of their ancestors. Their lives are indeed migratory. From the middle of May to the beginning of September, they reside on the waters of the Columbia, where they consider themselves perfectly secure from the Pawkees, who have never yet found their way to that retreat. During this time they subsist chiefly on salmon, and as that fish disappears on the approach of autumn, they are obliged to seek subsistence elsewhere. They then cross the ridge to the waters of the Missouri, down which they proceed slowly and cautiously, till they are joined near the three forks by other bands, either of their own nation or of the Flatheads, with whom they associate against the common enemy. Being now strong in numbers, they venture to hunt buffaloe in the plains eastward of the mountains, near which they spend the winter, till the return of the salmon invites them to the Columbia. But such is their terror of the Pawkees that as long as they can obtain the scantiest subsistence they do not leave the interior of the mountains, and as soon as they collect a large stock of dried meat they again retreat, and thus alternately obtaining their food at the hazard of their lives, and hiding themselves to consume it. In this loose and wandering existence they suffer the extremes of want; for two-thirds of the year they are forced to live in the mountains, passing whole weeks without meat, and with nothing to eat but a few fish and roots. Nor can anything be imagined more wretched than their condition at the present time, when the salmon is fast retiring, when roots are becoming scarce, and they have not yet acquired strength to hazard an encounter with their enemies. So insensible are they, however, to these calamities, that the Shoshonees are not only cheerful but even gay; and their character, which is more interesting than that of any Indians we have seen, has in it much of the dignity of misfortune. In their intercourse with strangers they are frank and communicative, in their dealings perfectly fair, nor have we had, during our stay with them, any reason to suspect that the display of all our new and valuable wealth has tempted them into a single act of dishonesty. While they have generally shared with us the little they possess, they have always abstained from begging anything from us. With their liveliness of temper, they are fond of gaudy dresses, and of all sorts of amusements, particularly of games of hazard; and, like most Indians, fond of boasting of their own warlike exploits, whether real or fictitious. In their conduct towards ourselves they were kind and obliging, and though on one occasion they seemed willing to neglect us, yet we scarcely knew how to blame the treatment by which we suffered, when we recollected how few civilized chiefs would have hazarded the comforts or the subsistence of their people for the sake of a few strangers. This manliness of character may cause or it may be formed by the nature of their government, which is perfectly free from any restraint. Each individual is his own master, and the only control to which his conduct is subjected is the advice of a chief supported by his influence over the opinions of the rest of the tribe. The chief himself is in fact no more than the most confidential person among the warriors, a rank neither distinguished by any external honor nor invested by any ceremony, but gradually acquired from the good wishes of his companions and by superior merit. Such an officer has, therefore, strictly no power; he may recommend or advise or influence, but his commands have no effect on those who incline to disobey, and who may at any time withdraw from their voluntary allegiance. His shadowy authority, which cannot survive the confidence which supports it, often decays with the personal vigour of the chief, or is transferred to some more fortunate or favourite hero.

In their domestic economy, the man is equally sovereign. The man is the sole proprietor of his wives and daughters, and can barter them away, or dispose of them in any manner he may think proper. The children are seldom corrected; the boys, particularly, soon become their own masters; they are never whipped, for they say that it breaks their spirit, and that after being flogged they never recover their independence of mind, even when they grow to manhood. A plurality of wives is very common; but these are not generally sisters, as among the Minnetarees and Mandans, but are purchased of different fathers. The infant daughters are often betrothed by the father to men who are grown, either for themselves or for their sons, for whom they are desirous of providing wives. The compensation to the father is usually made in horses or mules; and the girl remains with her parents till the age of puberty, which is thirteen or fourteen, when she is surrendered to her husband. At the same time the father often makes a present to the husband equal to what he had formerly received as the price of his daughter, though this return is optional with the parent. Sacajawea had been contracted in this way before she was taken prisoner, and when we brought her back, her betrothed was still living. Although he was double the age of Sacajawea, and had two other wives, he claimed her, but on finding that she had a child by her new husband, Chaboneau, he relinquished his pretensions and said he did not want her.

The chastity of the women does not appear to be held in much estimation. The husband will for a trifling present lend his wife for a night to a stranger, and the loan may be protracted by increasing the value of the present. Yet strange as it may seem, notwithstanding this facility, any connexion of this kind not authorized by the husband is considered highly offensive and quite as disgraceful to his character as the same licentiousness in civilized societies. The Shoshonees are not so importunate in volunteering the services of their wives as we found the Sioux were; and indeed we observed among them some women who appeared to be held in more respect than those of any nation we had seen. But the mass of the females are condemned, as among all savage nations, to the lowest and most laborious drudgery. When the tribe is stationary, they collect the roots, and cook; they build the huts, dress the skins, and make clothing; collect the wood, and assist in taking care of the horses on the route; they load the horses and have the charge of all the baggage. The only business of the man is to fight; he therefore takes on himself the care of his horse, the companion of his warfare; but he will descend to no other labour than to hunt and to fish. He would consider himself degraded by being compelled to walk any distance; and were he so poor as to possess only two horses, he would ride the best of them, and leave the other for his wives and children and their baggage; and if he has too many wives or too much baggage for the horse, the wives have no alternative but to follow him on foot; they are not, however, often reduced to those extremities, for their stock of horses is very ample. Notwithstanding their losses this spring they still have at least seven hundred, among which are about forty colts, and half that number of mules. There are no horses here which can be considered as wild; we have seen two only on this side of the Muscleshell river which were without owners, and even those, although shy, showed every mark of having been once in the possession of man. The original stock was procured from the Spaniards, but they now raise their own. The horses are generally very fine, of a good size, vigorous, and patient of fatigue as well as hunger. Each warrior has one or two tied to a stake near his hut both day and night, so as to be always prepared for action. The mules are obtained in the course of trade from the Spaniards, with whose brands several of them are marked, or stolen from them by the frontier Indians. They are the finest animals of that kind we have ever seen, and at this distance from the Spanish colonies are very highly valued. The worst are considered as worth the price of two horses, and a good mule cannot be obtained for less than three and sometimes four horses.

We also saw a bridle bit, stirrups, and several other articles which, like the mules, came from the Spanish colonies. The Shoshonees say that they can reach those settlements in ten days' march by the route of the Yellowstone river; but we readily perceive that the Spaniards are by no means favourites. They complain that the Spaniards refuse to let them have firearms, under pretence that these dangerous weapons will only induce them to kill each other. In the meantime, say the Shoshonees, we are left to the mercy of the Minnetarees, who, having arms, plunder them of their horses and put them to death without mercy. "But this should not be," said Cameahwait, fiercely, "if we had guns; instead of hiding ourselves in the mountains and living like the bears on roots and berries, we would then go down and live in the buffaloe country in spite of our enemies, whom we never fear when we meet on equal terms."

As war is the chief occupation, bravery is the first virtue among the Shoshonees. None can hope to be distinguished without having given proofs of it, nor can there be any preferment or influence among the nation without some warlike achievement. Those important events which give reputation to a warrior, and which entitle him to a new name, are killing a white bear, stealing individually the horses of the enemy, leading out a party who happen to be successful either in plundering horses or destroying the enemy, and lastly, scalping a warrior. These acts seem of nearly equal dignity, but the last, that of taking an enemy's scalp, is an honour quite independent of the act of vanquishing him. To kill your adversary is of no importance unless the scalp is brought from the field of battle, and were a warrior to slay any number of his enemies in action, and others were to obtain the scalps or first touch the dead, they would have all the honours, since they have borne off the trophy.

Although thus oppressed by the Minnetarees, the Shoshonees are still a very military people. Their cold and rugged country inures them to fatigue; their long abstinence makes them support the dangers of mountain warfare, and worn down, as we saw them, by want of sustenance, have a look of fierce and adventurous courage. The Shoshonee warrior always fights on horseback; he possesses a few bad guns, which are reserved exclusively for war, but his common arms are the bow and arrow, a shield, a lance, and a weapon called by the Chippeways, by whom it was formerly used, the poggamoggon. The bow is made of cedar or pine, covered on the outer side with sinews and glue. It is about two and a half feet long, and does not differ in shape from those used by the Sioux, Mandans, and Minnetarees. Sometimes, however, the bow is made of a single piece of the horn of an elk, covered on the back like those of wood with sinews and glue, and occasionally ornamented by a strand wrought of porcupine quills and sinews, which is wrapped round the horn near its two ends. The bows made of the horns of the bighorn are still more prized, and are formed by cementing with glue flat pieces of the horn together, covering the back with sinews and glue, and loading the whole with an unusual quantity of ornaments. The arrows resemble those of the other Indians, except in being more slender than any we have seen. They are contained, with the implements for striking fire, in a narrow quiver formed of different kinds of skin, though that of the otter seems to be preferred. It is just long enough to protect the arrows from the weather, and is worn on the back by means of a strap passing over the right shoulder and under the left arm. The shield is a circular piece of buffaloe hide about two feet four or five inches in diameter, ornamented with feathers, and a fringe round it of dressed leather, and adorned or deformed with paintings of strange figures. The buffaloe hide is perfectly proof against any arrow, but in the minds of the Shoshonees, its power to protect them is chiefly derived from the virtues which are communicated to it by the old men and jugglers. To make a shield is indeed one of their most important ceremonies. It begins by a feast, to which all the warriors, old men, and jugglers are invited. After the repast a hole is dug in the ground about eighteen inches in depth, and of the same diameter as the intended shield; into this hole red-hot stones are thrown and water poured over them, till they emit a very strong, hot steam. The buffaloe skin, which must be the entire hide of a male two years old, and never suffered to dry since it was taken from the animal, is now laid across the hole, with the fleshy side to the ground, and stretched in every direction by as many as can take hold of it. As the skin becomes heated, the hair separates and is taken off by the hand, till at last the skin is contracted into the compass designed for the shield. It is then taken off and placed on a hide prepared into parchment, and then pounded during the rest of the festival by the bare heels of those who are invited to it. This operation sometimes continues for several days, after which it is delivered to the proprietor, and declared by the old men and jugglers to be a security against arrows, and, provided the feast has been satisfactory, against even the bullets of their enemies. Such is the delusion that many of the Indians implicitly believe that this ceremony has given to the shield supernatural powers, and that they have no longer to fear any weapons of their enemies.

The paggamoggon is an instrument consisting of a handle twenty-two inches long, made of wood, covered with dressed leather about the size of a whip-handle; at one end is a thong of two inches in length, which is tied to a round stone weighing two pounds and held in a cover of leather; at the other end is a loop of the same material, which is passed round the wrist so as to secure the hold of the instrument, with which they strike a very severe blow.

Besides these, they have a kind of armour something like a coat of mail, which is formed by a great many folds of dressed antelope skins, united by means of a mixture of glue and sand. With this they cover their own bodies and those of their horses, and find it impervious to the arrow.

The caparison of their horses is a halter and a saddle; the first is either a rope of six or seven strands of buffaloe hair platted or twisted together, about the size of a man's finger, and of great strength; or merely a thong of raw hide, made pliant by pounding and rubbing, though the first kind is much preferred. The halter is very long, and is never taken from the neck of the horse when in constant use. One end of it is first tied round the neck in a knot, and then brought down to the under jaw, round which it is formed into a simple noose, passing through the mouth; it is then drawn up on the right side and held by the rider in his left hand, while the rest trails after him to some distance. At other times the knot is formed at a little distance from one of the ends, so as to let that end serve as a bridle, while the other trails on the ground. With these cords dangling alongside of them the horse is put to his full speed without fear of falling, and when he is turned to graze the noose is merely taken from his mouth. The saddle is formed, like the pack-saddles used by the French and Spaniards, of two flat thin boards which fit the sides of the horse, and are kept together by two cross pieces, one before and the other behind, which rise to a considerable height, ending sometimes in a flat point extending outwards, and always making the saddle deep and narrow. Under this a piece of buffaloe skin, with the hair on, is placed so as to prevent the rubbing of the boards, and when they mount they throw a piece of skin or robe over the saddle, which has no permanent cover. When stirrups are used, they consist of wood covered with leather; but stirrups and saddles are conveniences reserved for old men and women. The young warriors rarely use anything except a small leather pad stuffed with hair, and secured by a girth made of a leathern thong. In this way they ride with great expertness, and they have a particular dexterity in catching the horse when he is running at large. If he will not immediately submit when they wish to take him, they make a noose in the rope, and although the horse may be at a distance, or even running, rarely fail to fix it on his neck; and such is the docility of the animal that, however unruly he may seem, he surrenders as soon as he feels the rope on him. This cord is so useful in this way that it is never dispensed with, even when they use the Spanish bridle, which they prefer, and always procure when they have it in their power. The horse becomes almost an object of attachment; a favourite is frequently painted and his ears cut into various shapes; the mane and tail, which are never drawn nor trimmed, are decorated with feathers of birds, and sometimes a warrior suspends at the breast of his horse the finest ornaments he possesses.

Thus armed and mounted the Shoshonee is a formidable enemy, even with the feeble weapons which he is still obliged to use. When they attack at full speed they bend forward and cover their bodies with the shield, while with the right hand they shoot under the horse's neck.

The only articles of metal which the Shoshonees possess are a few bad knives, some brass kettles, some bracelets or arm-bands of iron and brass, a few buttons worn as ornaments in their hair, one or two spears about a foot in length, and some heads for arrows made of iron and brass. All these they had obtained in trading with the Crow or Rocky mountain Indians, who live on the Yellowstone. The few bridle-bits and stirrups they procured from the Spanish colonies.

The instrument which supplies the place of a knife among them is a piece of flint with no regular form, and the sharp part of it not more than one or two inches long; the edge of this is renewed, and the flint itself is formed into heads for arrows by means of the point of a deer or elk horn, an instrument which they use with great art and ingenuity. There are no axes or hatchets, all the wood being cut with flint or elk-horn, the latter of which is always used as a wedge in splitting wood. Their utensils consist, besides the brass kettles, of pots in the form of a jar, made either of earth, or of a stone found in the hills between Madison and Jefferson rivers, which, though soft and white in its natural state, becomes very hard and black after exposure to the fire. The horns of the buffaloe and the bighorn supply them with spoons.

The fire is always kindled by means of a blunt arrow and a piece of well-seasoned wood of a soft spongy kind, such as the willow or cottonwood.

The Shoshonees are of a diminutive stature, with thick flat feet and ankles, crooked legs, and are, generally speaking, worse formed than any nation of Indians we have seen. Their complexion resembles that of the Sioux, and is darker than that of the Minnetarees, Mandans, or Shawnees. The hair in both sexes is suffered to fall loosely over the face and down the shoulders; some men, however, divide it by means of thongs of dressed leather or otter skin into two equal queues, which hang over the ears and are drawn in front of the body; but at the present moment, when the nation is afflicted by the loss of so many relations killed in war, most of them have the hair cut quite short in the neck, and Cameahwait has the hair cut short all over his head, this being the customary mourning for a deceased kindred.

The dress of the men consists of a robe, a tippet, a shirt, long leggings, and moccasins. The robe is formed most commonly of the skins of antelope, bighorn, or deer, though when it can be procured the buffalo hide is preferred. Sometimes, too, they are made of beaver, moonax, and small wolves, and frequently during the summer of elk skin. These are dressed with the hair on, and reach about as low as the middle of the leg. They are worn loosely over the shoulders, the sides being at pleasure either left open or drawn together by the hand, and in cold weather kept close by a girdle round the waist. This robe answers the purpose of a cloak during the day, and at night is their only covering.

The tippet is the most elegant article of Indian dress we have ever seen. The neck or collar of it is a strip about four or five inches wide, cut from the back of the otter skin, the nose and eyes forming one extremity, and the tail another. This being dressed with the fur on, they attach to one edge of it from one hundred to two hundred and fifty little rolls of ermine skin, beginning at the ear, and proceeding towards the tail. These ermine skins are the same kind of narrow strips from the back of that animal, which are sewed round a small cord of twisted silkgrass thick enough to make the skin taper towards the tail which hangs from the end, and are generally about the size of a large quill. These are tied at the head into little bundles of two, three, or more, according

## UP THE MISSOURI 457

to the caprice of the wearer, and then suspended from the collar, and a broad fringe of ermine skin is fixed so as to cover the parts where they unite, which might have a coarse appearance. Little tassels of fringe of the same materials are also fastened to the extremities of the tail, so as to show its black colour to greater advantage. The centre of the collar is further ornamented with the shells of the pearl oyster. Thus adorned, the collar is worn close round the neck, and the little rolls fall down over the shoulders nearly to the waist, so as to form a sort of short cloak, which has a very handsome appearance. These tippets are very highly esteemed, and are given or disposed of on important occasions only. The ermine is the fur known to the northwest traders by the name of the white weasel, but is the genuine ermine, and by encouraging the Indians to take them, might no doubt be rendered a valuable branch of trade. These animals must be very abundant, for the tippets are in great numbers, and the construction of each requires at least one hundred skins.

The shirt is a covering of dressed skin without the hair, and formed of the hide of the antelope, deer, bighorn, or elk, though the last is more rarely used than any other for this purpose. It fits the body loosely, and reaches half way down the thigh. The aperture at the top is wide enough to admit the head, and has no collar, but is either left square, or most frequently terminates in the tail of the animal, which is left entire, so as to fold outwards, though sometimes the edges are cut into a fringe, and ornamented with quills of the porcupine. The seams of the shirt are on the sides, and are richly fringed and adorned with porcupine quills till within five or six inches of the sleeve, where it is left open, as is also the under side of the sleeve from the shoulder to the elbow, where it fits closely round the arm as low as the wrist, and has no fringe like the sides and the under part of the sleeve above the elbow. It is kept up by wide shoulder straps, on which the manufacturer displays his taste by the variety of figures wrought with porcupine quills of different colours, and sometimes by beads when they can be obtained. The lower end of the shirt retains the natural shape of the fore legs and neck of the skin, with the addition of a slight fringe; the hair, too, is left on the tail and near the hoofs, part of which last is retained and split into a fringe.

The leggings are generally made of antelope skins, dressed without the hair, and with the legs, tail, and neck hanging to them. Each legging is formed of a skin nearly entire, and reaches from the ankle to the upper part of the thigh, and the legs of the skin are tucked before and behind under a girdle round the waist. It fits closely to the leg, the tail being worn upwards, and the neck, highly ornamented with fringe and porcupine quills, drags on the ground behind the heels. As the legs of the animal are tied round the girdle, the wide part of the skin is drawn so high as to conceal the parts usually kept from view, in which respect their dress is much more decent than that of any nation of Indians on the Missouri. The seams of the leggings down the sides are also fringed and ornamented, and occasionally decorated with tufts of hair taken from enemies whom they have slain. In making all these dresses, their only thread is the sinew taken from the backs and loins of deer, elk, buffaloe, or any other animal.

The moccasin is of the deer, elk, or buffaloe skin, dressed without the hair, though in winter they use the buffaloe skin with the hairy side inward, as do most of the Indians who inhabit the buffaloe country. Like the Mandan moccasin, it is made with a single seam on the outer edge, and sewed up behind, a hole being left at the instep to admit the foot. It is variously ornamented with figures wrought with porcupine

## UP THE MISSOURI

quills, and sometimes the young men most fond of dress cover it with the skin of a polecat, and trail at their heels the tail of the animal.

The dress of the women consists of the same articles as that of their husbands. The robe, though smaller, is worn in the same way; the moccasins are precisely similar. The shirt or chemise reaches half way down the leg, is in the same form, except that there is no shoulder strap, the seam coming quite up to the shoulder, though for women who give suck both sides are open, almost down to the waist. It is also ornamented in the same way with the addition of little patches of red cloth, edged round with beads at the skirts. The chief ornament is over the breast, where there are curious figures made with the usual luxury of porcupine quills. Like the men they have a girdle round the waist, and when either sex wishes to disengage the arm, it is drawn up through the hole near the shoulder, and the lower part of the sleeve thrown behind the body.

Children alone wear beads round their necks; grown persons of both sexes prefer them suspended in little bunches from the ear, and sometimes intermixed with triangular pieces of the shell of the pearl oyster. Sometimes the men tie them in the same way to the hair of the fore part of the head, and increase the beauty of it by adding the wings and tails of birds, and particularly the feathers of the great eagle or calumet bird, of which they are extremely fond. The collars are formed either of sea shells procured from their relations to the southwest, or of the sweet-scented grass which grows in the neighbourhood, and which they twist or plait together to the thickness of a man's finger, and then cover with porcupine quills of various colours. The first of these is worn indiscriminately by both sexes, the second principally confined to the men, while a string of elk's tusks is a collar almost peculiar to the women and children. Another collar worn by the men is a string of round bones like the joints of a fish's back; but the collar most preferred, because most honourable, is one of the claws of the brown bear. To kill one of these animals is as distinguished an achievement as to have put to death an enemy, and in fact with their weapons is a more dangerous trial of courage. These claws are suspended on a thong of dressed leather, and, being ornamented with beads, are worn round the neck by the warriors with great pride. The men also frequently wear the skin of a fox, or a strip of otter skin, round the head in the form of a bandeau.

In short, the dress of the Shoshonees is as convenient and decent as that of any Indians we have seen.

They have many more children than might have been expected, considering their precarious means of support and their wandering life. This inconvenience is, however, balanced by the wonderful facility with which their females undergo the operations of child-birth. In the most advanced state of pregnancy they continue their usual occupations, which are scarcely interrupted longer than the mere time of bringing the child into the world.

The old men are few in number, and do not appear to be treated with much tenderness or respect.

The tobacco used by the Shoshonees is not cultivated among them, but obtained from the Indians of the Rocky mountains and from some of the bands of their own nation who live south of them; it is the same plant which is in use among the Minnetarees, Mandans, and Ricaras.

Their chief intercourse with other nations seems to consist in their association with other Snake Indians, and with the Flatheads when they go eastward to hunt buffaloe, and in the occasional visits made by the Flatheads to the waters of the Columbia for the purpose of fishing. Their intercourse with the Spaniards is much more rare, and it furnishes them with a few articles, such as mules, and some bridles and other ornaments for horses, which, as well as some of their kitchen utensils, are also furnished by the bands of Snake Indians from the Yellowstone. The pearl ornaments which they esteem so highly come from other bands, whom they represent as their friends and relations, living to the southwest beyond the barren plains on the other side of the mountains; these relations they say inhabit a good country, abounding with elk, deer, bear, and antelope, where horses and mules are much more abundant than they are here, or to use their own expression, as numerous as the grass of the plains.

The names of the Indians vary in the course of their life; originally given in childhood, from the mere necessity of distinguishing objects, or from some accidental resemblance to external objects, the young warrior is impatient to change it by some achievement of his own. Any important event, the stealing of horses, the scalping an enemy, or killing a brown bear, entitles him at once to a new name, which he then selects for himself, and it is confirmed by the nation. Sometimes the two names subsist together: thus, the chief Cameahwait, which means, "one who never walks," has the war name of Tooettecone, or "black gun," which he acquired when he first signalized himself. As each new action gives a warrior a right to change his name, many of them have had several in the course of their lives. To give to a friend his own name is an act of high courtesy, and a pledge, like that of pulling off the moccasin, of sincerity and hospitality. The chief in this way gave his name to Captain Clark when he first arrived, and he was afterwards known among the Shoshonees by the name of Cameahwait.

The diseases incident to this state of life may be supposed

to be few, and chiefly the result of accidents. We were particularly anxious to ascertain whether they had any knowledge of the venereal disorder. After inquiring by means of the interpreter and his wife, we learnt that they sometimes suffered from it, and that they most usually die with it; nor could we discover what was their remedy. It is possible that this disease may have reached them in their circuitous communications with the whites through the intermediate Indians; but the situation of the Shoshonees is so insulated that it is not probable that it could have reached them in that way, and the existence of such a disorder among the Rocky mountains seems rather a proof of its being aboriginal.

## CHAPTER XVII.

The party, after procuring horses from the Shoshonees, proceed on their journey through the mountains—The difficulties and dangers of the route—A council held with another band of the Shoshonees, of whom some account is given—They are reduced to the necessity of killing their horses for food—Captain Clark, with a small party, precedes the main body in quest of food, and is hospitably received by the Pierced-nose Indians—Arrival of the main body amongst this tribe, with whom a council is held—They resolve to perform the remainder of their journey in canoes—Sickness of the party—They descend the Kooskooskee to its junction with Lewis river, after passing several dangerous rapids—Short description of the manners and dress of the Pierced-nose Indians.

AUGUST 27. We were now occupied in determining our route and procuring horses from the Indians. The old guide who had been sent on by Captain Clark now confirmed, by means of our interpreter, what he had already asserted, of a road up Berry creek which would lead to Indian establishments on another branch of the Columbia; his reports, however, were contradicted by all the Shoshonees. This representation we ascribed to a wish on their part to keep us with them during the winter, as well for the protection we might afford against their enemies, as for the purpose of consuming our merchandise amongst them; and as the old man promised to conduct us himself, that route seemed to be the most eligible. We were able to procure some horses, though not enough for all our purposes. This traffic, and our inquiries and councils with the Indians, consumed the remainder of the day.

August 28. The purchase of horses was resumed, and our stock raised to twenty-two. Having now crossed more than once the country which separates the head waters of the

Missouri from those of the Columbia, we can designate the easiest and most expeditious route for a portage; it is as follows:

From the forks of the river north 60° west five miles to the point of a hill on the right; then south 80° west ten miles to a spot where the creek is ten miles wide, and the highlands approach within two hundred yards; southwest five miles to a narrow part of the bottom; then turning south 70° west two miles to a creek on the right; thence south 80° west three miles to a rocky point opposite to a thicket of pines on the left; from that place west three miles to the gap where is the fountain of the Missouri; on leaving this fountain south 80° west six miles across the dividing ridge, to a run from the right, passing several small streams north 80° west four miles, over hilly ground, to the east fork of Lewis's river, which is here forty yards wide.

Thursday, 29th. Captain Clark joined us this morning, and we continued our bargains for horses. The late misfortunes of the Shoshonees make the price higher than common, so that one horse cost a pistol, one hundred balls, some powder, and a knife; another was changed for a musket, and in this way we obtained twenty-nine. The horses themselves are young and vigorous, but they are very poor, and most of them have sore backs in consequence of the roughness of the Shoshonee saddle. We are therefore afraid of loading them too heavily, and are anxious to obtain one at least for each man, to carry the baggage, or the man himself, or in the last resource to serve as food; but with all our exertions we could not provide all our men with horses. We have, however, been fortunate in obtaining for the last three days a sufficient supply of flesh, our hunters having killed two or three deer every day.

Friday, 30th. The weather was fine, and having now made

all our purchases, we loaded our horses and prepared to start. The greater part of the band, who had delayed their journey on our account, were also ready to depart. We then took our leave of the Shoshonees, who set out on their visit to the Missouri at the same time that we, accompanied by the old guide, his four sons, and another Indian, began the descent of the river, along the same road which Captain Clark had previously pursued. After riding twelve miles we encamped on the south bank of the river, and as the hunters had brought in three deer early in the morning we did not feel the want of provisions.

Saturday, 31st. At sunrise we resumed our journey, and halted for three hours on Salmon creek to let the horses graze. We then proceeded to the stream called Berry creek, eighteen miles from the camp of last night; as we passed along, the vallies and prairies were on fire in several places, in order to collect the bands of the Shoshonees and the Flatheads for their journey to the Missouri. The weather was warm and sultry, but the only inconvenience which we apprehend is a dearth of food, of which we had to-day an abundance, having procured a deer, a goose, one duck, and a prairie fowl. On reaching Tower creek we left the former track of Captain Clark and began to explore the new route, which is our last hope of getting out of the mountains. For four miles the road, which is tolerably plain, led us along Berry creek to some old Indian lodges, where we encamped for the night. The next day,

Sunday, September 1, 1805, we followed the same road, which here left the creek and turned to the northwest across the hills. During all day we were riding over these hills, from which are many drains and small streams running into the river to the left, and at the distance of eighteen miles came to a large creek called Fish creek, emptying into the Columbia, which is about six miles from us. It had rained in the course of the day, and commenced raining again towards evening. We therefore determined not to leave the low grounds to-night, and after going up Fish creek four miles formed our encampment. The country over which we passed is well watered, but poor and rugged or stony, except the bottoms of Fish creek, and even these are narrow. Two men were sent to purchase fish of the Indians at the mouth of the creek, and with the dried fish which they obtained, and a deer and a few salmon killed by the party, we were still well supplied. Two bear also were wounded, but we could procure neither of them.

Monday, 2d. This morning all the Indians left us, except the old guide, who now conducted us up Fish creek; at one mile and a half we passed a branch of the river coming in through a low ground covered with pine on the left, and two and a half miles farther is a second branch from the right; after continuing our route along the hills covered with pine, and a low ground of the same growth, we arrived, at the distance of three and a half miles, at the forks of the creek. The road which we were following now turned up the east side of these forks, and as our guide informed us led to the Missouri. We were therefore left without any track, but as no time was to be lost we began to cut our road up the west branch of the creek. This we effected with much difficulty; the thickets of trees and brush through which we were obliged to cut our way required great labour; the road itself was over the steep and rocky sides of the hills where the horses could not move without danger of slipping down, while their feet were bruised by the rocks and stumps of trees. Accustomed as these animals were to this kind of life they suffered severely; several of them fell to some distance down the sides of the hills, some turned over with the baggage, one was crippled, and two gave out exhausted with fatigue. After crossing the creek several times we at last made five miles, with great fatigue and labour, and encamped on the left side of the creek in a small stony low ground. It was not, however, till after dark that the whole party was collected, and then, as it rained, and we killed nothing, we passed an uncomfortable night. The party had been too busily occupied with the horses to make any hunting excursion, and though as we came along Fish creek we saw many beaver dams, we saw none of the animals themselves. In the morning,

Tuesday, 3d, the horses were very stiff and weary. We sent back two men for the load of the horse which had been crippled yesterday, and which we had been forced to leave two miles behind. On their return we set out at eight o'clock and proceeded up the creek, making a passage through the brush and timber along its borders. The country is generally supplied with pine, and in the low grounds is a great abundance of fir trees and underbushes. The mountains are high and rugged, and those to the east of us covered with snow. With all our precautions the horses were very much injured in passing over the ridges and steep points of the hills, and to add to the difficulty, at the distance of eleven miles the high mountains closed the creek, so that we were obliged to leave the creek to the right and cross the mountain abruptly. The ascent was here so steep that several of the horses slipped and hurt themselves, but at last we succeeded in crossing the mountain, and encamped on a small branch of Fish creek. We had now made fourteen miles, in a direction nearly north from the river; but this distance, though short, was very fatiguing, and rendered still more disagreeable by the rain which began at three o'clock. At dusk it commenced snowing, and continued till the ground was covered to the depth of two inches, when it changed into a sleet.

We here met with a serious misfortune, the last of our thermometers being broken by accident. After making a scanty supper on a little corn and a few pheasants killed in the course of the day, we laid down to sleep, and next morning,

Wednesday, 4th, found everything frozen, and the ground covered with snow. We were obliged to wait some time in order to thaw the covers of the baggage, after which we began our journey at eight o'clock. We crossed a high mountain which forms the dividing ridge between the waters of the creek we had been ascending and those running to the north and west. We had not gone more than six miles over the snow when we reached the head of a stream from the right, which directed its course more to the westward. We descended the steep sides of the hills along its border, and at the distance of three miles found a small branch coming in from the eastward. We saw several of the argalia, but they were too shy to be killed, and we therefore made a dinner from a deer shot by one of the hunters. Then we pursued the course of the stream for three miles, till it emptied itself into a river from the east. In the wide valley at their junction, we discovered a large encampment of Indians; when we had reached them and alighted from our horses, we were received with great cordiality. A council was immediately assembled, white robes were thrown over our soldiers, and the pipe of peace introduced. After this ceremony, as it was too late to go any farther, we encamped, and continued smoking and conversing with the chiefs till a late hour. The next morning,

Thursday, 5th, we assembled the chiefs and warriors, and informed them who we were and the purpose for which we visited their country. All this was, however, conveyed to them through so many different languages that it was not comprehended without difficulty. We therefore proceeded to the more intelligible language of presents, and made four chiefs by giving a medal and a small quantity of tobacco to each. We received in turn from the principal chief a present consisting of the skins of a braro, an otter, and two antelopes, and were treated by the women to some dried roots and berries. We then began to traffic for horses, and succeeded in exchanging seven, purchasing eleven, for which we gave a few articles of merchandise.

This encampment consists of thirty-three tents, in which were about four hundred souls, among whom eighty were men. They are called Ootlashoots, and represent themselves as one band of a nation called Tushepaws, a numerous people of four hundred and fifty tents, residing on the heads of the Missouri and Columbia rivers, and some of them lower down the latter river. In person these Indians are stout, and their complexion lighter than that common among Indians. The hair of the men is worn in queues and otter skin, falling in front over the shoulders. A shirt of dressed skin covers the body to the knee, and on this is worn occasionally a robe. To these were added leggings and moccasins. The women suffer their hair to fall in disorder over the face and shoulders, and their chief article of covering is a long shirt of skin, reaching down to the ankles, and tied round the waist. In other respects, as also in the few ornaments which they possess, their appearance is similar to that of the Shoshonees; there is, however, a difference between the language of these people, which is still further increased by the very extraordinary pronunciation of the Ootlashoots. Their words have all a remarkably gutteral sound, and there is nothing which seems to represent the tone of their speaking more exactly than the clucking of a fowl or the noise of a parrot. This peculiarity renders their voices scarcely audible, except at a short distance, and when many of them are talking forms a strange confusion of sounds. The common conversation we overheard consisted of low gutteral sounds, occasionally broken by a loud word or two, after which it would relapse and scarcely be distinguished. They seem kind and friendly, and willingly shared with us berries and roots, which formed their only stock of provisions. Their only wealth is their horses, which are very fine, and so numerous that this party had with them at least five hundred.

Friday, 6th. We continued this morning with the Ootlashoots, from whom we purchased two more horses, and procured a vocabulary of their language. The Ootlashoots set off about two o'clock to join the different bands who were collecting at the three forks of the Missouri. We ourselves proceeded at the same time, and taking a direction N. 30W., crossed, within the distance of one mile and a half, a small river from the right and a creek coming in from the north. This river is the main stream, and when it reaches the end of the valley, where the mountains close in upon it, is joined by the river on which we encamped last evening, as well as by the creek just mentioned. To the river thus formed we gave the name of Captain Clark, he being the first white man who had ever visited its waters. At the end of five miles on this course we had crossed the valley and reached the top of a mountain covered with pine; this we descended along the steep sides and ravines for a mile and a half, when we came to a spot on the river where the Ootlashoots had encamped a few days before. We then followed the course of the river, which is from twenty-five to thirty yards wide, shallow, stony, and the low grounds on its borders narrow. Within the distance of three and a half miles we crossed it several times, and after passing a run on each side, encamped on its right bank, after making ten miles during the afternoon. The horses were turned out to graze, but those we had lately bought were secured and watched, lest they should escape or be stolen by their former owners. Our stock of flour was now exhausted and we had but little corn, and as our hunters had killed nothing except two pheasants, our supper consisted chiefly of berries.

Saturday, 7th. The greater part of the day the weather was dark and rainy; we continued through the narrow low grounds along the river, till at the distance of six miles we came to a large creek from the left, after which the bottoms widen. Four miles lower is another creek on the same side, and the valley now extends from one to three miles, the mountains on the left being high and bald, with snow on the summits, while the country to the right is open and hilly. Four miles beyond this is a creek running from the snow-topped mountains, and several runs on both sides of the river. Two miles from this last is another creek on the left. The afternoon was now far advanced, but not being able to find a fit place to encamp we continued six miles farther till after dark, when we halted for the night. The river here is still shallow and stony, but is increased to the width of fifty yards. The valley through which we passed is of a poor soil, and its fertility injured by the quantity of stone scattered over it. We met two horses which had strayed from the Indians and were now quite wild. No fish was to be seen in the river, but we obtained a very agreeable supply of two deer, two cranes, and two pheasants.

Sunday, 8th. We set out early; the snow-topped hills on the left approach the river near our camp, but we soon reached a valley four or five miles wide, through which we followed the course of the river in a direction due north. We passed three creeks on the right, and several runs emptying themselves into the opposite side of the river. At the distance of eleven miles the river turned more towards the west; we pursued it for twelve miles, and encamped near a large creek coming in from the right, which, from its being divided into four different channels, we called Scattering creek. The valley continues to be a poor, stony land, with scarcely any timber, except some pine trees along the waters and partially scattered on the hills to the right, which, as well as those on the left, have snow on them. The plant which forces itself most on our attention is a species of prickly pear very common on this part of the river; it grows in clusters, in an oval form about the size of a pigeon's egg, and its thorns are so strong and bearded that when it penetrates our feet it brings away the pear itself. We saw two mares and a colt, which, like the horses seen yesterday, seemed to have lost themselves and become wild. Our game to-day consisted of two deer, an elk, and a prairie fowl.

Monday, 9th. We resumed our journey through the valley, and leaving the road on our right crossed the Scattering creek, and halted at the distance of twelve miles on a small run from the east, where we breakfasted on the remains of yesterday's hunt. We here took a meridian altitude, which gave the latitude of 46° 41′ 38″ 9‴. We then continued, and at the distance of four miles passed over to the left bank of the river, where we found a large road through the valley. At this place is a handsome stream of very clear water, a hundred yards wide, with low banks, and a bed formed entirely of gravel; it has every appearance of being navigable, but as it contains no salmon we presume there must be some fall below which obstructs their passage. Our guide could not inform us where this river discharged its waters; he said that as far as he knew its course it ran along the mountains to the north, and that not far from our present position it was joined by another stream nearly as large as itself, which rises in the mountains to the east near the Missouri, and flows through an extensive valley or open prairie. Through this prairie is the great Indian road to the waters of the Missouri; and so direct is the route that in four days' journey from this place we might reach the Missouri about thirty miles above what we called the Gates of the Rocky mountains, or the spot where the valley of that river widens into an extensive plain on entering the chain of mountains. At ten miles from our camp is a small creek falling in from the eastward, five miles below which we halted at a large stream which empties itself on the west side of the river. It is a fine, bold creek of clear water about twenty yards wide, and we called it *Traveller's-rest* creek; for as our guide told us that we should here leave the river, we determined to remain for the purpose of making celestial observations and collecting some food, as the country through which we are to pass has no game for a great distance.

The valley of the river through which we have been passing is generally a prairie from five to six miles in width, and with a cold, gravelly, white soil. The timber which it possesses is almost exclusively pine, chiefly of the long-leafed kind, with some spruce, and a species of fir resembling the Scotch fir; near the water courses are also seen a few narrow-leafed cottonwood trees, and the only underbrush is the redwood, honeysuckle, and rosebushes. Our game was four deer, three geese, four ducks, and three prairie fowls; one of the hunters brought in a red-headed woodpecker of the large kind common in the United States, but the first of the kind we have seen since leaving the Illinois.

Tuesday, 10th. The morning being fair all the hunters were sent out, and the rest of the party employed in repairing their clothes. Two of them were sent to the junction of the river from the east, along which the Indians go to the Missouri; it is about seven miles below Traveller's-rest creek; the country at the forks is seven or eight miles wide, level and open, but with little timber; its course is to the north, and we incline to believe that this is the river which the Minnetarees had described to us as running from south to north along the west side of the Rocky mountains, not far from the sources of Medicine river; there is, moreover, reason to suppose that after going as far northward as the headwaters of that river it turns to the westward and joins the Tacootchetessee. Towards evening one of the hunters returned with three Indians whom he had met in his excursion up Traveller's-rest creek; as soon as they saw him they prepared to attack him with arrows, but he quieted them by laying down his gun and advancing towards them, and soon persuaded them to come to the camp. Our Shoshonee guide could not speak the language of these people, but by the universal language of signs and gesticulations, which is perfectly intelligible among the Indians, he found that these were three Tushepaw Flatheads in pursuit of two men, supposed to be Shoshonees, who had stolen twenty-three of their horses. We gave them some boiled venison and a few presents, such as a fishhook, a steel to strike fire, and a little powder; but they seemed better pleased with a piece of riband which we tied in the hair of each of them. They were, however, in such haste, lest their horses should be carried off, that two of them set off after sunset in quest of the robbers; the third, however, was persuaded to remain with us and conduct us to his relations; these he said were numerous, and resided on the Columbia in the plain below the mountains. From that place, he added, the river was navigable to the ocean; that some of his relations had been there last fall and seen an old white man who resided there by himself, and who gave them some handkerchiefs like those we have. The distance from this place is five sleeps or days' journey. When our hunters had all joined us we found our provisions consisted of four deer, a beaver, and three grouse.

The observation of to-day gave 46° 48' 28" as the latitude of Traveller's-rest creek.

Wednesday, 11th. Two of our horses having strayed away, we were detained all the morning before they were caught. In the meantime our Tushepaw Indian became impatient of the delay, and set out to return home alone. As usual we had dispatched four of our best hunters ahead, and as we hoped with their aid and our present stock of provisions to subsist on the route, we proceeded at three o'clock up the right side of the creek, and encamped under some old Indian huts at the distance of seven miles. The road was plain and good; the valley is, however, narrower than that which we left, and bordered by high and rugged hills to the right, while the mountains on the left were covered with snow. The day was fair and warm, the wind from the northwest.

Thursday, 12th. There was a white frost this morning. We proceeded at seven o'clock, and soon passed a stream falling in on the right, near which was an old Indian camp with a bath or sweating-house covered with earth. At two miles distance we ascended a high mountain, and thence continued through a hilly and thickly timbered country for nine miles, when we came to the forks of the creek, where the road branches up each fork. We followed the western route, and finding that the creek made a considerable bend at the distance of four miles, crossed a high mountain in order to avoid the circuit. The road had been very bad during the first part of the day, but the passage of the mountain, which was eight miles across, was very painful to the horses, as we were obliged to go over steep stony sides of hills and along the hollows and ravines, rendered more disagreeable by the fallen timber, chiefly pine, spruce pine, and fir. We at length reached the creek, having made twenty-three miles of a route so difficult that some of the party did not join us before ten o'clock. We found the account of the scantiness of game but too true, as we were not able to procure anything during the whole of yesterday, and to-day we killed only a single pheasant. Along the road we observed many of the pine trees peeled off, which is done by the Indians to procure the inner bark for food in the spring.

Friday, 13th. Two of the horses strayed away during the night, and one of them being Captain Lewis's, he remained with four men to search for them while we proceeded up the creek; at the distance of two miles we came to several springs issuing from large rocks of a coarse hard grit, and nearly boiling hot. These seem to be much frequented, as there are several paths made by elk, deer, and other animals, and near one of the springs a hole or Indian bath, and roads leading in different directions. These embarrassed our guide, who, mistaking the road, took us three miles out of the proper course over an exceedingly bad route. We then fell into the right road, and proceeded on very well, when, having made five miles, we stopped to refresh the horses. Captain Lewis here joined us, but not having been able to find his horse; two men were sent back to continue the search. We then proceeded along the same kind of country which we passed yesterday, and after crossing a mountain and leaving the sources of the Traveller's-rest creek on the left, reached, after five miles riding, a small creek which also came in from the left hand, passing through open glades, some of which were half a mile wide. The road, which had been as usual rugged and stony, became firm, plain, and level after quitting the head of Traveller's-rest. We followed the course of this new creek for two miles, and encamped at a spot where the mountains close on each side. Other mountains covered with snow are in view to the southeast and southwest. We were somewhat more fortunate to-day in killing a deer and several pheasants, which were of the common species, except that the tail was black.

Saturday, 14th. The day was very cloudy, with rain and hail in the vallies, while on the top of the mountains some snow fell. We proceeded early, and continuing along the right side of Glade creek crossed a high mountain, and at the distance of six miles reached the place where it is joined by another branch of equal size from the right. Near the forks the Tushepaws have had an encampment which is but recently abandoned, for the grass is entirely destroyed by horses, and two fish weirs across the creek are still remaining; no fish were, however, to be seen. We here passed over to the left side of the creek and began the ascent of a very high and steep mountain, nine miles across. On reaching the other side we found a large branch from the left, which seems to rise in the snowy mountains to the south and southeast. We continued along the creek two miles farther, when, night coming on, we encamped opposite a small island at the mouth of a branch on the right side of the river. The mountains which we crossed to-day were much more difficult than those of yesterday; the last was particularly fatiguing, being steep and stony, broken by fallen timber, and thickly overgrown by pine, spruce, fir, hacmatack, and tamarac. Although we had made only seventeen miles, we were all very weary. The whole stock of animal food was now exhausted, and we therefore killed a colt, on which we made a hearty supper. From this incident we called the last creek we had passed from the south Colt-killed creek. The river itself is eighty yards wide, with a swift current and a stony channel. Its Indian name is Kooskooskee.

Sunday, 15th. At an early hour we proceeded along the right side of the Kooskooskee, over steep rocky points of land, till at the distance of four miles we reached an old Indian fishing place; the road here turned to the right of the water, and began to ascend a mountain; but the fire and wind had prostrated or dried almost all the timber on the south side, and the ascents were so steep that we were forced to wind in every direction round the high knobs, which constantly impeded our progress. Several of the horses lost their foothold and slipped; one of them, which was loaded with a desk and small trunk, rolled over and over for forty yards, till his fall was stopped by a tree. The desk was broken, but the poor animal escaped without much injury. After clambering in this way for four miles, we came to a high snowy part of the mountain where was a spring of water, at which we halted two hours to refresh our horses.

On leaving the spring the road continued as bad as it was below, and the timber more abundant. At four miles we reached the top of the mountain, and foreseeing no chance of meeting with water, we encamped on the northern side of the mountain, near an old bank of snow, three feet deep. Some of this we melted, and supped on the remains of the colt killed yesterday. Our only game to-day was two pheasants, and the horses on which we calculated as a last resource begin to fail us, for two of them were so poor and worn out with fatigue, that we were obliged to leave them behind. All around us are high rugged mountains, among which is a lofty range from southeast to northwest, whose tops are without timber, and in some places covered with snow. The night was cloudy and very cold, and three hours before daybreak,

Monday, 16th, it began to snow, and continued all day, so that by evening it was six or eight inches deep. This covered the track so completely that we were obliged constantly to halt and examine, lest we should lose the route. In many places we had nothing to guide us except the branches of the trees, which, being low, have been rubbed by the burdens of the Indian horses. The road was, like that of yesterday, along steep hillsides, obstructed with fallen timber, and a growth of eight different species of pine, so thickly strewed that the snow falls from them as we pass and keeps us continually wet to the skin, and so cold that we are anxious lest our feet should be frozen, as we have only thin moccasins to defend them.

At noon we halted to let the horses feed on some long grass on the south side of the mountains, and endeavoured by making fires to keep ourselves warm. As soon as the horses were refreshed, Captain Clark went ahead with one man, and at the distance of six miles reached a stream from the right, and prepared fires by the time of our arrival at dusk. We here encamped in a piece of low ground, thickly timbered, but scarcely large enough to permit us to lie level. We had now made thirteen miles. We were all very wet, cold, and hungry; but although before setting out this morning we had seen four deer, yet we could not procure any of them, and were obliged to kill a second colt for our supper.

Tuesday, 17th. Our horses became so much scattered during the night that we were detained till one o'clock before they were all collected. We then continued our route, over high rough knobs and several drains and springs, and along a ridge of country separating the waters of two small rivers. The road was still difficult, and several of the horses fell and injured themselves very much, so that we were unable to advance more than ten miles to a small stream, on which we encamped.

We had killed a few pheasants, but these being insufficient for our subsistence we killed another of the colts. This want of provisions, and the extreme fatigue to which we were subjected, and the dreary prospects before us, began to dispirit the men. It was therefore agreed that Captain Clark should go on ahead with six hunters, and endeavour to kill something for the support of the party. He therefore set out,

Wednesday, 18th, early in the morning, in hopes of finding a level country from which he might send back some game. His route lay S. 85° W. along the same high dividing ridge, and the road was still very bad; but he moved on rapidly, and at the distance of twenty miles was rejoiced on discovering far off an extensive plain towards the west and southwest, bounded by a high mountain. He halted an hour to let the horses eat a little grass on the hillsides, and then went on twelve and a half miles till he reached a bold creek, running to the left, on which he encamped. To this stream he gave the very appropriate name of Hungry creek, for having procured no game, they had nothing to eat.

In the meantime we were detained till after eight o'clock by the loss of one of our horses, which had strayed away and could not be found. We then proceeded, but having soon finished the remainder of the colt killed yesterday, felt the want of provisions, which was more sensible from our meeting with no water, till towards nightfall we found some in a ravine among the hills. By pushing on our horses almost to their utmost strength we made eighteen miles.

We then melted some snow, and supped on a little portable soup, a few canisters of which, with about twenty weight of bear's oil, are our only remaining means of subsistence. Our guns are scarcely of any service, for there is no living creature in these mountains, except a few small pheasants, a small species of gray squirrel, and a blue bird of the vulture kind about the size of a turtle dove or jay, and even these are difficult to shoot.

Thursday, 19th. Captain Clark proceeded up the creek, along which the road was more steep and stony than any he had yet passed. At six miles distance he reached a small plain, in which he fortunately found a horse, on which he breakfasted, and hung the rest on a tree for the party in the rear. Two miles beyond this he left the creek and crossed three high mountains, rendered almost impassable from the steepness of the ascent and the quantity of fallen timber. After clambering over these ridges and mountains, and passing the heads of some branches of Hungry creek, he came to a large creek running westward. This he followed for four miles, then turned to the right down the mountain, till he came to a small creek to the left. Here he halted, having made twenty-two miles on his course, south 80° west, though the winding route over the mountains almost doubled the distance. On descending the last mountain, the heat became much more sensible after the extreme cold he had experienced for several days past. Besides the breakfast in the morning, two pheasants were their only food during the day, and the only kinds of birds they saw were the blue jay, a small white-headed hawk, a larger hawk, crows, and ravens.

We followed soon after sunrise. At six miles the ridge terminated, and we had before us the cheering prospect of the large plain to the southwest. On leaving the ridge we again ascended and went down several mountains, and six miles farther came to Hungry creek where it was fifteen yards wide and received the waters of a branch from the north. We went up it on a course nearly due west, and at three miles crossed a second branch flowing from the same quarter. The country is thickly covered with pine timber, of which we have enumerated eight distinct species. Three miles beyond this last branch of Hungry creek we encamped, after a fatiguing route of eighteen miles. The road along the creek is a narrow, rocky path near the borders of very high precipices, from which a fall seems almost inevitable destruction. One of our horses slipped, and rolling over with his load down the hillside, which was nearly perpendicular and strewed with large irregular rocks, nearly a hundred yards, and did not stop till he fell into the creek; we all expected he was killed, but to our astonishment, on taking off his load he rose, and seemed but little injured, and in twenty minutes proceeded with his load. Having no other provision we took some portable soup, our only refreshment during the day. This abstinence, joined with fatigue, has a visible effect on our health. The men are growing weak and losing their flesh very fast; several are afflicted with the dysentery, and eruptions of the skin are very common.

Friday, 20th. Captain Clark went on through a country as rugged as usual, till on passing a low mountain he came, at the distance of four miles, to the forks of a large creek. Down this he kept on a course south 60° west for two miles, then turning to the right, continued over a dividing ridge where were the heads of several little streams, and at twelve miles distance descended the last of the Rocky mountains and reached the level country. A beautiful open plain, partially supplied with pine, now presented itself. He continued for five miles, when he discovered three Indian boys, who, on observing the party, ran off and hid themselves in the grass. Captain Clark immediately alighted, and giving his horse and gun to one of the men went after the boys. He soon relieved their apprehensions and sent them forward to the village, about a mile off, with presents of small pieces of riband. Soon after the boys had reached home a man came out to meet the party, with great caution, but he conducted them to a large tent in the village, and all the inhabitants gathered round to view with a mixture of fear and pleasure these wonderful strangers. The conductor now informed Captain Clark, by signs, that the spacious tent was the residence of the great chief, who had set out three days ago with all the warriors to attack some of their enemies towards the southwest; that he would not return before fifteen or eighteen days, and that in the meantime there were only a few men left to guard the women and children. They now set before them a small piece of buffaloe meat, some dried salmon, berries, and several kinds of roots. Among these last is one which is round and much like an onion in appearance, and sweet to the taste; it is called quamash, and is eaten either in its natural state, or boiled into a kind of soup, or made into a cake which is then called pasheco. After the long abstinence this was a sumptuous treat; we returned the kindness of the people by a few small presents, and then went on, in company with one of the chiefs, to a second village in the same plain, at the distance of two miles. Here the party was treated with great kindness, and passed the night. The hunters were sent out, but though they saw some tracks of deer, were not able to procure anything.

We were detained till ten o'clock before we could collect our scattered horses; we then proceeded for two miles, when to our great joy we found the horse which Captain Clark had killed, and a note apprising us of his intention of going to the plains towards the southwest, and collect provisions by the time we reached him. At one o'clock we halted on a small stream, and made a hearty meal of horse flesh. On examination it now appeared that one of the horses was missing, and the man in whose charge he had been was directed to return and search for him. He came back in about two hours without having been able to find the horse; but as the load was too valuable to be lost, two of the best woodsmen were directed to continue the search while we proceeded. Our general course was south 25° west through a thick forest of large pine, which has fallen in many places and very much obstructs the road. After making about fifteen miles we encamped on a ridge, where we could find but little grass and no water. We succeeded, however, in procuring a little from a distance, and supped on the remainder of the horse.

On descending the heights of the mountains the soil becomes gradually more fertile, and the land through which we passed this evening is of an excellent quality. It has a dark gray soil, though very broken, and with large masses of gray free-stone above the ground in many places. Among the vegetable productions we distinguished the alder, honeysuckle, and huckleberry, common in the United States, and a species of honeysuckle, known only westward of the Rocky mountains, which rises to the height of about four feet, and bears a white berry. There is also a plant resembling the chokecherry, which grows in thick clumps eight or ten feet high, and bears a black berry, with a single stone, of a sweetish taste. The arbor vitæ, too, is very common, and grows to a great size, being from two to six feet in diameter.

Saturday, 21st. The free use of food, to which he had not been accustomed, made Captain Clark very sick both yesterday evening and during the whole of to-day. He therefore sent out all the hunters and remained himself at the village, as well on account of his sickness as for the purpose of avoiding suspicion and collecting information from the Indians as to the route.

The two villages consist of about thirty double tents, and the inhabitants call themselves Chopunnish or Pierced-nose. The chief drew a chart of the river, and explained that a greater chief than himself, who governed this village and was called the Twisted-hair, was now fishing at the distance of half a day's ride down the river; his chart made the Kooskooskee fork a little below his camp; a second fork below; still farther on a large branch flowed in on each side, below which the river passed the mountains; here was a great fall of water, near which lived white people, from whom were procured the white beads and brass ornaments worn by the women.

A chief of another band made a visit this morning, and smoked with Captain Clark. The hunters returned without having been able to kill anything. Captain Clark purchased as much dried salmon, roots, and berries as he could with the few articles he chanced to have in his pockets, and, having sent them by one of the men and a hired Indian back to Captain Lewis, he went on towards the camp of the Twisted-hair. It was four o'clock before he set out, and the night soon came on; but having met an Indian coming from the river, they engaged him, by a present of a neckcloth, to guide them to the Twisted-hair's camp. For twelve miles they proceeded through the plain before they reached the river hills, which are very high and steep. The whole valley from these hills to the Rocky mountain is a beautiful level country, with a rich soil covered with grass; there is, however, but little timber, and the ground is badly watered; the plain is so much lower than the surrounding hills, or so much sheltered by them, that the weather is quite warm, while the cold of the mountains was extreme. From the top of the river hills they proceeded down for three miles till they reached the water side, between eleven and twelve o'clock at night; here we found a small camp of five squaws and three children, the chief himself being encamped, with two others, on a small island in the river; the guide called to him, and he soon came over. Captain Clark gave him a medal, and they smoked together till one o'clock.

We could not set out till eleven o'clock, because, being obliged in the evening to loosen our horses to enable them to find subsistence, it is always difficult to collect them in the morning. At that hour we continued along the ridge on which we had slept, and at a mile and a half reached a large creek running to our left, just above its junction with one of its branches. We proceeded down the low grounds of this creek, which are level, wide, and heavily timbered, but turned to the right at the distance of two and a half miles, and began to pass the broken and hilly country; but the thick timber had fallen in so many places that we could scarcely make our way. After going five miles we passed the creek on which Captain Clark had encamped during the night of the 19th, and continued five miles farther over the same kind of road, till we came to the forks of a large creek. We crossed the northern branch of this stream, and proceeded down it on the west side for a mile; here we found a small plain where there was tolerable grass for the horses, and therefore remained during the night, having made fifteen miles on a course S. 30° W.

The arbor vitæ increases in size and quantity as we advance, some of the trees we passed to-day being capable of forming periogues at least forty-five feet in length. We were so fortunate, also, as to kill a few pheasants and a prairie wolf, which, with the remainder of the horse, supplied us with one meal, the last of our provisions, our food for the morrow being wholly dependent on the chance of our guns.

Sunday, 22d. Captain Clark passed over to the island with the Twisted-hair, who seemed to be cheerful and sincere in his conduct. The river at this place is about one hundred and sixty yards wide, but interrupted by shoals, and the low grounds on its borders are narrow. The hunters brought in three deer; after which Captain Clark left his party, and, accompanied by the Twisted-hair and his son, rode back to the village, where he arrived about sunset; they then walked up together to the second village, where we had just arrived. We had intended to set out early, but one of the men having neglected to hobble his horse he strayed away, and we were obliged to wait till nearly twelve o'clock. We then proceeded on a western course for two and a half miles, when we met the hunters sent by Captain Clark from the village, seven and a half miles distant, with provisions. This supply was most seasonable, as we had tasted nothing since last night, and the fish and roots and berries, in addition to a crow which we killed on the route, completely satisfied our hunger. After this refreshment we proceeded in much better spirits, and at a few miles were overtaken by the two men who had been sent back after a horse on the 20th. They were perfectly exhausted with the fatigue of walking and the want of food; but as we had two spare horses they were mounted and brought on to the village.

They had set out about three o'clock in the afternoon of the 20th, with one horse between them; after crossing the mountain they came to the place where we had eaten the horse. Here they encamped, and having no food made a fire and roasted the head of the horse, which even our appetites had spared, and supped on the ears, skin, lips, etc., of the animal. The next morning, 21st, they found the track of the horse, and pursuing it recovered the saddle-bags, and at length, about eleven o'clock, the horse himself. Being now both mounted, they set out to return and slept at a small stream. During the day they had nothing at all except two pheasants, which were so torn to pieces by the shot that the head and legs were the only parts fit for food. In this situation they found the next morning, 22d, that during the night their horses had run away from them or been stolen by the Indians. They searched for them until nine o'clock, when, seeing that they could not recover them, and fearful of starving if they remained where they were, they set out on foot to join us, carrying the saddle-bags alternately. They walked as fast as they could during the day, till they reached us in a deplorable state of weakness and inanition.

As we approached the village, most of the women, though apprised of our being expected, fled with their children into the neighbouring woods. The men, however, received us without any apprehension, and gave us a plentiful supply of provisions. The plains were now crowded with Indians, who came to see the persons of the whites and the strange things they brought with them; but as our guide was perfectly a stranger to their language, we could converse by signs only. Our inquiries were chiefly directed to the situation of the country, the courses of the rivers, and the Indian villages, of all which we received information from several of the Indians, and as their accounts varied but little from each other we were induced to place confidence in them. Among others, the Twisted-hair drew a chart of the river on a white elk-skin. According to this, the Kooskooskee forks a few miles from this place; two days towards the south is another and larger fork, on which the Shoshonee or Snake Indians fish; five days' journey farther is a large river from the northwest into which Clark's river empties itself; from the mouth of that river to the falls is five days' journey farther; on all the forks, as well as on the main river, great numbers of Indians reside, and at the falls are establishments of whites. This was the story of the Twisted-hair.

Monday, 23d. The chiefs and warriors were all assembled

this morning, and we explained to them where we came from, the objects of our visiting them, and our pacific intentions towards all the Indians. This, being conveyed by signs, might not have been perfectly comprehended, but appeared to give perfect satisfaction. We now gave a medal to two of the chiefs, a shirt in addition to the medal already received by the Twisted-hair, and delivered a flag and a handkerchief for the grand chief on his return. To these were added a knife, a handkerchief, and a small piece of tobacco for each chief. The inhabitants did not give us any provisions gratuitously. We therefore purchased a quantity of fish, berries (chiefly red haws), and roots, and in the afternoon went on to the second village. The Twisted-hair introduced us into his own tent, which consisted, however, of nothing more than pine bushes and bark, and gave us some dried salmon boiled. We continued our purchases, and obtained as much provision as our horses could carry, in their present weak condition, as far as the river. The men exchanged a few old canisters for dressed elk-skins, of which they made shirts; great crowds of the natives are round us all night, but we have not yet missed anything except a knife and a few other articles stolen yesterday from a shot-pouch. At dark we had a hard wind from the southwest, accompanied with rain, which lasted half an hour, but in the morning,

Tuesday, 24th, the weather was fair. We sent back Colter in search of the horses lost on the mountains, and having collected the rest, set out at ten o'clock along the same route already passed by Captain Clark towards the river. All round the village the women are busily employed in gathering and dressing the pasheco root, of which large quantities are heaped up in piles over the plain. We now felt severely the consequence of eating heartily after our late privations; Captain Lewis and two of the men were taken very ill last evening, and to-day he could scarcely sit on his horse, while others were obliged to be put on horseback, and some, from extreme weakness and pain, were forced to lie down alongside of the road for some time. At sunset we reached the island where the hunters had been left on the 22d. They had been unsuccessful, having killed only two deer since that time, and two of them are very sick. A little below this island is a larger one, on which we encamped and administered Rush's pills to the sick.

Wednesday, 25th. The weather was very hot and oppressive to the party, most of whom are now complaining of sickness. Our situation, indeed, rendered it necessary to husband our remaining strength, and it was determined to proceed down the river in canoes. Captain Clark, therefore, set out with the Twisted-hair and two young men, in quest of timber for canoes. As he went down the river he crossed at the distance of a mile a creek from the right, which, from the rocks that obstructed its passage, he called Rockdam river. The hills along the river are high and steep, the low grounds are narrow, and the navigation of the river embarrassed by two rapids. At the distance of three miles farther he reached two nearly equal forks of the river, one of which flowed in from the north. Here he rested for an hour, and cooked a few salmon which one of the Indians caught with a gig. Here, too, he was joined by two canoes of Indians from below; they were long, steady, and loaded with the furniture and provisions of two families. He now crossed the south fork and returned to the camp on the south side, through a narrow pine bottom the greater part of the way, in which was found much fine timber for canoes. One of the Indian boats, with two men, set out at the same time, and such was their dexterity in managing the pole that they reached camp within fifteen minutes after him, although they had to drag the canoe over three rapids. He found Captain Lewis and several of the men still very sick, and distributed to such as were in need of it salts and tartar emetic.

Thursday, 26th. Having resolved to go down to some spot calculated for building canoes, we set out early this morning and proceeded five miles, and encamped on low ground on the south, opposite the forks of the river. But so weak were the men that several were taken sick in coming down, the weather being oppressively hot. Two chiefs and their families followed us, and encamped with a great number of horses near us; and soon after our arrival we were joined by two Indians, who came down the north fork on a raft. We purchased some fresh salmon, and having distributed axes, and portioned off the labour of the party, began,

Friday, 27th, at an early hour, the preparations for making five canoes. But few of the men, however, were able to work, and of these several were soon taken ill, as the day proved very hot. The hunters too, returned without any game, and seriously indisposed, so that nearly the whole party was now ill. We procured some fresh salmon; and Colter, who now returned with one of the horses, brought half a deer, which was very nourishing to the invalids. Several Indians from a camp below came up to see us.

Saturday, 28th. The men continue ill, though some of those first attacked are recovering. Their general complaint is a heaviness at the stomach, and a lax, which is rendered more painful by the heat of the weather, and the diet of fish and roots, to which they are confined, as no game is to be procured. A number of Indians collect about us in the course of the day to gaze at the strange appearance of everything belonging to us.

Sunday, 29th. The morning was cool, the wind from the southwest; but in the afternoon the heat returned. The men continue ill, but all those who are able to work are occupied at the canoes. The spirits of the party were much recruited by three deer brought in by the hunters; and the next day,

Monday, 30th, the sick began to recruit their strength, the morning being fair and pleasant. The Indians pass in great numbers up and down the river, and we observe large quantities of small duck going down this morning.

Tuesday, October 1, 1805. The morning was cool, the wind easterly, but the latter part of the day was warm. We were visited by several Indians from the tribes below, and others from the main south fork. To two of the most distinguished men we made presents of a ring and brooch, and to five others a piece of riband, a little tobacco, and the fifth part of a neckcloth. We now dried our clothes and other articles, and selected some articles such as the Indians admire in order to purchase some provisions, as we have nothing left except a little dried fish, which operates as a complete purgative.

Wednesday, 2d. The day is very warm. Two men were sent to the village with a quantity of these articles to purchase food. We are now reduced to roots, which produce violent pains in the stomach. Our work continued as usual, and many of the party are convalescent. The hunters returned in the afternoon with nothing but a small prairie-wolf, so that, our provisions being exhausted, we killed one of the horses to eat, and provide soup for the sick.

Thursday, 3d. The fine cool morning and easterly wind had an agreeable effect upon the party, most of whom are now able to work. The Indians from below left us, and we were visited by others from different quarters.

Friday, 4th. Again we had a cool east wind from the mountains. The men were now much better, and Captain Lewis himself so far recovered as to walk about a little.

Three Indians arrived to-day from the Great river to the south. The two men also returned from the village with roots and fish, and as the flesh of the horse killed yesterday was exhausted, we were confined to that diet, although unwholesome as well as unpleasant. The afternoon was warm.

Saturday, 5th. The wind easterly, and the weather cool. The canoes being nearly finished it became necessary to dispose of our horses. They were therefore collected to the number of thirty-eight, and being branded and marked, were delivered to three Indians, the two brothers and the son of a chief who promises to accompany us down the river. To each of these men we gave a knife and some small articles, and they agreed to take good care of the horses till our return. The hunters, with all their diligence, are unable to kill anything, the hills being high and rugged, and the woods too dry to hunt deer, which is the only game in the country. We therefore continue to eat dried fish and roots, which are purchased from the squaws by means of small presents, but chiefly white beads, of which they are extravagantly fond. Some of these roots seem to possess very active properties, for after supping on them this evening we were swelled to such a degree as to be scarcely able to breathe for several hours. Towards night we launched two canoes, which proved to be very good.

Sunday, 6th. This morning is again cool, and the wind easterly. The general course of the winds seems to resemble that which we observed on the east side of the mountain. While on the head waters of the Missouri, we had every morning a cool wind from the west. At this place a cool breeze springs up during the latter part of the night, or near daybreak, and continues till seven or eight o'clock, when it subsides, and the latter part of the day is warm. Captain Lewis is not so well as he was, and Captain Clark was also taken ill. We had all our saddles buried in a cache near the river, about half a mile below, and deposited at the same time a canister of powder and a bag of balls. The time which could be spared from our labours on the canoes was devoted to some astronomical observations. The latitude of our camp as deduced from the mean of two observations is 46° 34' 56"3'" north.

Monday, 7th. This morning all the canoes were put in the water and loaded, the oars fixed, and every preparation made for setting out, but when we were all ready the two chiefs who had promised to accompany us were not to be found, and at the same time we missed a pipe tomahawk. We therefore proceeded without them. Below the forks this river is called the Kooskooskee, and is a clear, rapid stream, with a number of shoals and difficult places. For some miles the hills are steep, the low grounds narrow, but then succeeds an open country with a few trees scattered along the river. At the distance of nine miles is a small creek on the left. We passed in the course of the day ten rapids, in descending which one of the canoes struck a rock and sprung a leak; we, however, continued for nineteen miles, and encamped on the left side of the river, opposite to the mouth of a small run. Here the canoe was unloaded and repaired, and two lead canisters of powder deposited; several camps of Indians were on the sides of the river, but we had little intercourse with any of them.

Tuesday, 8th. We set out at nine o'clock. At eight and a half miles we passed an island; four and a half miles lower a second island, opposite a small creek on the left side of the river. Five miles lower is another island on the left, a mile and a half below which is a fourth. At a short distance from this is a large creek from the right, to which we gave the name of Colter's creek, from Colter, one of the men. We had left this creek about a mile and a half, and were passing the last of fifteen rapids which we had been fortunate enough to escape, when one of the canoes struck, and a hole being made in her side, she immediately filled and sank. The men, several of whom could not swim, clung to the boat till one of our canoes could be unloaded, and with the assistance of an Indian boat they were all brought to shore. All the goods were so much wet that we were obliged to halt for the night and spread them out to dry. While all this was exhibited, it was necessary to place two sentinels over the merchandise, for we found that the Indians, though kind and disposed to give us every aid during our distress, could not resist the temptation of pilfering some of the small articles. We passed, during our route of twenty miles to-day, several encampments of Indians on the islands and near the rapids, which places are chosen as most convenient for taking salmon. At one of these camps we found our two chiefs, who, after promising to descend the river with us, had left us; they, however, willingly came on board after we had gone through the ceremony of smoking.

Wednesday, 9th. The morning was, as usual, cool; but as the weather both yesterday and to-day was cloudy, our merchandise dried but slowly. The boat, though much injured, was repaired by ten o'clock so as to be perfectly fit for service; but we were obliged to remain during the day till the articles were sufficiently dry to be reloaded; the interval we employed in purchasing fish for the voyage and conversing with the Indians. In the afternoon we were surprised at hearing that our old Shoshonee guide and his son had left us, and been seen running up the river several miles above. As he had never given any notice of his intention, nor had even received his pay for guiding us, we could not imagine the cause of his desertion, nor did he ever return to explain his conduct. We requested the chief to send a horseman after him to request that he would return and receive what we owed him. From this, however, he dissuaded us, and said very frankly that his nation, the Chopunnish, would take from the old man any presents that he might have on passing their camp.

The Indians came about our camp at night, and were very gay and good-humoured with the men. Among other exhibitions was that of a squaw who appeared to be crazy; she sang in a wild, incoherent manner, and would offer to the spectators all the little articles she possessed, scarifying herself in a horrid manner if anyone refused her present; she seemed to be an object of pity among the Indians, who suffered her to do as she pleased without interruption.

Thursday, 10th. A fine morning. We loaded the canoes, and set off at seven o'clock. At the distance of two and a half miles we had passed three islands, the last of which is opposite to a small stream on the right. Within the following three and a half miles is another island and a creek on the left, with wide, low grounds, containing willow and cottonwood trees, on which were three tents of Indians. Two miles lower is the head of a large island, and six and a half miles farther we halted at an encampment of eight lodges on the left, in order to view a rapid before us; we had already passed eight, and some of them difficult; but this was worse than any of them, being a very hazardous ripple strewed with rocks; we here purchased roots and dined with the Indians. Among them was a man from the falls who says that he saw white people at that place, and is very desirous of going down with us, an offer which, however, we declined. Just above this camp we had passed a tent, near which was an Indian bathing himself in a small pond or hole of water, warmed by throwing in hot stones. After finishing our meal we descended the rapid, with no injury except to one of our boats, which ran against a rock, but in the course of an hour was brought off with only a small split in her side. This ripple, from its appearance and difficulty, we named the Rugged rapid. We went on over five other rapids of a less dangerous kind, and at the distance of five miles reached a large fork of the river from the south; and after coming twenty miles, halted below the junction on the right side of the river. Our arrival soon attracted the attention of the Indians, who flocked in all directions to see us. In the evening the Indian from the falls, whom we had seen at the Rugged rapid, joined us with his son, in a small canoe, and insisted on accompanying us to the falls. Being again reduced to fish and roots, we made an experiment to vary our food by purchasing a few dogs, and after having been accustomed to horse-flesh, felt no disrelish to this new dish. The Chopunnish have great numbers of dogs, which they employ for domestic purposes, but never eat; and our using the flesh of that animal soon brought us into ridicule as dog-eaters.

The country at the junction of the two rivers is an open plain on all sides, broken towards the left by a distant ridge of highland, thinly covered with timber; this is the only body of timber which the country possesses, for at the forks there is not a tree to be seen, and during almost the whole descent of sixty miles down the Kooskooskee from its forks there are very few. This southern branch is, in fact, the main stream of Lewis's river, on which we encamped when among the Shoshonees. The Indians inform us that it is navigable for sixty miles; that not far from its mouth it receives a branch from the south; and a second and larger branch, two days' march up, and nearly parallel to the first Chopunnish villages we met near the mountains. This branch is called Pawnashte, and is the residence of a chief who, according to their expression, has more horses than he can count. The river has many rapids, near which are situated many fishing camps, there being ten establishments of this before reaching the first southern branch: one on that stream, five between that and the Pawnashte, one on that river, and two above it; besides many other Indians who reside high up on the more distant waters of this river. All these Indians belong to the Chopunnish nation, and live in tents of an oblong form, covered with flat roofs.

At its mouth Lewis's river is about two hundred and fifty yards wide, and its water is of a greenish blue colour. The Kooskooskee, whose waters are clear as crystal, is one hundred and fifty yards in width, and after the union the river enlarges to the space of three hundred yards; at the point of the union is an Indian cabin, and in Lewis's river a small island.

The Chopunnish, or Pierced-nose nation, who reside on the Kooskooskee and Lewis's rivers, are in person stout, portly, well-looking men; the women are small, with good features, and generally handsome, though the complexion of both sexes is darker than that of the Tushepaws. In dress they resemble that nation, being fond of displaying their ornaments. The buffaloe or elk-skin robe decorated with beads, sea-shells, chiefly mother-of-pearl, attached to an otter-skin collar and hung in the hair, which falls in front in two queues; feathers, paints of different kinds, principally white, green, and light blue, all of which they find in their own country: these are the chief ornaments they use. In the winter they wear a short shirt of dressed skins, long painted leggings and moccasins, and a plait of twisted grass round the neck.

The dress of the women is more simple, consisting of a long shirt of argalia or ibex skin, reaching down to the ankles without a girdle; to this are tied little pieces of brass and shells and other small articles; but the head is not at all ornamented. The dress of the female is indeed more modest, and more studiously so, than any we have observed, though the other sex is careless of the indelicacy of exposure.

The Chopunnish have very few amusements, for their life is painful and laborious, and all their exertions are necessary to earn even their precarious subsistence. During the summer and autumn they are busily occupied in fishing for salmon and collecting their winter store of roots. In the winter they hunt the deer on snowshoes over the plains, and towards spring cross the mountains to the Missouri for the purpose of trafficking for buffaloe robes. The inconveniences of that comfortless life are increased by frequent encounters with their enemies from the west, who drive them over the mountains, with the loss of their horses and sometimes the lives of many of the nation. Though originally the same people, their dialect varies very perceptibly from that of the Tushepaws; their treatment to us differed much from the kind and disinterested services of the Shoshonees; they are indeed selfish and avaricious; they part very reluctantly with every article of food or clothing; and while they expect a recompense for every service, however small, do not concern themselves about reciprocating any presents we may give them.

They are generally healthy, the only disorders which we have had occasion to remark being of a scrophulous kind, and for these, as well as for the amusement of those who are in good health, hot and cold bathing is very commonly used.

The soil of these prairies is of a light yellow clay, intermixed with small smooth grass; it is barren, and produces little more than a bearded grass about three inches high, and a prickly pear, of which we now found three species. The first is of the broad-leafed kind, common to the Missouri. The second has the leaf of a globular form, and is also frequent on the upper part of the Missouri, particularly after it enters the Rocky mountains. The third is peculiar to this country, and is much more inconvenient than the other two; it consists of small thick leaves of a circular form, which grow from the margin of each other as in the broad-leafed pear of the Missouri; these leaves are armed with a greater number of thorns, which are stronger, and appear to be barbed; and as the leaf itself is very slightly attached to the stem, as soon as one thorn touches the moccasin it adheres and brings with it the leaf, which is accompanied by a reënforcement of thorns.

END OF VOLUME I.

# Now back in print
## Original Journals of Lewis and Clark
### Edited by Reuben Gold Thwaites

Now back in print "Original Journals of The Lewis and Clark Expedition" as published by Dodd Mead in 1904/1905. In preparation for the Bicentennial commemoration of the historic Lewis and Clark Expedition, Digital Scanning, Inc. (DSI) announces the release of their digital reprint edition. The 1903-04 set of **"Original Journals of the Lewis and Clark Expedition"** have been described as the most accurate, work on the expedition. Edited and including an introduction and index by Reuben Gold Thwaites, this set is considered a valuable resource for historians, students and history buffs. This set includes 7 two-part volumes and the Atlas. Illustrated throughout by Karl Bodmer.

| Trade Paper Editions | Hardcover Editions |
|---|---|
| 8 Volume Trade Paper Set 1582186510 | 8 Volume Hardcover Ser 158218660X |
| ISBN | ISBN |
| TP Volume 1  1582186529 | HC Volume 1  1582186618 |
| TP Volume 2  1582186537 | HC Volume 2  1582186626 |
| TP Volume 3  1582186545 | HC Volume 3  1582186634 |
| TP Volume 4  1582186553 | HC Volume 4  1582186642 |
| TP Volume 5  1582186561 | HC Volume 5  1582186650 |
| TP Volume 6  158218657X | HC Volume 6  1582186669 |
| TP Volume 7  1582186588 | HC Volume 7  1582186677 |
| TP Volume 8  1582186596 | HC Volume 8  1582186685 |

Additional Information is available at http://www.Digitalscanning.com or http://www.PDFLibrary.com.

# "The Trail of Lewis and Clark" by Olin D. Wheeler
## "As Published in 1904"

A Century after Lewis and Clark explored the newly purchased lands west of the Mississippi, Olin D. Wheeler set out on his own epic journey. Using the explorer's journals as a guide, he followed the old trail and recorded the changes to the land and landscape. Wheeler traveled by train, steamboat and pack train accompanied by a photographer. The 2 Volume set contains hundreds of photographs, sketches and maps.

| Volume 1 | ISBN | Volume 2 | ISBN |
| --- | --- | --- | --- |
| Individually | | Individually | |
| Trade paper | 1582187258 | Trade paper | 1582187266 |
| Hardcover | 1582187274 | Hardcover | 1582187282 |
| 2 Vol. Sets | | 2 Vol. Sets | |
| Trade paper | 1582187290 | Hardcover | 1582187304 |

# First Across The Continent By Noah Brooks
## As Published in 1901

*First Across the Continent: The Story of the Lewis and Clark Expedition* is presented as a captivating tale. It is drawn from the original journals of the explorers. Noah Brooks uses extensive, carefully selected excerpts from the journals to entice the reader, and then sends the armchair adventurer along on the trek with Lewis and Clark. The detailed description and faithful narratives immerse you in one of the most amazing journeys in history. Originally published in 1901.

Noah Brooks (1830-1903 was a political confidant and personal friend of Abraham Lincoln. A journalist for the *Sacramento Union* during the Lincoln presidency, Brooks was a frequent guest at the White House. After Lincoln's assassination, Brooks moved to the east coast and wrote for other newspapers, including the *New York Tribune* and the *New York Times*.

A great introduction for the young reader, audience 10 to adult.

ISBN TP 1582186820  HC 1582186839 eBook 1582186812

Information and samples available at:
http://www.digitalscanning.com and http://www.PDFlibrary.com

# Other Explorers titles offered by *Digital Scanning, Inc.*

**The Life of Dr. Elisha Kent Kane and Other Distinguished American Explorers,**
by Samuel M. Smucker
As Published in 1858.
TP: 1582182663 ($19.95)
HC: 1582182671 ($34.95)

**The Louisiana Purchase and the Exploration, Early History and Building of the West,**
by Ripley Hitchcock
As Published in 1903.
TP: 1582182361 ($19.95)
HC: 158218237X ($34.95)

**Our Lost Explorers: The Narrative of the Jeannette Arctic Expedition,**
by Raymond Lee Newcomb
As Published in 1888.
TP: 1582182825 ($24.95)
HC: 1582182833 ($39.95)

**In the Lena Delta: The Search for Lt. Commander DeLong, etc.,**
by George W. Melville
As Published in 1884.
TP: 1582183783 ($24.95)
HC: 1582183791 ($39.95)

**Crooked Trails,**
by Frederic Remington
As Published in 1899.
TP: 1582182981 ($14.95)
HC: 158218299X ($27.95)

**Pioneer Life and Frontier Adventures of Kit Carson and his Companions,**
by DeWitt C. Peters
As Published in 1881.
TP: 1582182248 ($29.95)
HC: 1582182256 ($45.95)

**Two Years Before the Mast,**
by Richard Henry Dana, Jr.
As Published in 1840.
TP: 158218285X ($24.95)
HC: 1582182868 ($34.95)

**Trails of the Pathfinders,**
by George Bird Grinnell
As Published in 1912.
TP: 1582185964 ($22.95)
HC: 1582185972 ($36.95)

**The Making of the Ohio Valley States,**
by Samuel A. Drake
As Published in 1894.
TP: 1582184224 ($14.95)
HC: 1582184232 ($27.95)

**The Making of the Great West,**
by Samuel A. Drake
As Published in 1894.
TP: 1582184380 ($17.95)
HC: 1582184399 ($29.95)

**The "Teddy" Expedition (Among the Ice Flows of Greenland),**
by Kai R. Dahl
As Published in 1925.
TP: 1582184623 ($15.95)
HC: 1582184631 ($29.95)

**The Spanish Pioneers,**
by Charles F. Lummis
As Published in 1899.
TP: 1582186243 ($14.95)
HC: 1582186251 ($27.95)

**Our Arctic Province (Alaska and the Seal Islands),**
by Henry W. Elliott
As Published in 1886.
TP: 1582184585 ($24.95)
HC: 1582184593 ($39.95)

**The Story of the Railroad,**
by Cy Warman
As Published in 1898.
TP: 1582186324 ($14.95)
HC: 1582186332 ($27.95)

**On the Border with Crook,**
by John G. Bourke
As Published in 1896.
TP: 1582184461 ($24.95)
HC: 158218447X ($39.95)

**Original Journals of the Lewis and Clark Expedition *(8 Volumes, 15 Parts including the Atlas),***
Edited, with Introduction, Notes and Index by Reuben Gold Thwaites
As Published in 1904 & 1905.
TP Set: 1582186510 ($175.00)
HC Set: 158218660X ($275.00)

# Other Explorers titles offered by Digital Scanning, Inc. (cont.):

**On the Storied Ohio,**
*by Reuben Gold Thwaites*
As Published in 1903.
TP: 1582182914 ($19.95)
HC: 1582182922 ($34.95)

**The Expedition of Lewis and Clark**
*(2 Volume Set),*
*by James K. Hosmer*
As Published in 1903.
Volumes 1:
TP: 1582186987 ($27.95)
HC: 1582186995 ($39.95)
Volumes 2:
TP: 1582187029 ($27.95)
HC: 1582187037 ($39.95)

**The Mississippi Basin,**
*by Justin Winsor*
As Published in 1895.
TP: 1582186448 ($22.95)
HC: 1582186456 ($34.95)

**The Adventures of Christopher Hawkins,**
*by Charles I. Bushnell*
As Published in 1864.
TP: 1582184542 ($17.95)
HC: 1582184550 ($29.95)

**The Old Northwest,**
*by B. A. Hinsdale*
As Published in 1888.
TP: 1582186782 ($21.95)
HC: 1582186790 ($34.95)

**First Across the Continent,**
*by Noah Brooks*
As Published in 1901.
TP: 1582186820 ($17.95)
HC: 1582186839 ($31.95)

**Life Explorations and Public Services of John Charles Fremont,**
*by Charles Wentworth Upham*
As Published in 1856.
TP: 1582183945 ($17.95)
HC: 1582183953 ($34.95)

**The Westward Movement,**
*by Justin Winsor*
As Published in 1897.
TP: 1582186480 ($27.95)
HC: 1582186499 ($39.95)

**Lasalle and the Discovery of the Great West,**
*by Francis Parkman*
As Published in 1889.
TP: 1582184909 ($24.95)
HC: 1582184917 ($39.95)

## To order any of the above titles:

*Contact your local bookstore and order through *Ingram Books*.
*Contact the publisher directly
 (for general information or special event purchases):

Digital Scanning, Inc.
344 Gannett Rd., Scituate, MA 02066
Phone: (781) 545-2100

email: books@digitalscanning.com
www.digitalscanning.com

Printed in the United States
203297BV00005B/10-12/A